STAR WARE

THE AMATEUR ASTRONOMER'S ULTIMATE GUIDE TO CHOOSING, BUYING, AND USING TELESCOPES AND ACCESSORIES

Philip S. Harrington

John Wiley & Sons, Inc.

New York ● Chichester ● Brisbane ● Toronto ● Singapore

On the cover: (Left) The Starsplitter 20 Newtonian reflector. (Upper right) The Celestron 8 Schmidt-Cassegrain telescope (left) accompanied by the Questar 3.5 Maksutov telescope. (Lower right) A bevy of astronomical accessories, including a pair of finderscopes, a sun filter, Telrad telescope-aiming device, and assorted color filters.

This text is printed on acid-free paper.

This publication is designed to provide accurate and authoritative information in regard to the subject matter covered. It is sold with the understanding that the publisher is not engaged in rendering legal, accounting, or other professional services. If legal advice or other expert assistance is required, the services of a competent professional person should be sought.

Library of Congress Cataloging-in-Publication Data

Harrington, Philip S.
    Star ware : the amateur astronomer's ultimate guide to choosing, buying, and using telescopes and accessories / Philip S. Harrington.
        p.   cm.
    Includes index.
    ISBN 0-471-57671-9 (pbk.)
    1. Telescopes—Purchasing—Guidebooks. 2. Telescopes—Amateurs' manuals. I. Title.
QB88.H37   1994
681'.412'0297—dc20   93-40226

Printed in the United States of America
10 9 8 7 6 5 4 3 2 1

For my daughter, Helen, the star of my life

Contents

	Preface	*vii*
1	*Parlez Vous* Telescope?	*1*
2	In the Beginning . . .	*11*
3	So You Want to Buy a Telescope!	*25*
4	Attention, Shoppers!	*53*
5	Dealer Options Extra	*105*
6	The Right Stuff	*127*
7	The Homemade Astronomer	*183*
8	Till Death Do You Part	*217*
9	A Few Tricks of the Trade	*241*
10	It's Time to Solo	*267*
	Appendices	
A	The Astronomical Yellow Pages	*339*
B	An Astronomer's Survival Guide	*349*
C	Upcoming Eclipses, 1994–2000	*351*
D	Visibility of the Planets, 1994–2000	*353*
E	The Messier Catalogue Plus	*357*
F	The Constellations	*363*
G	English/Metric Conversion	*367*
	Index	*369*

Preface

> If the pure and elevated pleasure to be derived from the possession and use of a good telescope . . . were generally known, I am certain that no instrument of science would be more commonly found in the homes of intelligent people.
>
> There is only one way in which you can be sure of getting a good telescope. First, decide how large a glass you are to have, then go to a maker of established reputation, fix upon the price you are willing to pay—remembering that good work is never cheap—and finally see that the instrument furnished to you answers the proper tests for telescopes of its size. There are telescopes and there are telescopes . . .

With these words of advice, Garrett Serviss opened his classic work *Pleasures of the Telescope*. Upon its publication in 1901, this book inspired many an armchair astronomer to change from being merely a spectator to a participant, actively observing the universe instead of just reading about it. In many ways, that book was an inspiration for the volume you hold before you.

The telescope market today is radically different than it was in the days of Serviss. Back then, amateur astronomy was an activity of the wealthy. The selection of commercially made telescopes was restricted to only one type of instrument—the refractor, and these sold for many times what their modern progenitors cost today (after correcting for inflation). By contrast, we live in an age that thrives on choice. Amateur astronomers must now wade through an ocean of literature and propaganda before being able to select a telescope intelligently. For many a budding astronomer, this chore appears overwhelming.

That is where this book comes in. You and I are going hunting for telescopes. After opening chapters that explain telescope jargon and history, today's astronomical marketplace is dissected and explored. Where is the best place to buy a telescope? Is there one telescope that does everything well? How should a telescope be cared for? What accessories are needed? The list of questions goes on and on. Happily, so do the answers. Although there is no single set of answers that are right for everybody, all of the available options will be explored so that you can make an educated decision.

Not all of the best astronomical equipment is available for sale, however; some of it has to be made at home. Ten homemade projects are outlined further in the book. These range in complexity from the simplicity of a dew cap to the complexity of a complete observatory. The book concludes with a discussion of how to use a telescope, and finally, some suggestions of what to look for in the night sky.

Yes, the telescope marketplace has certainly changed in the past century (in the past decade!), and so has the universe. The amateur astronomer has grown with these changes to explore the depths of space in ways that our ancestors could not have even imagined.

Acknowledgments

To put together and complete a book of this sort could not have been possible were it not for the support of many other players. I would be an irresponsible author if I relied solely on my own humble opinions about astronomical equipment. To compile the section reviewing telescopes, eyepieces, and accessories, I solicited input from amateur astronomers around the world by placing announcements in leading astronomical periodicals and announcing my survey at various astronomical conventions and in astronomical computer bulletin boards. The responses I received were very revealing and immensely helpful. Unfortunately, space does not permit me to list everyone here; to all who contributed, you have my heartfelt thanks. I would, however, like to acknowledge Thomas Back, Thomas Barkume, David Hasenaurer, Mike Harvey, Andrew Jaffe, and Sherrill Shaffer, all of whom went above and beyond in providing useful information. I also wish to acknowledge the contributions of the companies and dealers who provided me with their latest information, references, and other vital data. Brad Berger of Berger Brother Camera Exchange in Amityville, New York, deserves special recognition for allowing me to borrow and test telescopes and accessories.

As you will see, Chapter 7 is a selection of build-at-home projects for amateur astronomers. Several were invented and constructed by amateur astronomers who were looking to enhance their enjoyment of the hobby. They were kind enough to supply me with information, drawings, and photographs so that I could pass these projects along to you. For their invaluable contributions, I wish to thank Gerry Atkinson, Chris Bayus, Jerry Burns, Greg Bohemier, Bob Deen, Randy Hammock, Dave Kratz, Carl Lancaster, John Stanbury, and George Viscome.

All the celestial photographs that adorn this book were taken by amateur astronomers. Astrophotography is not easy, and so I must thank those accomplished photographers who graciously allowed me to use some of their work. These marvelous illustrations were photographed by Brian Kennedy, Richard Sanderson, Gregory Terrance, and George Viscome.

I wish to pass on my sincere appreciation to my proofreaders Dave Kratz, Alan MacRobert, Jack Megas, Cal Powell, and Richard Sanderson. I am especially indebted to them for submitting constructive suggestions while massaging my sensitive ego. My thanks as well to Eric Hilton, Susan Ring of *Southern Sky* magazine, and Warren Voegelin for their assistance behind the scenes. Many thanks also to Kate Bradford of John Wiley & Sons for her diligent guidance and help. Finally, my deepest thanks and appreciation go to my ever-patient family, my wife, Wendy, and daughter, Helen. They have continually provided me with boundless love and encouragement over the years. Were it not for their understanding my need to go out at three in the morning or drive an hour or more from home just to look at the stars, this book could not exist. I love them both dearly for that.

You, the reader, have a stake in all this, too. This book is not meant to be written, read, and forgotten about. It is meant to change, just as the hobby of astronomy changes. As you read through this occasionally opinionated book (did I say "occasionally"?), there may be a passage or two to which you take exception. Or maybe you own a telescope or something else astronomical that you are either happy or unhappy with. If so, great! This book is meant to kindle emotion. Drop me a line and tell me about it. *I want to know*. Please address all correspondence to me in care of John Wiley & Sons, Inc., 605 Third Avenue, New York, New York 10158. I shall try to answer all letters, but in case I miss yours, thank you in advance!

1

Parlez-Vous
Telescope?

Before the telescope, ours was a mysterious universe. Events occurred nightly that struck both awe and dread into the hearts and minds of early stargazers. Was the firmament populated with powerful gods who looked down upon the pitiful Earth? Would the world be destroyed if one of these deities became displeased? Eons passed without an answer.

The invention of the telescope was the key that unlocked the vault of the cosmos. Though it is still rich with intrigue, the universe of today is no longer one to be feared. Instead, we sense that it is our destiny to study, explore, and embrace the heavens. From our backyards we are now able to spot incredibly distant phenomena that could not have been imagined just a generation ago. Such is the marvel of the modern telescope.

Today's amateur astronomers have a wide and varied selection of equipment from which to choose. To the novice stargazer, it all appears very enticing but very complicated. One of the most confusing aspects of amateur astronomy is telescope vernacular—terms whose meanings seem shrouded in mystery. "Do astronomers speak a language all their own?" is the cry frequently echoed by newcomers to the hobby. The answer is yes, but it is a language that, unlike some foreign tongues, is easy to learn. Here is your first lesson.

Many different kinds of telescopes have been developed over the years. Even though their variations in design are great, all fall into one of three broad categories according to how they gather and focus light. *Refractors*, shown in Figure 1.1a, have a large lens (the *objective*) mounted in the front of the tube to perform this task, whereas *reflectors*, shown in Figure 1.1b, use a large mirror (the *primary mirror*) at the tube's bottom. The third class of telescope, called *catadioptrics* (Figure 1.1c), places a lens (here called a *corrector plate*) in front of the primary mirror. In each instance, the telescope's *prime optic* (objective lens or primary mirror) brings the incoming light to a *focus* and then directs

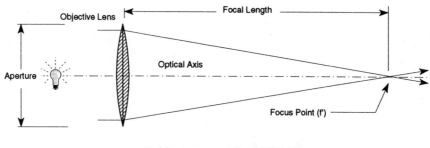

(a) Lens system (Refractor)

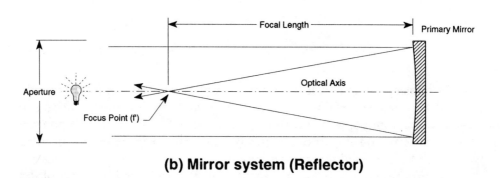

(b) Mirror system (Reflector)

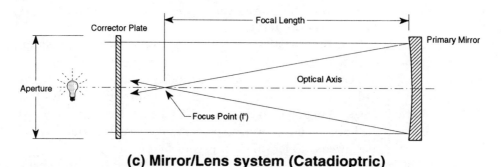

(c) Mirror/Lens system (Catadioptric)

Figure 1.1   *The basic principles of the telescope. Using either a lens (a), a mirror (b), or a combination (c), a telescope bends parallel rays of light to a focus point, or prime focus.*

that light through an *eyepiece* to the observer's waiting eye. Although Chapter 2 addresses the history and development of these grand instruments, we will begin here by exploring the many facets and terms that all telescopes share. As you read through the following discussion, be sure to pause and refer to the diagrams found in Chapter 2. This way, you can see how individual terms relate to the various types of telescopes.

Aperture

Let's begin with the basics. When we refer to the size of a telescope, we speak of its *aperture*. The aperture is simply the diameter (usually expressed in inches, centimeters, or millimeters) of the instrument's prime optic. In the case of a refractor, the diameter of the objective lens is cited; whereas in reflectors and catadioptric instruments, the diameters of their primary mirrors are specified. For instance, the objective lens in Galileo's first refractor was about 1.5 inches in diameter; it is therefore designated a 1.5-inch refractor. Sir Isaac Newton's first reflecting telescope employed a 1.33-inch mirror and would be referred to today as a 1.33-inch Newtonian reflector.

Many amateur astronomers consider aperture to be the most important criterion when selecting a telescope. In general (and there are exceptions to this, as pointed out in Chapter 3), the larger a telescope's aperture, the brighter and clearer the image it will produce. And that is the name of the game: sharp, vivid views of the universe.

Focal Length

The *focal length* is the distance from the objective lens or primary mirror to the *focal point* or *prime focus*, which is where the light rays converge. In a reflector and a catadioptric, this distance depends on the curvature of the telescope's mirrors, with a deeper curve resulting in a shorter focal length. The focal length of a refractor is dictated by the curves of the objective lens as well as by the type of glass used to manufacture the lens.

As with aperture, focal length is commonly expressed in either inches, centimeters, or millimeters.

Focal Ratio

Looking through astronomical books and magazines, it's not unusual to see a telescope specified as, say, an 8-inch f/10 or a 14-inch f/4.5. This f-number is the instrument's *focal ratio*, which is simply the focal length divided by the aperture. Therefore, an 8-inch telescope with a focal length of 56 inches would have a focal ratio (f-ratio) of f/7, because $56 \div 8 = 7$. Likewise, by turning the expression around, we know that a 6-inch f/8 telescope has a focal length of 48 inches, because $6 \times 8 = 48$.

Readers familiar with photography may already be used to referring to lenses by their focal ratios. In the case of cameras, a lens with a faster focal ratio (that is, a smaller f-number) will produce brighter images on film, thereby allowing shorter exposures when shooting dimly lit subjects. The same is true for telescopes. Instruments with faster focal ratios will produce brighter images on film, reducing the exposure times needed to record faint objects. However, a telescope with a fast focal ratio will *not* produce brighter images when used visually. The view of a particular object through, say, an 8-inch f/5 and an 8-inch f/10 will be identical when both are used at the same magnification. How bright an object appears to the eye depends only on telescope aperture and magnification.

Magnification

Many people, especially those new to telescopes, are under the false impression that the higher the magnification, the better the telescope. How wrong they are! It's true that as the power of a telescope increases, the apparent size of whatever is in view grows larger, but what most people fail to realize is that at the same time, the images become fainter and fuzzier. Finally, as the magnification climbs even higher, image quality becomes so poor that less detail will be seen than at lower powers.

It's easy to figure out the magnification of a telescope. If you look at the barrel of any eyepiece, you will notice a number followed by *mm*. It might be 26 mm, 12 mm, or 7 mm, among others; this is the focal length of that particular eyepiece expressed in millimeters. Magnification is calculated by dividing the telescope's focal length by the eyepiece's focal length. Remember to first convert the two focal lengths into the same units of measure—that is, both in inches or both in millimeters. A helpful hint: There are 25.4 millimeters in an inch.

For example, let's figure out the magnification of an 8-inch f/10 telescope with a 26-mm eyepiece. The telescope's 80-inch focal length equals 2,032 mm (80 × 25.4 = 2,032). Dividing 2,032 by the eyepiece's 26-mm focal length tells us that this telescope/eyepiece combination yields a magnification of 78× (read *78 power*), because 2,032 ÷ 26 = 78.

Most books and articles state that magnification should not exceed 60× per inch of aperture. This is under *ideal* conditions, something most observers rarely enjoy. Due to atmospheric turbulence (what astronomers call *poor seeing*), interference from artificial lighting, and other sources, many experienced observers seldom exceed 40× per inch. Some add a caveat to this: Never exceed 300× even if the telescope's aperture permits it. Others insist there is nothing wrong with using more than 60× per inch, as long as the sky conditions and optics are good enough. As you can see, the issue of magnification is always a hot topic of debate. My advice for the moment is to use the lowest magnification required to see what you want to see, but we are not done with the subject just yet. Magnification will be spoken of again in Chapter 5.

Light-gathering Ability

The human eye is a wondrous optical device, but its usefulness is severely limited in dim lighting conditions. When fully dilated under the darkest circumstances, the pupils of our eyes expand to about a quarter of an inch, or

7 mm, although this varies from person to person—the older you get, the less your pupils will dilate. In effect, we are born with a pair of quarter-inch refractors.

Telescopes effectively expand our pupils from fractions of an inch to many inches in diameter. The heavens now unfold before us with unexpected glory. A telescope's ability to reveal faint objects depends primarily on the diameter of either its objective lens or primary mirror (in other words, its aperture), not on magnification; quite simply, the larger the aperture, the more light gathered. Doubling a telescope's diameter increases light-gathering power by a factor of four, tripling its aperture expands it by nine times, and so on.

A telescope's *limiting magnitude* is a measure of how faint a star the instrument will show. Table 1.1 lists the faintest stars that can be seen through some popular telescope sizes. Trying to quantify limiting magnitude is anything but precise due to a large number of variables. Apart from aperture, other factors affecting this value include the quality of the telescope's optics, meteorological conditions, light pollution, excessive magnification, apparent size of the target, and the observer's vision and experience. These numbers are conservative estimates; experienced observers under dark, crystalline skies can better these by perhaps half a magnitude or more.

Resolving Power

A telescope's *resolving power* is its ability to reveal fine detail in whatever it is aimed at. Though resolving power plays a big part in everything we look at, it is especially important when viewing subtle planetary features, small surface markings on the Moon, or searching for close-set double stars.

A telescope's ability to resolve fine detail is always expressed in *arc-seconds*. You may remember this term from high-school geometry. Recall that in

Table 1.1   **Limiting Magnitudes**

Telescope in.	Aperture mm	Faintest Magnitude
2	51	10.3
3	76	11.2
4	102	11.8
6	152	12.7
8	203	13.3
10	254	13.8
12.5	318	14.3
14	356	14.5
16	406	14.8
18	457	15.1
20	508	15.3
24	610	15.7
30	762	16.2

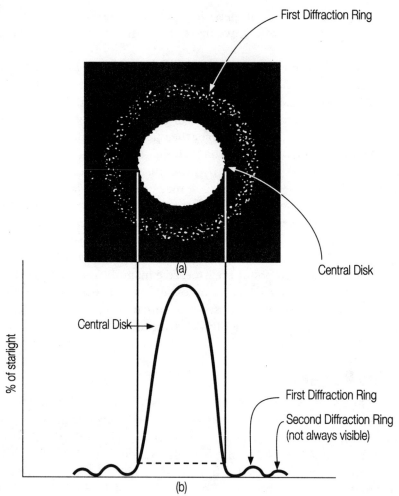

Figure 1.2 *The Airy disk (a) as it appears through a highly-magnified telescope and (b) graphically showing the distribution of light.*

the sky there are 90 degrees from horizon to the overhead point, or zenith, and 360 degrees around the horizon. Each one of those degrees may be broken into 60 equal parts called *arc-minutes*. For example, the apparent diameter of the Moon in our sky may be referred to as either 0.5° or 30 arc-minutes, each one of which may be further broken down into 60 arc-seconds. Therefore, the Moon may also be sized as 1,800 arc-seconds.

Regardless of the size, quality, or location of a telescope, stars will never appear as perfectly sharp points. This is partially due to atmospheric interference and partially due to the fact that light consists of slightly fuzzy waves rather than mathematically straight lines. Even with perfect atmospheric conditions, what we see is a blob, technically called the *Airy disk* (named in honor of its discoverer, Sir George Airy, Britain's Astronomer Royal from 1835 to 1892). Because light is composed of waves, rays from different parts of a telescope's prime optic (be it a mirror or lens) alternately interfere with and enhance each other, producing a series of dark and bright concentric rings around the Airy disk (Figure 1.2a). The whole display is known as a *diffraction pattern*. Ideally, through a telescope without a central obstruction (that is, without a secondary mirror), 84% of the starlight remains concentrated in the central disk, 7% in the first bright ring, and 3% in the second bright ring, with the rest distributed among progressively fainter rings.

Figure 1.2b graphically presents a typical diffraction pattern. The central peak represents the bright central disk, and the smaller humps show the successively fainter rings.

The apparent diameter of the Airy disk plays a direct role in determining an instrument's resolving power. This becomes especially critical for observations of close-set double stars. How large an Airy disk will a given telescope produce? Table 1.2 summarizes the results for most common amateur-size telescopes.

Although these values would appear to indicate the resolving power of the given apertures, some telescopes can actually exceed these bounds. The nineteenth-century English astronomer William Dawes found experimentally that the closest a pair of 6th-magnitude yellow stars can be to each other and still be distinguishable as two points can be estimated by dividing 4.56 by the telescope's aperture. This is called *Dawes' Limit* (Figure 1.3). Table 1.3 lists Dawes' Limit for some common telescope sizes.

When using telescopes of less than 6-inch aperture, some amateurs can readily exceed Dawes' Limit, while others will never reach it. Does this mean that they are doomed to be failures as observers? Not at all! Remember that Dawes' Limit was developed under very precise conditions that may have been far different than your own. Just as with limiting magnitude, reaching Dawes' Limit can be adversely affected by many factors, such as turbulence in our

Table 1.2 **Resolving Power**

Telescope in.	Aperture mm	Diameter of Airy Disk (theoretical) Arc-seconds
2	51	2.7
3	76	1.8
4	102	1.40
6	152	0.91
8	203	0.68
10	254	0.55
12.5	318	0.44
14	356	0.39
16	406	0.34
18	457	0.30
20	508	0.27
24	610	0.23
30	762	0.18

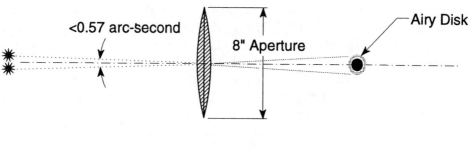

(a) Not Resolved

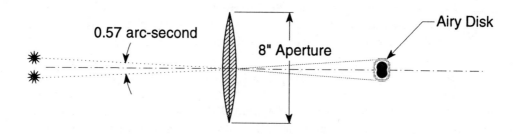

(b) Barely Resolved (Dawes Limit for an 8-inch)

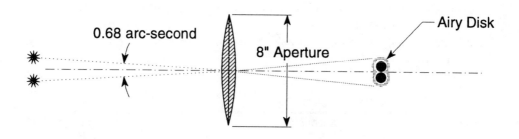

(c) Fully Resolved

Figure 1.3 *The resolving power of an 8-inch telescope: (a) not resolved, (b) barely resolved, and the Dawes' Limit for the aperture, (c) fully resolved.*

atmosphere, a great disparity in the test stars' colors and/or magnitudes, mis-aligned or poor-quality optics, and the observer's visual acuity.

Rarely will a large-aperture telescope—that is, one greater than about 10 inches—resolve to its Dawes' Limit. Even the largest backyard instruments can almost never show detail finer than between 0.5 arc-seconds (abbreviated 0.5″) and 1 arc-second (1″). In other words, a 16- to 18-inch telescope will offer little additional detail compared to an 8- to 10-inch one when used under most

Table 1.3   ***Dawes' Limit***

Telescope in.	Aperture mm	Limit of resolution Arc-seconds
2	51	2.3
3	76	1.5
4	102	1.1
6	152	0.76
8	203	0.57
10	254	0.46
12.5	318	0.36
14	356	0.33
16	406	0.29
18	457	0.25
20	508	0.23
24	610	0.19
30	762	0.15

observing conditions. Interpret Dawes' Limit as a telescope's equivalent to the projected gas mileage of an automobile: "These are test results only—your actual numbers may vary."

We have just begun to digest a few of the multitude of telescope terms that are out there. Others will be introduced in the succeeding chapters as they come along, but for now, the ones we have learned will provide enough of a foundation for us to begin our journey.

2

In the Beginning . . .

To appreciate the grandeur of the modern telescope, we must first understand its history and development. It is a rich history, indeed. Since its invention, the telescope has captured the curiosity and commanded the respect of princes and paupers, scientists and laypersons. Peering through a telescope renews the sense of wonder we all had as children. In short, it is a tool that sparks the imagination in us all.

Who is responsible for this marvelous creation? Ask this question of most people and they probably will answer, "Galileo." Galileo Galilei did, in fact, usher in the age of telescopic astronomy when he first turned his telescope, illustrated in Figure 2.1, toward the night sky. With it, he became the first person in human history to witness craters on the Moon, the phases of Venus, four of the moons orbiting Jupiter, and many other hitherto unknown heavenly sights. Though he was ridiculed by his contemporaries and persecuted for heresy, Galileo's observations changed humankind's view of the universe as no single individual's ever had before or has since. But he did not make the first telescope.

So who did? The truth is that no one knows for certain just who came up with the idea, or even when. Many knowledgeable historians tell us that it was Jan Lippershey, a spectacle maker from Middelburg, Holland. Records indicate that in 1608 he first held two lenses in line and noticed that they seemed to bring distant scenes closer. Subsequently, Lippershey sold many of his telescopes to his government, which recognized the military importance of such a tool. In fact, many of his instruments were sold in pairs, thus creating the first field glasses.

Other evidence may imply a much earlier origin for the telescope. Archaeologists have unearthed glass in Egypt that dates to about 3500 B.C., while primitive lenses have been found in Turkey and Crete that are thought to be

11

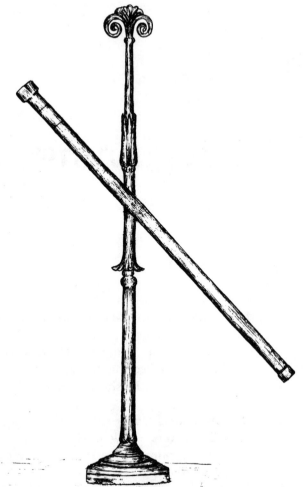

Figure 2.1 *Artist's rendition of Galileo's first telescope. Artwork by David Gallup.*

4,000 years old! In the third century B.C., Euclid wrote about the reflection and refraction of light. Four hundred years later, the Roman writer Seneca referred to the magnifying power of a glass sphere filled with water.

Although it is unknown if any of these independent works led to the creation of a telescope, the English scientist Roger Bacon wrote of an amazing observation made in the thirteenth century: ". . . Thus from an incredible distance we may read the smallest letters. . . . the Sun, Moon and stars may be made to descend hither in appearance . . ." Might he have been referring to the view through a telescope? We may never know.

Refracting Telescopes

Though its inventor may be lost to history, this early kind of telescope is called a *Galilean* or *simple* refractor. The Galilean refractor consists of two lenses: a convex (curved outward) lens held in front of a concave (curved inward) lens a certain distance away. As you know, the telescope's front lens is called the objective, while the other is referred to as the eyepiece, or *ocular*. The Galilean refractor placed the concave eyepiece *before* the objective's prime focus; this produced an upright, extremely narrow field of view, like today's inexpensive opera glasses.

Not long after Galileo made his first telescope, Johannes Kepler improved on the idea by simply swapping the concave eyepiece for a double convex lens, placing it behind the prime focus. The *Keplerian refractor* proved to be far superior to Galileo's instrument. The modern refracting telescope continues to be based on Kepler's design. The fact that the view is upside down is of little consequence to astronomers because there is no up and down in space; for terrestrial viewing, extra lenses may be added to flip the image a second time, reinverting the scene.

Unfortunately, both the Galilean and the Keplerian designs have several optical deficiencies. Chief among these is *chromatic aberration* (Figure 2.2). As you may know, when we look at any white-light source, we are not actually looking at a single wavelength of light but rather a collection of wavelengths mixed together. To prove this for yourself, shine sunlight through a prism. The light going in is refracted within the prism, exiting not as a unit but instead broken up, forming a rainbow-like spectrum. Each color of the spectrum has its own unique wavelength.

If you use a lens instead of a prism, each color will focus at a slightly different point. The net result is a zone of focus, rather than a point. Through such a telescope, everything appears blurry and surrounded by halos of color. This effect is called chromatic aberration.

Another problem of simple refractors is *spherical aberration* (Figure 2.3). In this instance, the curvature of the objective lens causes the rays of light entering around its edges to focus at a slightly different place than those striking the center. Once again, the light focuses within a range rather than at a single point, making the telescope incapable of producing a clear, razor-sharp image.

Modifying the inner and outer curves of the lens proved somewhat helpful. Experiments showed that both defects could be reduced (but not totally eliminated) by increasing the focal length—that is, decreasing the curvature—of

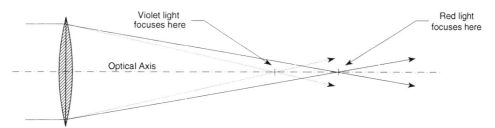

Figure 2.2 *Chromatic aberration, the result of a simple lens focusing different wavelengths of light at different distances.*

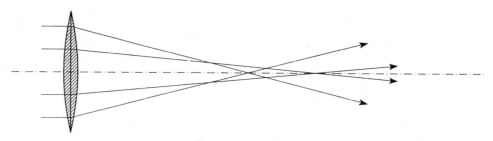

(a) Lens-induced spherical aberration

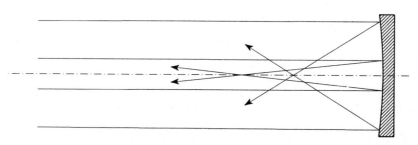

(b) Mirror-induced spherical aberration

Figure 2.3 *Spherical aberration. Both (a) lens-induced and (b) mirror-induced spherical aberration are caused by incorrectly figured optics.*

the objective lens. And so, in an effort to improve image quality, the refractor became longer . . . and longer . . . and even longer! The longest refractor on record was constructed by Johannes Hevelius in Denmark during the latter part of the seventeenth century; it measured about 150 feet from objective to eyepiece and required a complex sling system suspended high above the ground on a wooden mast to hold it in place! Can you imagine the effort it must have taken to swing around such a monster just to look at the Moon or a bright planet? Surely, there had to be a better way.

In an effort to combat these imperfections, Chester Hall developed a two-element *achromatic lens* in 1733. Hall learned that by using two matching lenses made of different types of glass, aberrations could be greatly reduced. In an achromatic lens, the outer element is usually made of crown glass, while the inner element is typically flint glass. Crown glass has a lower dispersion effect, and therefore bends light rays less than flint glass, which has a higher dispersion. The convergence of light passing through the crown-glass lens is compensated by its divergence through the flint-glass lens, resulting in greatly dampened aberrations. Ironically, though Hall made several telescopes using this arrangement, the idea of an achromatic objective did not catch on for another quarter century.

In 1758, John Dollond reacquainted the scientific community with Hall's idea when he was granted a patent for a two-element aberration-suppressing lens. Though quality glass was hard to come by for both of these pioneers, it appears that Dollond was more successful at producing a high-quality instrument. Perhaps that is why history records John Dollond, rather than Chester Hall, as the father of the modern refractor.

Regardless of who first devised it, this new and improved design has come to be called the *achromatic refractor* (Figure 2.4a), with the compound objective simply labelled an *achromat*. Though the methodology for improving the refractor was now known, the problem of getting high-quality glass (especially flint glass) persisted. In 1780, Pierre Louis Guinard, a Swiss bell maker, began experimenting with various casting techniques in an attempt to improve the glass-making process. It took him close to 20 years, but Guinard's efforts ultimately paid off, for he learned the secret of producing flawless optical disks as big as roughly six inches in diameter.

Later, Guinard was to team up with Joseph von Fraunhofer, inventor of the spectroscope. While studying under Guinard's guidance, Fraunhofer experimented by slightly modifying the lens curves suggested by Dollond, which resulted in the highest-quality objective yet created. In Fraunhofer's design, the front surface is strongly convex. The two central surfaces differ slightly from each other, requiring a narrow air space between the elements, while the innermost surface is almost perfectly flat. These innovations bring two wavelengths of light across the lens's full diameter to a common focus, thereby greatly reducing chromatic and spherical aberration.

The world's largest refractor is the 40-inch f/19 telescope at Yerkes Observatory in Williams Bay, Wisconsin. This mighty instrument was constructed by Alvan Clark and Sons, Inc., America's premier telescope maker of the nineteenth century. Other examples of the Clarks's exceptional skill include the 36-inch at Lick Observatory in California, the 26-inch at the U.S. Naval Observatory in Washington, D.C., and many smaller refractors at universities and colleges worldwide. Even today, Clark refractors are considered to be among the finest available.

The most advanced modern refractors offer features that the Clarks could not have imagined. *Apochromatic* refractors effectively eliminate just about all aberrations common to their Galilean, Keplerian, and achromatic cousins. More about these when we examine consumer considerations in Chapter 3.

Reflecting Telescopes

But there is more than one way to skin a cat. The second general type of telescope utilizes a large mirror, rather than a lens, to focus light to a point—not just any mirror, mind you, but a mirror with a precisely figured surface. To understand how a mirror-based telescope works, we must first reflect on how mirrors work (sorry about that). Take a look at a mirror in your home. Chances

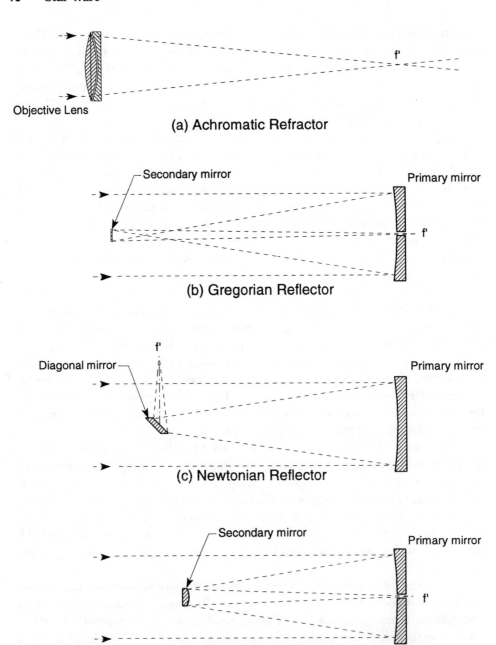

Objective Lens

(a) Achromatic Refractor

(b) Gregorian Reflector

(c) Newtonian Reflector

(d) Cassegrain Reflector

Figure 2.4 *Telescopes come in all different shapes and sizes: (a) achromatic refractor, (b) Gregorian reflector, (c) Newtonian reflector, (d) Cassegrain reflector, (e) Schmidt catadioptric telescope, (f) Maksutov-Cassegrain telescope, and (g) Schmidt-Cassegrain telescope.*

are it is flat, as shown in Figure 2.5a. Light that is cast onto the mirror's polished surface in parallel rays is reflected back in parallel rays. If the mirror is convex (Figure 2.5b), the light diverges after it strikes the surface. But if the mirror is concave (Figure 2.5c), then the rays converge toward a common point, or focus. (It should be pointed out here that household mirrors are *second-*

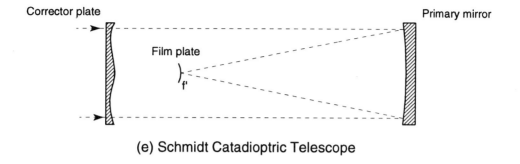

(e) Schmidt Catadioptric Telescope

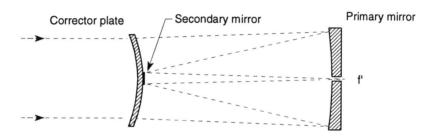

(f) Maksutov-Cassegrain Catadioptric Telescope

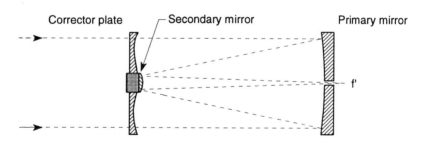

(g) Schmidt-Cassegrain Catadioptric Telescope

Figure 2.4 continued

surface mirrors; that is, their reflective coating is applied onto the back surface. Reflecting telescopes use *front-surface* mirrors, coated on the front.)

The first reflecting telescope was designed by James Gregory in 1663. His system centered around a concave mirror (called the *primary mirror*). The primary mirror reflected light to a smaller concave *secondary mirror*, which, in turn, bounced the light back through a central hole in the primary and out to the eyepiece. The *Gregorian reflector* (Figure 2.4b) had the benefit of yielding an upright image, but its optical curves proved difficult for Gregory and his contemporaries to fabricate.

A second design was later conceived by Sir Isaac Newton in 1672 (Figure 2.6). Like Gregory, Newton realized that a concave mirror would reflect and

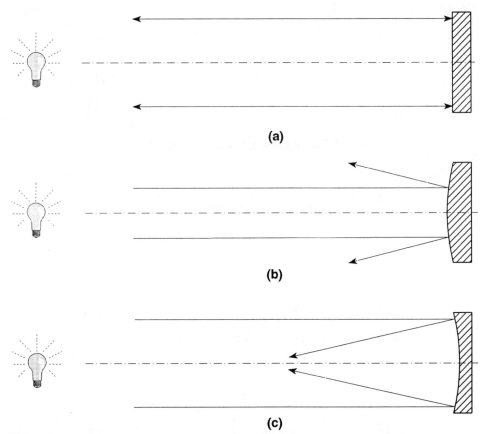

Figure 2.5 *Three mirrors, each with a different front-surface curve, reflect light differently. A flat mirror (a) reflects light straight back to the source, a convex mirror (b) causes light to diffuse, and a concave mirror (c) focuses light.*

focus light back along the optical axis to a point called the prime focus. Here an observer could view a magnified image through an eyepiece. Quickly realizing that his head got in the way, Newton inserted a flat mirror at a 45° angle some distance in front of the primary. The secondary, or *diagonal*, mirror acted to bounce the light at a 90° angle out through a hole in the side of the telescope's tube. This arrangement has since become known as the *Newtonian reflector* (Figure 2.4c).

That same year, the French sculptor Sieur Cassegrain announced a third variation of the reflecting telescope. His system is strongly reminiscent of Gregory's original design. The biggest difference between a *Cassegrain reflector* (Figure 2.4d) and a Gregorian reflector is the curve of the secondary mirror's surface. The Gregorian design uses a concave secondary mirror positioned outside the main focus, whereas the Cassegrain uses a convex secondary mirror inside the main focus.

Figure 2.6 *Newton's first reflecting telescope. From* Great Astronomers *by Sir Robert S. Ball, London, 1912.*

Both Newton and Cassegrain received acclaim for their independent inventions, but neither telescope saw further development for more than half a century. It seems that good mirrors were just too difficult to come by. One of the greatest difficulties to overcome was the lack of information on suitable materials for their construction. Newton, for instance, made his mirrors out of bell metal whitened with arsenic. Others chose speculum metal, an amalgam consisting of copper, tin, and arsenic.

Another complication faced by makers of early reflecting telescopes was generating accurately figured mirrors. In order for all the light striking its surface to focus to a point, a primary mirror's concave surface must be a parabola precisely shaped to within a few millionths of an inch—a fraction of the wavelength of light. Unfortunately, the first reflectors were outfitted with spherically figured mirrors. In this case, rays striking the mirror's edge come to a different focus than the rays striking its center. The net result: spherical aberration.

The first reflector to use a parabolic mirror was constructed by Englishman John Hadley in 1722. The primary mirror of his Newtonian measured about six inches across and had a focal length of 62⅝ inches. But whereas Newton and the others had failed to generate mirrors with accurate parabolic concave curves, Hadley succeeded. Extensive tests were performed on Hadley's reflector after he presented it to the Royal Society. In direct comparison between it and the society's 123-foot-focal- length refractor of the same diameter, the reflector performed equally well and was immeasurably simpler to use.

A second success story for the early reflecting telescope was that of James Short, another English craftsman. Short created several fine Newtonian and Gregorian instruments in his optical shop from the 1730s through the 1760s. He placed many of his telescopes on a special type of support that permitted easier tracking of sky objects (what is today termed an *equatorial mount*—see Chapter 3). Today, the popularity of the Gregorian reflector has long since faded away, though it is interesting to note that NASA chose that design for its highly successful Solar Max mission of 1980.

Sir William Herschel, a musician who became interested in astronomy when he was given a telescope in 1722, ground some of the finest mirrors of his day. As his interest in telescopes grew, Herschel continued to refine the reflector by devising his own system. The *Herschelian* design called for the primary mirror to be tilted slightly, thereby casting the reflection toward the front rim of the oversized tube, where the eyepiece would be mounted. The biggest advantage to this arrangement is that with no secondary mirror to block the incoming light, the telescope's aperture is unobstructed by a second mirror; disadvantages included image distortion due to the tilted optics and heat from the observer's head. Herschel's largest telescope was completed in 1789. The metal speculum around which it was based measured 48 inches across and had a focal length of 40 feet. Records indicate that it weighed something in excess of one ton.

Even this great instrument was to be eclipsed in 1845, when Lord Rosse completed the largest speculum ever made. It measured 72 inches in diameter and weighed in at an incredible 8,380 pounds. This telescope (Figure 2.7), mounted in Parsonstown, Ireland, is famous in the annals of astronomical history as the first to reveal spiral structure in what were then thought to be nebulae and are now known to be spiral galaxies.

The poor reflective qualities of speculum metal, coupled with its rapid tarnishing, made it imperative to develop a new mirror-making process. That evolutionary step was taken in the following decade. The first reflector to use a glass mirror instead of a metal speculum was constructed in 1856 by Dr. Karl Steinheil of Germany. The mirror, which measured four inches across, was coated with a very thin layer of silver; the procedure for chemically bonding silver to glass had been developed by Justus von Liebig about 1840. Though it apparently produced a very good image, Steinheil's attempt received very little attention from the scientific community. The following year, Jean Foucault (creator of the Foucault pendulum and the Foucault mirror test procedure, among others) independently developed a silvered mirror for his astronomical telescope. He brought his instrument before the French Academy of Sciences, which immediately made his findings known to all. Foucault's methods of working glass and testing the results elevated the reflector to new heights of excellence and availability.

Although silver-on-glass specula proved far superior to the earlier metal versions, this new development was still not without flaws. For one thing, silver

Figure 2.7   *Lord Rosse's 72-inch reflecting telescope. From* Elements of Descriptive Astronomy *by Herbert A. Howe, New York, 1897.*

tarnished quite rapidly, although not as fast as speculum metal. The twentieth century dawned with experiments aimed to remedy the situation, which ultimately led to the process used today of evaporating a thin film of aluminum onto glass in a vacuum chamber. Even though aluminum is not quite as highly reflective as silver, its longer useful lifespan more than makes up for that slight difference.

Although reflectors do not suffer from the refractor's chromatic aberration, they are anything but flawless. We have already seen how spherical aberration can destroy image integrity, but other problems must be dealt with as well. These include *coma,* which describes objects away from the center of view appearing like tiny comets, with their tails aimed outward from the center; *astigmatism,* resulting in star images that focus to crosses rather than points; and *light loss,* which is caused by obstruction by the secondary mirror and the fact that no reflective surface returns 100% of the light striking it.

Today, there exist many variations of the reflecting telescope's design. While the venerable Newtonian has remained popular among amateur astronomers, the Gregorian is all but forgotten. In addition to the classical Cassegrain, we find two modified versions: the Dall-Kirkham and the Ritchey-Chretien. The former employs simpler mirror curves than a true Cassegrain and is therefore favored by amateur telescope makers. The latter is the best of the three at correcting aberrations but is quite difficult to make. Finally, for the true student of the reflector, there are several lesser-known instruments, such as the tri-schiefspiegler (a three-mirror telescope with tilted optics).

Like the refractor, today's reflectors enjoy the benefit of advanced materials and optical coatings. Although they are a far cry from the first telescopes of Newton, Gregory, and Cassegrain, we must still pause a moment to consider how different our understanding of the universe might be if it were not for these and other early optical pioneers.

Catadioptric Telescopes

Earlier this century, some comparative newcomers launched a whole new breed of telescope: the catadioptric. These telescopes combine attributes of both refractors and reflectors into one instrument. They can produce wide fields with few aberrations. Many declare that this genre is (at least potentially) the perfect telescope; others see it as a collection of compromises.

The first catadioptric was devised in 1930 by German astronomer Bernhard Schmidt. The *Schmidt telescope* (Figure 2.4e) passes starlight through a corrector plate *before* it strikes the spherical primary mirror. The curves of the corrector plate eliminate the spherical aberration that would result if the mirror were used alone. One of the chief advantages of the Schmidt is its fast f-ratio, typically f/1.5 or less. However, due to the fast optics, the Schmidt's prime focus point is inaccessible to an eyepiece, restricting the instrument to photographic applications only. To photograph through a Schmidt, film is placed in a special curved holder (to accommodate a slightly curved focal plane) at the instrument's prime focus, not far in front of the main mirror.

The second type of catadioptric instrument to be developed was the *Maksutov telescope*. By rights, the Maksutov telescope should probably be called the Bouwers telescope, after A. Bouwers of Amsterdam, Holland. Bouwers developed the idea for a photovisual catadioptric telescope in February 1941. Eight months later, D. Maksutov, an optical scientist working independently in Moscow, came up with the exact same design. Like the Schmidt, the Maksutov combines features of both refractors and reflectors. The most distinctive trait of the Maksutov is its deep-dish front corrector plate, or *meniscus*, which is placed inside the spherical primary mirror's radius of curvature. Light passes through the corrector plate to the primary and then to a convex secondary mirror.

Most Maksutovs resemble a Cassegrain in design and are therefore referred to as *Maksutov-Cassegrains* (Figure 2.4f). In these, the secondary mirror returns the light toward the primary mirror, passing through a central hole and out to the eyepiece. This layout allows a long focal length to be crammed into the shortest tube possible.

In 1957, John Gregory, an optical engineer working for Perkin-Elmer Corporation in Connecticut, modified the original Maksutov-Cassegrain scheme to improve its overall performance. The main difference in the *Gregory-Maksutov* telescope is that instead of a separate secondary mirror, a small central spot on the interior of the corrector is aluminized to reflect light to the eyepiece.

Though not as common, a Maksutov telescope may also be constructed in a Newtonian configuration. In this scheme, the secondary mirror is tilted at 45°. As in the classical Newtonian reflector, light from the target then passes through a hole in the side of the telescope's tube to the waiting eyepiece. The greatest advantage of the Maksutov-Newtonian over the traditional Newtonian is the availability of a short focal length (and therefore a wide field of view) with greatly reduced coma and astigmatism.

Finally, two hybrids of the Schmidt camera have also been developed: the *Schmidt-Newtonian* and the *Schmidt-Cassegrain* (Figure 2.4g). The Newtonian hybrid remains mostly in the realm of the amateur telescope maker, but since its introduction in the 1960s, the Schmidt-Cassegrain has grown to become the most popular type of telescope sold today. It combines a short-focal-length spherical mirror with an elliptical-figured secondary mirror and a Schmidt-like corrector plate. The net result is a large-aperture telescope that fits into a comparatively small package. Is the Schmidt-Cassegrain the right telescope for you? Only you can answer that question—with a little help from the next chapter, that is.

The telescope has certainly come a long way in its nearly 400-year history, but that history is by no means finished. The age of orbiting observatories, such as the Hubble Space Telescope, has just dawned with untold possibilities. Back here on the ground, newly designed giant telescopes, like the Keck reflector in Hawaii, using segmented mirrors—and even some whose exact curves are controlled and varied by computers to compensate for atmospheric conditions (so-called *adaptive optics*)—are now being aimed toward the universe. New advanced materials, construction techniques, and accessories are coming into use. All this means that the future will see even more diversity in this already diverse field. Stay tuned!

3

So You Want to
Buy a Telescope

So you want to buy a telescope? That's wonderful! A telescope will let you visit places that most people are not even aware exist. With it, you can soar over the stark surface of the Moon, travel to the other worlds in our solar system, and plunge into the dark void of deep space to survey clusters of jewel-like stars, huge interstellar clouds, and remote galaxies. You will witness firsthand exciting celestial objects that were unknown to astronomers only a generation ago. You can become a citizen of the universe without ever leaving your backyard.

Just as a pilot needs the right aircraft to fly from one point to another, so too must an amateur astronomer have the right instrument to journey into the cosmos. As we have seen already, many different types of telescopes have been devised in the past four centuries. Some remain popular today, while others are of interest from a historical viewpoint only.

Which telescope is right for you? Had I written this book back in the 1950s or 1960s, there would have been one answer: a 6-inch f/8 Newtonian reflector. Just about every amateur either owned one or knew someone who did. Though many different companies made this type of instrument, the most popular one was the RV6 Dynascope by Criterion Manufacturing Company of Hartford, Connecticut, which for years retailed for $194.95. The RV6 was to telescopes what the Volkswagen Beetle was to cars—a triumph of simplicity and durability at a great price!

Times have changed, the world has grown more complicated, and the hobby of amateur astronomy has become more complex. The venerable RV6 is no longer manufactured, although some can still be found in classified advertisements. Today, looking through astronomical product literature, we find sophisticated Schmidt-Cassegrains, mammoth Newtonian reflectors, and state-of-the-art refractors. With such a variety from which to choose, it is hard to know where to begin.

Optical Quality

Before examining specific types of telescopes, a few terms used to rate the caliber of telescope lenses and mirrors must be defined and discussed. In the everyday world, when we want to express the accuracy of something, we usually write it in fractions of an inch, centimeter, or millimeter. For instance, when building a house, a carpenter might call for a piece of wood that is, say, 4 feet long plus or minus one-eighth of an inch. In other words, as long as the piece of wood is within an eighth of an inch of 4 feet, it is close enough to be used.

In the optical world, however, close is not always close enough. Because the curves of a lens or mirror must be made to such tight tolerances, it is not practical to refer to optical quality in everyday measurements. Instead, it is usually expressed in fractions of the wavelength of light. Each color in the spectrum has a different wavelength, so opticians use the color that the human eye is most sensitive to: yellow-green. Yellow-green, in the middle of the visible spectrum, has a wavelength of 550 nanometers (that's 0.00055 mm, or 0.00002 inch).

For a lens or a mirror to be accurate to, say, ⅛ wave (a value frequently quoted by telescope manufacturers), its surface shape cannot deviate from perfection by more than 0.000069 mm, or 0.000003 inch! This means that none of the little irregularities (commonly called *hills* and *valleys*) on the optical surface exceed a height or depth of ⅛ of the wavelength of yellow light. As you can see, the smaller the fraction, the better the optics. Given the same aperture and conditions, telescope A with a ⅛-wave prime optic (lens or mirror) should outperform telescope B with a ¼-wave lens or mirror, while both should be exceeded by telescope C with a ¹⁄₂₀-wave prime optic.

Stop right there. Companies are quick to boast about the quality of their primary mirrors and objective lenses, but in reality, we should be concerned with the *final wavefront* reaching the observer's eye, double the wave error of the prime optic alone. For instance, a reflecting telescope with a ⅛-wave mirror has a final wavefront of ¼ wave. This value is known as *Rayleigh's Criterion* and is usually considered the lowest quality level that will produce acceptable images. Clearly, an instrument with a ⅛ to ¹⁄₁₀ final wavefront is very good. However, even these figures must be taken loosely because there is no industrywide method of testing.

Due to increasing consumer dissatisfaction with the quality of commercial telescopes, both *Sky & Telescope* and *Astronomy* magazines have begun to purchase instruments for testing and evaluation, subsequently publishing the results. Talk about a shot heard around the world! Both organizations quickly found out that the claims made by some manufacturers (particularly a few producers of Newtonian reflectors and Schmidt-Cassegrains) were a bit, shall I say, inflated.

In light of this shake-up, many companies have dropped claims of their optics' wavefront, referring to them instead as being *diffraction limited*, meaning that the optics are so good that performance is limited only by the wave properties of light itself and not by any flaws in optical accuracy. In general, to be diffraction limited, an instrument's final wavefront must be at least ¼ wave, the Rayleigh Criterion. Once again, however, this can prove to be a subjective claim.

Telescope Point-Counterpoint

So which telescope would I recommend for you? None of them . . . or all of them! Actually, the answer is that there is no one answer anymore. It all depends on what you want to use the telescope for, how much money you can afford to spend, and many other considerations. To help sort all this out, you and I are about to go telescope hunting together. We will begin by looking at each type of telescope that is commercially available. The chapter's second section will examine the many different mounting systems used to hold a telescope in place. Finally, all considerations will be weighed together to let *you* decide which telescope is right for you.

Binoculars

What are binoculars doing in a book about telescopes? The fact of the matter is that every amateur astronomer should own a pair of good quality binoculars regardless of his or her other telescopic equipment. In fact, if you are limited in budget or are just starting out in the hobby, do NOT even consider buying an inexpensive telescope. Spend your money wisely by purchasing a good pair of binoculars plus a star atlas and a few of the books listed in Chapter 6.

Binoculars (Figure 3.1) may be thought of as two refracting telescopes strapped together. Light from a target enters a pair of objective lenses, bounces through two identical sets of prisms, and exits through the eyepieces. Modern binoculars are available in two basic styles depending on the type of prisms used: *roof-prism* and *porro-prism*. Which should you consider? All other things being equal, porro-prism binoculars will yield brighter, sharper images than roof-prism glasses. Although the difference may not be noticeable in terrestrial applications, the effect can become quite pronounced when viewing the night sky.

All binoculars are labeled with two numbers, such as 7×35 or 10×50. The first refers to the pair's magnification, while the second specifies the diameter (in millimeters) of the two front lenses. Typically, values range from 7 power ($7\times$) to 20 power ($20\times$), with objectives measuring between 35 mm (1.5 inches) and 125 mm (5 inches).

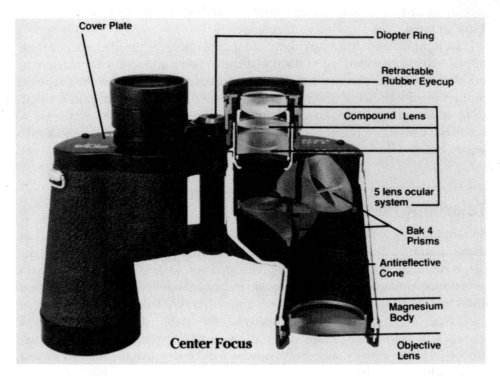

Figure 3.1 *Cross-sectional view of porro-prism binoculars. Photo courtesy of Swift, Inc.*

To be perfectly matched for wide-field nighttime skywatching, the diameter of the beam of light leaving the binoculars' eyepieces (the *exit pupil*) should match the clear diameter of the observer's pupils. Although the diameter of everybody's pupil can vary (especially with age), values typically range from about 2.5 mm in the brightest lighting conditions to about 7 mm under the dimmest conditions. If the binoculars' exit pupil is much smaller, then we lose the binocular's *rich-field* viewing capability, while too large an exit pupil will waste some of the light the binoculars collect. (For a complete discussion on what exactly an exit pupil is, fast forward to Chapter 5. Go ahead, I'll wait.)

To find the exit pupil for any pair of binoculars, simply divide the size of the aperture by the magnification. If you are young and plan on doing most of your observing from a rural setting, then you would do best with a pair of binoculars that yield a 7 mm exit pupil (such as 7×50 or 10×70). However, if you are older or are a captive of a light-polluted city or suburb, then you may do better with binoculars yielding a 4 mm or 5 mm exit pupil (7×35, 10×50, and so on).

When shopping for a pair of astronomical binoculars, look at several brands and models side by side if possible. Does the manufacturer state that the lenses are coated? Optical coatings improve light transmission and reduce scattering. An uncoated lens reflects about 4% of the light hitting it. By applying a thin layer of magnesium fluoride onto both surfaces of the lens, reflection is reduced to 1.5%. Top-of-the-line binoculars receive multiple antireflection coatings, reducing reflection to less than 0.5%. Be sure to insist on fully coated, or better yet, fully multicoated optics.

Another feature to look for is the type of glass used to make the porro prisms. Better binoculars use prisms made from BaK-4 (barium crown) glass, while less-expensive binoculars utilize prisms of BK-7 (borosilicate). BaK-4 glass yields slightly brighter, sharper images because it passes practically all of the light that enters (what optical experts call *total internal reflection*). BK-7 prisms do not have total internal reflection, causing light falloff and, consequently, somewhat dimmer images.

Most manufacturers will state "BaK-4 prisms" right on the binoculars, but if not, you can always check for yourself. Hold the binoculars at arm's length and look at the circle of light floating, as it were, behind the eyepieces. This is the previously mentioned exit pupil, which will appear perfectly circular if the prisms are made from BaK-4 glass but somewhat diamond-shaped with gray-edged shadows (because of the light falloff) with BK-7 prisms.

The ultimate in portability, binoculars offer unparalleled views of rich Milky Way starfields thanks to their low power and wide fields of view. As much as this is an advantage to the deep-sky observer, it is a serious drawback to those interested in looking for fine detail on the planets, where higher powers are required. In these cases, the hobbyist has no choice but to purchase a telescope.

Refracting Telescopes

After many years of being all but ignored by the amateur community, the astronomical refractor (Figure 3.2) is making a strong comeback. Hobbyists are rediscovering the exquisite images seen through well-made refractors. Crisp views of the Moon, razor-sharp planetary vistas, and pinpoint stars are all possible through the refracting telescope.

Achromatic refractors. As mentioned in Chapter 2, many refractors of yesteryear were plagued with a wide and varied assortment of aberrations and image imperfections. The most difficult of these faults to correct are chromatic aberration and spherical aberration.

Achromatic objective lenses, in which a convex crown lens is paired with a concave flint element, go a long way in suppressing chromatic aberration. Indeed, at f/15 or greater, chromatic aberration is effectively eliminated. Even at focal ratios down to f/10, chromatic aberration is frequently not too offensive if the objective elements are *well made*. High-quality achromatic refractors sold today range in size from 2.4 inch (6 cm) up to 6 inch (15 cm). Even though chromatic aberration can be dealt with effectively, a lingering bluish or purplish glow will frequently be seen around brighter stars and planets. This glow is known as *secondary spectrum* and is almost always present in achromatic refractors.

Figure 3.2   *The 4-inch Tele Vue Genesis-SDF apochromatic refractor, one of the finest refractors for the amateur astronomer. Photo courtesy of Tele Vue, Inc.*

Another point in favor of the refractor is that its aperture is *clear*—that is, nothing blocks any part of the light as it travels from the objective to the eyepiece. As you can tell from looking at the diagrams in Chapter 2, this is not the case for reflector and catadioptric instruments. As soon as a secondary mirror interferes with the path of the light, some loss of contrast and image degradation are inevitable.

In addition to sharp images, the achromatic refractor is also famous for its portability and ruggedness. If constructed properly, a refractor should deliver years of service without its optics needing to be realigned (recollimated). The sealed-tube design means that dust and dirt are prevented from infiltrating the optical system, and contaminants can be kept off the objective's exterior simply by using a lens cap.

On the minus side of the achromatic refractor is its small aperture. Although this is of less concern to lunar, solar, and planetary observers, the instrument's small light-gathering area means that faint objects such as nebulae and galaxies will appear dimmer than larger-but-cheaper reflectors. In addition, the long tubes of achromatic refractors can make them difficult to store and transport to dark, rural skies.

Another problem common to refractors is their inability to provide comfortable viewing angles at all elevations above the horizon. The long tube and short tripod typically provided can work against the observer in some cases. For instance, if the mounting is set at the proper height to view near the zenith, the eyepiece will swing high off the ground as soon as the telescope is aimed toward the horizon. This disadvantage can be partially offset by using a star diagonal between the telescope's drawtube and eyepiece, but doing so has disadvantages of its own. Most star diagonals use either a prism or a flat mirror to bounce the light at a right angle, flipping the field left to right and making it difficult to compare the view with star charts.

Apochromatic refractors. For the true connoisseur who will settle for nothing but the best, there are apochromatic refractors. While an achromat brings two wavelengths of light at opposite ends of the spectrum to a common focus, it still leaves a secondary spectrum along the optical axis. Though not as distracting as chromatic aberration from a single lens, secondary spectrum can still contaminate critical viewing and photography.

Apos, as they are affectionately known to many owners, greatly reduce chromatic aberration and secondary spectrum, allowing manufacturers to increase aperture and decrease focal length. First popularized in the 1980s, apochromatic refractors use either two-, three-, or four-element objective lenses with one or more elements of an unusual glass type—often fluorite or ED (short for *extra-low dispersion*) glass. All apochromats minimize dispersion of light by bringing all wavelengths to just about the same focus, reducing chromatic aberration and secondary spectrum dramatically and thereby permitting shorter, more manageable focal lengths.

Much has been written about the pros and cons of fluorite (monocrystalline calcium fluoride) lenses. The most popular myth is that they do not stand the test of time. Some so-called experts claim that fluorite absorbs moisture and/or fractures more easily than other types of glass. This is simply not true. Fluorite objectives work very well and are just as durable as conventional lenses. Like all lenses, they will last a lifetime if given a little care. (Besides, fluorite is not normally used as an outer element.)

Although durability is not a problem, there are a couple of hitches to fluorite refractors. One problem that is not popularly known is fluorite's high thermal expansion. This means that the fluorite element will require relatively more time to adjust to ambient temperature; telescope optics change shape slightly as they cool or warm, and this tendency is more pronounced in fluorite than in other materials.

Other hindrances are shared by all apos. Like most commercially sold achromatic refractors, apochromatic refractors are limited to smaller apertures, usually somewhere in the range from 3 to 7 inches or so. This is not because of unleashed aberrations at larger apertures; it's simply a question of economics, which brings us to their second (and biggest) stumbling block: Apochromats are not cheap! When we compare dollars per inch of aperture, it soon becomes apparent that apochromatic refractors are *the* most expensive telescopes. Given the same type of mounting and accessories, an apochromatic refractor can retail for more than twice the price of a comparable achromat. That's a big difference, but the difference in image quality can be even bigger. For a first telescope, an achromatic refractor is just fine, but if this going to be your ultimate dream telescope, then you ought to consider an apochromat.

Reflecting Telescopes

Reflectors offer an alternative to the small apertures and big prices of refractors. Let's compare. Each of the two or more elements in a refractor's objective lens must be accurately figured and made of high-quality, homogeneous glass. By contrast, the single optical surfaces of a reflector's primary and secondary mirrors favor construction of large apertures at comparatively modest prices.

Another big advantage that reflectors enjoy over refractors is complete freedom from chromatic aberration, which is a property of light refraction but not reflection. This means that only the true colors of whatever a reflecting telescope is aiming at will come shining through. Of course, the eyepieces used to magnify the image for our eyes use lenses, so we are not completely out of the woods.

These two important pluses are frequently enough to sway amateurs in favor of a reflector. They feel that although there are drawbacks to the designs, these are outweighed by the many strong points. But just what are the problems with reflecting telescopes? Some are peculiar to certain breeds, while others affect them all.

One problem common to all telescopes of this genre is the simple fact that mirrors do not reflect all the light that strikes them. Just how much light is lost depends on the kind of reflective coating used. For instance, most telescope mirrors are coated with a thin layer of aluminum and overcoated with a clear layer of silicon monoxide for added protection against scratches and pitting. This combination reflects about 89% of visible light. But consider this: Given primary and secondary mirrors with standard aluminum coatings, the combined reflectivity is only 79% of the light striking the primary! That's why special enhanced coatings have become so popular in recent years. Enhanced coatings increase overall system reflectivity to about 90%—a noticeable improvement.

Reflectors also lose some light and image contrast because of obstruction by the secondary mirror. The degree of blockage depends on the size of the secondary, which in turn depends on the focal length of the primary mirror. Generally speaking, the shorter the focal length of the primary, the larger its secondary must be to bounce all of the light toward the eyepiece. For primaries with very short focal lengths, this value can exceed 10% of the primary's total light-gathering area. The only reflectors that do not suffer from this ailment are Herschelians and members of the Schiefspeigler family of instruments. Nevertheless, their availability is severely limited because of economics and practicality.

Since the idea of a telescope that uses mirrors to focus light was first conceived in 1663, different schemes have come and gone. Today, two designs continue to stand the test of time: the Newtonian reflector and the Cassegrain reflector. Each shall be examined separately.

Newtonian reflectors. For sheer brute-force light-gathering ability, Newtonian reflectors rate as a best buy. No other type of telescope will give you as large an aperture for the money. Given a similar style mounting, we could buy an 8-inch Newtonian reflector for the same amount of money needed for a 4-inch achromatic refractor.

Newtonians (Figure 3.3) are famous for their panoramic views of star fields, making them especially attractive to deep-sky fans, but they also can be

Figure 3.3 *The Meade Starfinder 8-inch f/6 Newtonian reflector, an excellent telescope for those just starting out in the hobby. Photo courtesy of Meade Instruments Corporation.*

equally adept at moderate-to-high–powered glimpses of the Moon and planets. These highly versatile instruments come in a wide variety of styles. Commercial models range from 3 inches to more than 2 *feet* in diameter, with focal ratios stretching between f/3.5 and about f/10. Of course, not all apertures are available at all focal ratios. Can you imagine climbing more than 20 feet to the eyepiece of a 24-inch f/10?

For the sake of discussion, I have divided Newtonians into two groups based on focal ratio. Those of f/6 and less have been broadly classified as *rich-field* telescopes, or RFTs for short. Newtonian reflectors with focal lengths greater than f/6 here will be called *normal-field* telescopes, or simply NFTs.

Let's examine normal-field telescopes first. Pardon my bias, but NFTs have always been my favorite type of telescope. They are capable of delivering clear views of the Moon, Sun, and other members of the solar system as well as thousands of deep-sky objects. NFTs with apertures between 3 inches (8 cm) and 8 inches (20 cm) are usually small enough to be moved from home to observing site and set up quickly with little trouble. Once the viewing starts, most amateurs happily find that looking through both the eyepiece and small finderscope is effortless because the telescope's height closely matches their eye level.

A 6-inch f/8 Newtonian is still one of the best all-around telescopes for those new to astronomy. It is compact enough so as not to be a burden to transport and assemble, yet it is large enough to provide years of fascination and is reasonably priced—maybe not $194.95 like the old RV6, but it's still not a bad deal. Better yet is an 8-inch f/7 to f/9 Newtonian. The increased aperture permits even finer views of nighttime targets. Keep in mind, however, that as aperture grows, so grows a telescope's size and weight. Unless you live in the country and can store your telescope where it is easily accessible, an NFT larger than an 8-inch might be difficult to manage.

Most experienced visual observers agree that NFT Newtonian reflectors are tough to beat. In fact, an optimized Newtonian reflector can deliver views of the Moon and planets that eclipse those possible through a Schmidt-Cassegrain telescope and compare favorably with a refractor of similar size, but at a fraction of the refractor's cost. Though the commercial telescope market now offers a wide range of superb refractors, it has yet to embrace the long-focus reflector fully. Why? NFT Newtonians were quite popular back in the 1950s and 1960s; us old-timers still remember 12.5-inch f/8s.

The dawn of the 1970s saw both reflectors and refractors taking a back seat to the incredibly popular Schmidt-Cassegrain telescope. The reason was simple. Long-focus reflectors and refractors can be much more difficult to store and transport, so manufacturers, fearing the loss of customers, dropped them from their lines. But this is the 1990s, and with a widespread renewal of interest in observing and photographing the planets, amateurs are rediscovering the virtues of these fine instruments. Several companies have recently reintroduced 6- and 8-inch Newtonians in the f/7 to f/8 range that are sure to be a hit among amateurs.

While NFTs provide fine views of the planets and deep-sky objects up to about 0.5° across, rich-field telescopes (RFTs) return outstanding vistas of extended deep-sky objects such as widespread open clusters and diffuse nebulae. When combined with one of the wide-field eyepieces described in Chapter 5, these instruments give panoramic views of Milky Way starfields that are beyond written description. A few companies sell small-aperture RFT Newtonians, but most sold today are 10 inches across and larger (in some cases, much larger).

Most large RFTs use thin-section primary mirrors of short focal length. Traditionally, primary mirrors have a diameter-to-thickness ratio of 6 : 1. This means that a 12-inch mirror measures a full 2 inches thick. That is one heavy piece of glass to support. Thin-section mirrors cut this ratio to 12 : 1 or 13 : 1, slashing the weight by 50%. This sounds good at first, but practice shows that large, thin mirrors tend to sag under their own weight (thicker mirrors are more rigid), thereby distorting the parabolic curve, when held in a conventional three-point mirror cell. To prevent mirror sag, a new support system was devised to support the primary at nine (or more) evenly spaced points across its back surface. These cells are frequently called *mirror flotation systems*, as they do not clamp down around the mirror's rim, thereby preventing possible edge distortions by pinching.

If big apertures mean bright images, why not buy the biggest aperture available? Actually, there are several reasons not to. For one thing, unless they are made very well, Newtonians (especially those with short focal lengths) are susceptible to a number of optical aberrations, including *spherical aberration* and *astigmatism*. Spherical aberration results when light rays near the edge of an improperly made mirror (or lens) focus to a slightly different point than those from the optic's center. Astigmatism is due to a mirror (or lens, once again) that was not symmetrically ground around its center. The result: elongated star images that appear to flip their orientation by 90° when the eyepiece is brought from one side of the focus point to the other. Coma, especially apparent in short-focal-length RFTs, is evident when stars near the edge of the field of view distort into tiny blobs resembling comets, while stars at the center appear as sharp points. With any or all of these present, resolution suffers greatly. (Note that coma can be, for all purposes, eliminated using a *coma corrector*—see the discussion in Chapter 5.)

Furthermore, if you observe from a light-polluted area, large apertures will likely produce results inferior to instruments with smaller apertures. While they gather more starlight, larger mirrors also gather more sky glow, washing out the field of view. In these cases, you probably will do best by sticking with a telescope no larger than 8 to 10 inches in aperture.

Both NFT and RFT Newtonians share many other pitfalls as well. One of the more troublesome is that of all the different types of telescopes, Newtonians are among the most susceptible to collimation problems. If either or both of the mirrors are not aligned correctly, image quality will suffer greatly, possibly to the point of making the telescope worthless. Sadly, many commercial reflectors are delivered with misaligned mirrors. The new owner, perhaps not knowing better, immediately condemns his or her telescope's poor performance as a case of bad optics. In reality, however, the optics may be fine, just a little out of alignment. Chapter 8 details how to examine and adjust a telescope's collimation, a procedure that should be repeated frequently. The need for precise alignment grows more critical as the primary's focal ratio shrinks, making it especially important to double-check collimation at the start of every observing session if your telescope is f/6 or less.

There are cases where no matter how well aligned the optics are, image quality is still lacking. If this is the case, then the fault undoubtedly lies with one or both of the mirrors themselves. As the saying goes, you get what you pay for, and that is as true with telescopes as with anything else. Clearly, manufacturers of low-cost models must cut their expenses somewhere in order to underbid their competition. These cuts are usually found in the nominal-quality standard equipment supplied with the instrument but may also sometimes affect optical testing procedures and quality control.

Cassegrain reflectors. Though they have never attained the widespread following among amateur astronomers that Newtonians continue to enjoy, Cassegrain reflectors (Figure 3.4) have always been considered highly competent instruments. Cassegrains are characterized by long focal lengths, making them ideally suited for high-power, high-resolution applications such as solar, lunar, and planetary studies. While Newtonians also may be constructed with these focal ratios, observers would have to go to great lengths to reach their eyepieces! This is not the case with the Cassegrain, where the eyepiece is conveniently located along the optical axis behind the backside of the primary mirror.

The Cassegrain's long focal length is created not by the primary mirror (which typically ranges around f/4) but rather by the convex, hyperbolic secondary mirror. As it reflects the light from the primary back toward the eyepiece, the convex secondary actually magnifies the image, thereby stretching the telescope's effective focal ratio to between f/10 and f/15. The net result is a telescope that is much more compact and easier to manage than a Newtonian of equivalent aperture and focal length.

Unfortunately, while the convex secondary mirror gives the Cassegrain its great compactness, it also contributes to many of the telescope's biggest disadvantages. First, in order to reflect all the light from the primary back toward the eyepiece, the secondary mirror must be placed quite close to the primary. This requires its diameter to be quite large, noticeably larger than the flat diagonal of a Newtonian. With the secondary blocking more light, image brightness, clarity, and contrast all suffer. Secondly, the convex secondary combined with the short-focus primary mirror make alignment critical to the Cassegrain's proper function and at the same time cause the telescope to be more difficult to collimate than a similar Newtonian. Finally, Cassegrains are prone to coma just like RFT Newtonians, making it impossible to achieve sharp focus around the edge of the field of view.

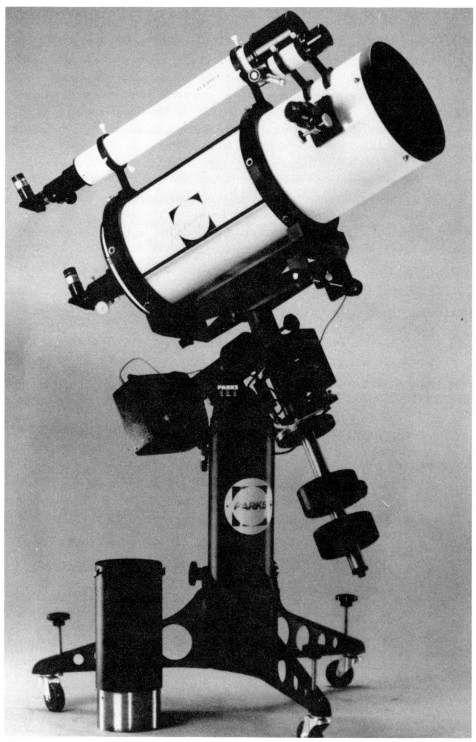

Figure 3.4 *The Parks H.I.T. telescope, a fine Cassegrain-Newtonian reflector. Photo courtesy of Parks Optical.*

The advantage of the eyepiece's placement along the optical axis can also work against the instrument's performance. The most obvious objection will become painfully apparent the first time an observer aims a Cassegrain near the zenith and tries to look through the eyepiece. That can be a real pain in the neck, although the use of a star diagonal will help alleviate the problem. Another problem that may not be quite as apparent involves a very localized case of light pollution, caused by extraneous light passing around the secondary and flooding the field of view. To combat this, manufacturers invariably install a long baffle tube protruding in front of the primary. The size of the baffle is critical, as it must shield the eyepiece field from all sources of incidental light while allowing the full intensity of the target to shine through.

Though Cassegrains remain the most common type of telescope in professional observatories, their popularity among today's amateur astronomers is low. So it should come as no surprise to find that so few companies offer complete Cassegrain systems for the hobbyist.

Catadioptric Telescopes

Most amateur astronomers who desire a compact telescope now favor hybrid designs that combine some of the best attributes of the reflector with some from the refractor, creating a completely different kind of beast: the catadioptric. Catadioptric telescopes (also known as *compound telescopes*) are comparative johnny-come-latelies on the amateur scene. Yet in only a few decades, they have developed a loyal following of backyard astronomers who staunchly defend them as the ultimate telescopes.

Most lovers of catadioptrics fall into one, two, or possibly all three of the following categories:

1. They are urban or suburban astronomers who prefer to travel to remote observing sites.
2. They enjoy astrophotography (or aspire to at least try it).
3. They just love gadgets.

If any or all of these profiles fit you, then a catadioptric telescope might just be the one for you.

Catadioptric telescopes for visual use may be constructed in either Newtonian or Cassegrain configurations. Only two catadioptrics have made lasting impacts on the world of amateur astronomy: the Schmidt-Cassegrain and the Maksutov-Cassegrain. For our purposes here, the discussion will be confined to these two designs.

Schmidt-Cassegrain telescopes. Take a look through practically any astronomy magazine published just about anywhere in the world and you are bound to find at least one advertisement for a Schmidt-Cassegrain telescope (also known as a *Schmidt-Cas* or an *SCT*). As your eyes digest the ads chock-full of mouth-watering celestial photographs that have been taken through these instruments, you suddenly get the irresistible urge to run right out and buy one. Don't worry—you would not be the first to find these telescopes so appealing. In the last few decades, sales of Schmidt-Cassegrains have outpaced both refractors and reflectors to become the most popular serious telescope among amateur astronomers. Though SCTs are available in apertures from 4 to 14 inches, the favorite size of all is the 8-inch model.

Is the Schmidt-Cassegrain (Figure 3.5) the perfect telescope? Admittedly, it can be attractive. By far, its greatest asset has to be the compact design. No other telescope can fit as large an aperture and as long a focal length into such a short tube assembly as a Schmidt-Cas; they are usually only about twice as

Figure 3.5 *The Meade 8-inch LX200 Schmidt-Cassegrain telescope, one of the most sophisticated instruments on the market today. Photo courtesy of Meade Instruments Corporation.*

long as the aperture. If storing and transporting the telescope are major concerns for you, then this will be an especially important benefit.

Here is another point in their favor. Nothing can end an observing session quicker than a fatigued observer. For instance, owning a Newtonian reflector, its eyepiece positioned at the front end of the tube, usually means having to remain standing—sometimes even on a stool or a ladder—just to take a peek. Compare this to a Schmidt-Cassegrain telescope, which allows the observer to enjoy comfortable, seated viewing of just about all points in the sky. Your back and legs will certainly thank you! The eyepiece is difficult to reach only when the telescope is aimed close to the zenith. As with a refractor and Cassegrain, a right-angle star diagonal placed between the telescope and eyepiece will help a little, but these have their drawbacks, too. Most annoying of all is that a diagonal will flip everything right-to-left, creating a mirror image that makes the view difficult to compare with star charts.

All commercially made Schmidt-Cassegrain telescopes look pretty much the same *at a quick glance*, but then again, so do many products to the uninitiated. Only after closer scrutiny will the features unique to individual models come shining through. Standard-equipment levels vary greatly, as reflected in the wide price range of SCTs. Some basic models come with an undersized finderscope, one eyepiece, maybe a couple of other bare-bones accessories, and some cardboard boxes for storage, whereas top-of-the-line instruments are supplied with foamed-lined footlockers, advanced eyepieces, large finders, and a multitude of electronic gadgets. (As I mentioned before, if you love widgets and whatchamacallits, then the Schmidt-Cassegrain will certainly appeal to you.) Most amateurs can find happiness with a model somewhere between these two extremes.

Another big plus of the Schmidt-Cassegrain is its sealed tube. The front corrector plate acts as a shield to keep dirt, dust, and other foreign contaminants off the primary and secondary mirrors. This is especially handy if you travel a lot with your telescope and are constantly taking it in and out of its carrying case. A sealed tube can also help extend the useful life of the mirrors' aluminized coatings by sealing well against the elements. (Always make sure the mirrors are dry before storing the telescope to prevent the onset of mold and mildew.)

While the corrector seals the two mirrors against dust contamination, it also can act as a dew collector when you are observing. Depending on local weather conditions, correctors can fog over in a matter of hours or even minutes, or they may remain clear all night. To help fight the onslaught of dew, manufacturers sell *dew caps* or *dew shields*. Dew caps are a must-have accessory for all Cassegrain-based catadioptrics. Consult Chapter 6 for more information, or see Chapter 7 for hints on how to build your own.

Many of the accessories for SCTs revolve around astrophotography, an activity enjoyed by many amateur astronomers. Here again, the SCT pulls ahead of the crowd. Because of their comparatively short, lightweight tubes, Schmidt-Cassegrains permit easy tracking of the night sky. Just about all are held on fork-style equatorial mounts complete with motorized clock drives. Once the equatorial mount is properly aligned to the celestial pole (a tedious activity at times—see Chapter 9), you can turn on the drive motor, and the telescope will track the stars by compensating for Earth's rotation. With various accessories (many of which are intended to be used only with SCTs), the amateur is now ready to photograph the universe.

What about optical performance? Here is where the Schmidt-Cassegrain telescope begins to teeter. Due to the comparatively large secondary mirrors required to reflect light back toward their eyepieces, SCTs produce images that are fainter and with less contrast than other telescope designs of the same aperture size. This can prove especially critical when searching for fine planetary detail or hunting for faint deep-sky objects at the threshold of visibility. One way to help the situation is to use enhanced optical coatings. As mentioned earlier in this chapter, these coatings improve light transmission and reduce scattering. They can make the difference between seeing a marginally visible object and missing it, and they are an absolute must for all Schmidt-Cassegrains.

Most 8-inch SCTs operate at f/10, while a few work at f/6.3. What's the difference? On the outside, they both look the same, the only difference being in the secondary mirrors. Are there pluses to using one over the other? Yes and no. If the telescopes are used visually (that is, if you are just going to look through them), then there should be negligible difference between the performance of an f/10 telescope and an f/6.3 telescope *when operated at the same magnification*. Image brightness is controlled by clear aperture, not by f-ratio.

The faster focal ratio may actually work against the observer. To achieve an f/6.3 instrument, a larger secondary mirror is required (3.5 inches across, compared to between 2.75 inches to 3 inches across in an f/10 instrument). The larger central obstruction in an f/6.3 SCT causes a decrease in contrast, making them less useful for planetary observation than their f/10 brethren. If you really want to split hairs, there is also a slight difference in image brightness. As Dennis di Cicco pointed out in his extensive review of SCTs in the December 1989 issue of *Sky & Telescope* magazine (p. 582), "When used at the same magnification, there is no question that images in the f/6.3 . . . will be *fainter* than those in . . . f/10 scopes. Why? Because the larger central obstruction in the [f/6.3 scope] actually reduces the light-gathering power. . . ." But di Cicco quickly adds that the difference is negligible—only about 5%. (Of course, in an 8-inch f/7 Newtonian reflector, the secondary blocks only 1.5 inches of the full aperture, which helps to explain their superior performance.)

The biggest advantage to using an f/6.3 SCT is enjoyed by astrophotographers. When set up for prime-focus photography, with the camera body coupled directly to the eyepiece-less telescope, exposure time can be cut by a factor of 2.5 to get the same image brightness as an f/10. Of course, image size is going to be reduced at the same time, but for many deep-sky objects, this is usually not a problem. (See also the discussion in Chapter 5 about focal-length reducers for SCTs.)

Image sharpness in a Schmidt-Cassegrain is not as precise as through a refractor or a reflector. Perhaps this is due to the loss of contrast mentioned above or because of optical misalignment, another problem of the Schmidt-Cassegrain. In any telescope, optical misalignment will play havoc with image quality. What should you do if the optics of a Schmidt-Cassegrain are out of alignment? If only the secondary is off, then you may follow the procedure outlined in Chapter 8, but if the primary is out, then manufacturers suggest that the telescope be returned to the factory. That is good advice. Remember—although just about anyone can take a telescope apart, not everyone can put it back together!

Finally, aiming an SCT can sometimes prove to be a frustrating experience. This is not the fault of the telescope but is due instead to the low position of the *finderscope*, a small auxiliary telescope mounted sidesaddle and used to aim the main instrument. Traditionally, most SCTs are supplied with right-angle finders. Although these greatly reduce back fatigue, they introduce a whole cauldron of problems of their own. See Chapter 6 for more about right-angle finders and why you shouldn't use them.

In general, Schmidt-Cassegrain telescopes represent good values for the money. They offer acceptable views of the Sun, Moon, planets, and deep-sky objects and work reasonably well for astrophotography. But for exacting views of celestial objects, SCTs are outperformed by other types of telescopes. For observations of solar system members, it is hard to beat an NFT Newtonian (especially f/10 or higher) or an apochromatic refractor, while the myriad faint deep-sky objects are best seen with large-aperture Newtonians. You might think of Schmidt-Cassegrain telescopes as jack-of-all-trades-but-master-of-none telescopes.

Maksutov-Cassegrain telescopes. The final stop on our telescope world tour is the Maksutov-Cassegrain catadioptric. Many people feel that Maksutovs are the finest telescopes of all. And why not? They offer all the advantages inherent in the Cassegrain and Schmidt-Cassegrain in an even smaller parcel while effectively eliminating coma and other aberrations. Maks, as they are called by some, provide views of the Moon, Sun, and planets that rival those possible with the best refractors and long-focus reflectors, and they are easily adaptable for astrophotography (though their high focal ratios mean longer exposures than needed with other telescopes of similar aperture). And traveling with them is a breeze.

Is there a downside to the Maksutov? Unfortunately, yes—a big one. Unlike Schmidt-Cassegrains, for which manufacturers have developed methods of mass-producing optical components of consistent quality while holding prices down, Maksutovs require precise handcrafting. In other words, they cost a lot. A second restriction of Maksutovs is aperture—or lack thereof. Even the smallest Maks cost more per inch of aperture than nearly any other type of telescope.

To help digest all this, take a look at Table 3.1, which summarizes all the pros and cons mentioned above. Use it to compare the good points and the bad between the more popular types of telescopes sold today.

Support Your Local Telescope

The telescope itself is only half of the story. Can you imagine trying to hold a telescope *by hand* while trying to look through it? If the instrument's weight did not get you first, surely every little shake would be magnified into a visual earthquake! To use a true astronomical telescope, we have no choice but to support it on some kind of external mounting. For small spotting scopes, this might be a simple tabletop tripod, whereas the most elaborate telescopes come equipped with equally elaborate support systems.

Selecting the proper mount is *just as important* as picking the right telescope. A good mount must be strong enough to support the telescope's weight while minimizing any vibrations induced by the observer (such as during focusing) and the environment (from wind gusts or even nearby road traffic). Indeed, without a sturdy mount to support the telescope, even the finest instrument will produce only blurry, wobbly images. A mounting also must allow for smooth motions when moving the telescope from one object to the next and permit easy access to any part of the sky.

Though Figure 3.6 shows many different types of telescope-mounting systems, all fall into one of two broad categories based on their construction: *altitude-azimuth* and *equatorial*. We shall examine both.

Altitude-Azimuth Mounts

Frequently referred to as either *alt-azimuth* or *alt-az*, these are the simplest types of telescope support available. As their name implies, alt-az systems move both in azimuth (horizontally) and in altitude (vertically). All camera-tripod heads, for instance, are alt-az systems.

This is the type of mounting most frequently supplied with smaller, less-expensive refractors and Newtonian reflectors. It allows the instrument to be aimed with ease toward any part of the sky. Once pointed in the proper direction, the mount's two axes (that is, the altitude axis and the azimuth axis) can be locked in place. Better alt-azimuth mounts are outfitted with *slow-motion controls*, one for each axis; together, they permit fine adjustment of the telescope's aiming simply by twisting one or both of the control knobs.

In the past 20 years, a variation of the alt-azimuth mount called the *Dobsonian mount* has become extremely popular among hobbyists. Dobsonian mounts are named for John Dobson, an amateur telescope maker and astronomy popularizer from the San Francisco area. Back in the 1970s, Dobson began to build large-aperture Newtonian reflectors in order to see the "real universe," as he put it. With the optical assembly complete, he faced the difficult challenge of designing a mount strong enough to support the instrument's girth yet sim-

Table 3.1 **Telescope Point-Counterpoint: A Summary**

	Point	Counterpoint
A. Binoculars Typically 1.4″ to 4″ apertures	• Most are comparatively inexpensive • Extremely portable • Wide field makes them ideal for scanning	• Low power makes them unsuitable for objects requiring high magnification • Small aperture restricts magnitude limit
B. Achromatic Refractors Typically 2.4″ to 5″ aperture, f/10 and above	• Portable in smaller apertures • Sharp images • Moderate price vs. aperture • Good for Moon, Sun, planets, double stars, and bright astrophotography	• Small apertures • Mounts may be shaky (attention, department-store shoppers!) • Possible chromatic aberration
C. Apochromatic Refractors Typically 3″ to 8″ aperture, f/5 and above	• Portable in smaller apertures • Very sharp images of high contrast • Excellent for Moon, Sun, planets, double stars, and astrophotography	• Very high cost vs. aperture
D. Normal-field Newtonian Reflectors (NFTs) Typically 4″ and larger apertures, f/7 and above	• Best all-around telescope • Low cost vs. aperture • Easy to collimate • Good for Moon, Sun, planets (especially at f/10 and above), deep-sky objects, and astrophotography	• Bulky/heavy over 8″ aperture • Collimation must be checked often • Open tube end permits dirt and dust contamination
E. Rich-field Newtonian Reflectors (RFTs) Typically 4″ and larger apertures, below f/7	• Very low cost vs. aperture • Wide fields of view • Large apertures mean maximum magnitude penetration • Easy to collimate • Good for both bright and faint deep-sky objects (solar system objects okay, but usually inferior to longer-focal-length instruments)	• Heavy • Larger apertures may require a ladder to reach the eyepiece • Low cost may indicate compromise in quality • Mounting does not track the stars • Collimation is critical (must be checked before each use) • Open tube end permits dirt and dust contamination • Susceptible to light pollution
F. Cassegrain reflector Typically 6″ and larger apertures, f/12 and above	• Portability • Convenient eyepiece position • Good for Moon, Sun, planets, and smaller deep-sky objects (e.g., double stars, planetary nebulae, and some galaxies)	• Large secondary • Moderate-to-high price vs. aperture • Narrow fields • Offered by few companies

Continued on next page

Table 3.1 continued

	Point	Counterpoint
G. Schmidt-Cassegrain (Catadioptric) Typically 4″ to 14″ apertures, f/6.3 or f/10	• Moderate cost vs. aperture • Portability • Convenient eyepiece position • Wide range of accessories • Easily adaptable to astrophotography • Good for viewing Moon, Sun, planets, bright deep-sky objects, and especially astrophotography	• Large secondary • Image quality not as good as refractors or reflectors (though optics with enhanced coatings help) • Slow f-ratio means longer exposures than faster Newtonians and refractors • Corrector plates are prone to dewing over • Potentially difficult to find objects without using an auxiliary finder or setting circles • Mirror shift (see Chapter 4)
H. Maksutov-Cassegrain (Catadioptric) 3.5″ to 12″ apertures, f/12 to f/15	• Sharp images • Convenient eyepiece position • Easily adaptable for astrophotography • Good for Moon, Sun, planets, and bright deep-sky photography	• Very high price vs. aperture • Some models use threaded eyepieces, making an adapter necessary to use other brand oculars. • Difficult to collimate • Slow f-ratio means longer exposures than faster Newtonians and refractors

ple enough to be constructed from common materials using hand tools. What resulted was an offshoot of the alt-az mount.

Using plywood, Formica, and Teflon, along with some glue and nails, Dobson devised a telescope mount that was capable of holding steady his huge Newtonians. Plywood is an ideal material for a telescope mount, as it has incredible strength as well as a terrific vibration-damping ability. Formica and Teflon together create smooth bearing surfaces, allowing the telescope to flow across the sky. No wonder Dobsonian mounts have become so popular.

Though both traditional alt-az mounts as well as Dobsonian mounts are wonderfully simple to use, they also possess some drawbacks. Perhaps the most obvious is caused not by the mounts but by the Earth itself! If an alt-azimuth mount is used to support a terrestrial spotting scope, then the fact that it moves horizontally and vertically plays in its favor. However, the sky is always moving due to the Earth's rotation. Therefore, to study or photograph celestial objects for extended periods without interruption, our telescopes have to move right along with them. If we were located exactly at either the North or South Pole, the stars would appear to trace arcs parallel to the horizon as they move around the sky. In these two cases, tracking the stars would be a simple matter with an alt-az mounting; one would simply tilt the telescope up at the desired target, lock the altitude axis in place, and slowly move the azimuth axis with the sky.

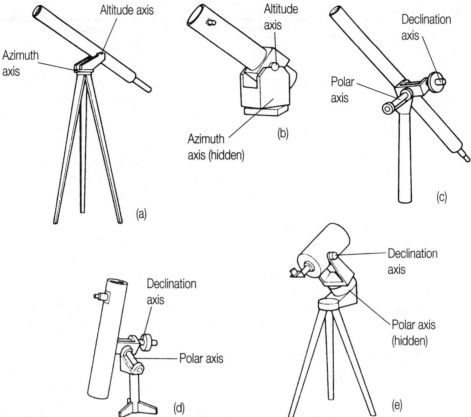

Figure 3.6 *A variety of modern telescope mountings. (a) A simple alt-azimuth mounting for a small refractor, (b) a Dobsonian alt-azimuth mounting for a Newtonian, (c) a German equatorial mounting for a refractor, (d) a German equatorial for a Newtonian, and (e) a fork equatorial mounting for a catadioptric telescope. Illustration from* Norton's 2000.0 Star Atlas and Reference Handbook, *edited by Ian Ridpath, Longman Scientific & Technical.*

Once we leave the poles, however, the tilt of the Earth's axis causes the stars to follow long, curved paths in the sky, causing most to rise diagonally in the east and set diagonally in the west. With an alt-azimuth mount, it now becomes necessary to nudge the telescope both horizontally and vertically in a steplike fashion to keep up with the sky. This is decidedly less convenient than the single motion enjoyed by an equatorial mount, a second way of supporting a telescope.

Equatorial Mounts

"If you can't raise the bridge, lower the river," so the saying goes. This is the philosophy of the equatorial mount. Because nothing can be done about the stars' apparent motion across the sky, the telescope's mounting method must accommodate it. An equatorial mount may be thought of as an altitude-azimuth mount tilted at an angle that matches your location's latitude.

Like its simpler sibling, an equatorial mount is made up of two perpendicular axes, but instead of referring to them as altitude and azimuth axes, we use the terms *right ascension* (or *polar*) axis and *declination* axis. (No doubt you have encountered these terms before, but to refresh your memory, you might want to turn to the discussion of *celestial coordinates* found in Chapter 9.) In order for an equatorial mount to track the stars, its polar axis must be aligned with the celestial pole, a procedure also detailed in Chapter 9.

There are many benefits to using an equatorial mount, the greatest of which is the ability to attach a motor drive onto the right ascension axis so the telescope follows the sky automatically and (almost) effortlessly. But there are more reasons favoring an equatorial mount. One is that once aligned to the pole, an equatorial will make finding objects in the sky much easier by simplifying hopping from one object to the next using a star chart, as well as by permitting the use of setting circles.

On the minus side of equatorial mounts, however, is that they are almost always larger, heavier, more expensive, and more cumbersome than alt-azimuth mounts. This is why the simple Dobsonian alt-az design is so popular for supporting large Newtonians. An equatorial large enough to support, say, a 12- to 14-inch f/4 reflector would probably tip the scales at close to 200 pounds, while a plywood Dobsonian mount would weigh under 50 pounds.

Just as there are many kinds of telescopes, so too are there many kinds of equatorial mounts. Some are quite extravagant, while others are simple to use and understand. We will examine the two most common styles.

German equatorial mounts. For years, this was the most popular type of mount among amateur astronomers. The German equatorial is shaped like a tilted letter T, with the polar axis representing the long leg and the declination axis marking the letter's cross bar. The telescope is mounted to one end of the cross bar, while a weight for counterbalance is secured to the opposite end.

The simplicity and sturdiness of German equatorials have made them the perennial favorite for supporting refractors and reflectors as well as some catadioptrics. They allow free access to just about any part of the sky (as with all equatorial mounts, things get a little tough around the poles), are easily outfitted with a clock drive, and may be held by either a tripod or a pedestal base. To help make polar alignment easier, some German equatorials have small alignment scopes built right into their right ascension axes—a big hit among astrophotographers.

Of course, as with everything in life, there are some flaws to the German mount as well. One strike against the design is that it cannot sweep continuously from east to west. Instead, when the telescope nears the meridian, the user must move it away from whatever was in view, swing the instrument around to the other side of the mounting, and re-aim it back to the target. Inconvenient as this is for the visual observer, it is disastrous for the astrophotographer caught in the middle of a long exposure because as the telescope is spun around, the orientation of the field of view is also rotated.

A second burden imposed by the German-style mount is a heavy one to bear: They can weigh a lot, especially for telescopes larger than 8 inches. Most of their weight comes from the axes (typically made of solid steel) as well as the counterweight used to offset the telescope. At the same time, I must quickly point out that weight does *not* necessarily beget sturdiness. For instance, some heavy German equatorials are so poorly designed that they could not even steadily support telescopes half as large as those with which they are sold. (More about checking a mount's rigidity later in this chapter.)

If you are looking at a telescope that comes with a German mount, pay especially close attention to the diameter of the right ascension and declination shafts. On well-designed mounts, each shaft is *at least* ⅛ of the telescope's aperture. For additional solidity, superior mounts use tapered shafts instead of straight shafts for the polar and declination axes. The latter carry the weight of the telescope more uniformly, thereby giving steadier support. It is easy to tell at a glance if a mount has tapered axes or not by looking at an equatorial's T-housing. If the mount has tapered shafts, then the housing will look like two truncated cones joined together at right angles; otherwise, it will look like two long, thin cylinders.

Finally, if you must travel with your telescope to a dark-sky site, moving a large German equatorial mount can be a tiring exercise. First the telescope must be disconnected from the mounting. Next, the equatorial mount (or *head*) must frequently be separated from its tripod or pedestal. Last, all three pieces (along with all eyepieces, charts, and other accessories) must be carefully stored away. The reverse sequence occurs when setting up at the site, and the whole thing happens all over again when it is time to go home. All this can add to the burden of observing, something we always try to minimize.

Fork equatorial mounts. Although German mounts are preferred for telescopes with long tubes, fork equatorial mounts are usually supplied with more compact instruments such as Schmidt-Cassegrains and Maksutov-Cassegrains. Fork mounts support their telescopes on bearings set between two short *tines*, or *prongs*, that permit full movement in declination. The tines typically extend from a rotatable circular base, which, in turn, acts as the right ascension axis when tilted at the proper angle.

Perhaps the biggest plus to the fork mount is its light weight. Unlike its Bavarian cousin, a fork equatorial usually does not require counterweighting to achieve balance; instead, the telescope is balanced by placing its center of gravity within the prongs, sort of like a seesaw. This is an especially nice feature for Cassegrain-style telescopes, as it permits convenient access to the eyepiece regardless of where the telescope is aimed . . . that is, except when it is aimed near the celestial pole. In this position, the eyepiece can be notoriously difficult to get to, usually requiring the observer to lean over the mounting without bumping into it.

The fork mounts that come with SCTs and Maksutovs are designed for maximum convenience and portability. They are compact enough to remain attached to their telescopes and fit together into their cases for easy transporting. Once at the observing site, the fork quickly secures to its tripod using thumb screws. It can't get much better than that, especially when compared with the German alternative.

Fork mounts quickly become impractical, however, for long-tubed telescopes such as Newtonians and refractors. In order for a fork-mounted telescope to be able to point toward any spot in the sky, the mount's two prongs must be long enough to let the ends of the instrument swing through without colliding with any other part of the mounting. To satisfy this requirement, the prongs must grow in length as the telescope becomes longer. At the same time, the fork tines must also grow in girth to maintain rigidity; otherwise, if the fork arms are undersized, they will transmit every little vibration to the telescope. (In those cases, maybe they ought to be called tuning fork mounts.)

One way around the need for longer fork prongs is to shift the telescope's center of gravity by adding counterweights onto the tube. Either way, however, the total weight will increase. In fact, in the end the fork-mounted telescope might weigh more than if it was held on an equally strong German equatorial.

A Telescope Pop Quiz

Now let's put all this newly acquired knowledge of yours to work. Here are eight questions to help you focus on which telescope might be best for you. Respond to each question as honestly and realistically as you can; remember, there is no right or wrong answer. Once completed, add up the scores that are listed in brackets after each response. By comparing your total score with those found in Table 3.2 at the end of the quiz, you will get a good idea of which telescopes are best suited for your needs, but use the results only as a guide, not as an absolute. And no fair peeking at your neighbor's answers!

1. Which statement best describes your level of astronomical expertise?
    a. Casual observer   [1]
    b. Enthusiastic beginner   [4]
    c. Intermediate space cadet   [6]
    d. Advanced amateur   [10]
2. Will this be your first telescope or binoculars?
    a. Yes   [4]
    b. No   [8]
3. If not, what other instrument(s) do you already own? (If you own more than one, select only the one that you use most often.)
    a. Binoculars   [1]
    b. Achromatic Refractor   [4]
    c. Apochromatic Refractor   [11]
    d. Newtonian Reflector (2″ to 4″ aperture)   [4]
    e. Newtonian Reflector (6″ to 10″ aperture on equatorial mount)   [6]

f.   Newtonian Reflector (>10″ aperture on equatorial mount)   [11]
g.   Newtonian Reflector (<12″ aperture on Dobsonian mount)   [5]
h.   Newtonian Reflector (12″ to 14″ aperture on Dobsonian mount)   [7]
i.   Newtonian Reflector (16″+ aperture on Dobsonian mount)   [10]
j.   Cassegrain Reflector (<10″ aperture)   [7]
k.   Cassegrain Reflector (10″ and larger aperture)   [11]
l.   Schmidt-Cassegrain catadioptric   [9]
m.   Maksutov   [11]

4.   What do you want to use the telescope for primarily?
   Choose only one.
   a.   Casual scan of the sky   [1]
   b.   Informal lunar/solar/planetary observing   [4]
   c.   Estimating magnitudes of variable stars   [6]
   d.   Comet hunting   [5]
   e.   Detailed study of solar system objects   [10]
   f.   Bright deep-sky objects (star clusters, nebulae, galaxies)   [6]
   g.   Faint deep-sky objects   [8]
   h.   Astrophotography of bright objects (Moon, Sun, etc.)   [4]
   i.   Astrophotography of faint objects (deep sky, etc.)   [9]

5.   How much money can you afford to spend on this telescope?
   (Be conservative; remember that you might want to buy some
   accessories for it—see Chapters 5 and 6.)
   a. $100 or less   [1]
   b. $200 to $400   [3]
   c. $400 to $800   [5]
   d. $800 to $1,200   [7]
   e. $1,200 to $1,600   [9]
   f. $1,600 to $2,000   [11]
   g. $2,000 to $3,000   [13]
   h.   As much as it takes (are you looking to adopt an older son?)   [15]

6.   Which of the following scenarios best describes your particular
   situation?
   a.   I live in the city and will use my telescope in the city.   [1]
   b.   I live in the city but will use my telescope in the suburbs.   [6]
   c.   I live in the city but will use my telescope in the country.   [8]
   d.   I live in the suburbs and will use my telescope in the suburbs.   [9]
   e.   I live in the suburbs but will use my telescope in the country.   [10]
   f.   I live in the country and will use my telescope in the country.   [12]
   g.   I live in the country but will use my telescope in the city.
         (Just kidding)   [-5]

7.   Which of the following best describes your observing site?
   a.   A beach   [4]
   b.   A rural open field or meadow away from any body of water.   [8]
   c.   A suburban park near a lake or river.   [5]

  d. A suburban site with a few trees and a few lights but away
     from any water. [6]
  e. An urban yard with a few trees and a lot of lights. [4]
  f. A rural hilltop far from all civilization. [13]
  g. A desert. [11]
  h. A rural yard with a few trees and no lights. [9]
8. Where will you store your telescope?
  a. In a room on the ground floor of my house. [6]
  b. In a room in my ground-floor apartment/co-op/condominium. [5]
  c. Upstairs. [4]
  d. In my (sometimes damp) basement [4]
  e. In a closet on the ground floor. [5]
  f. I'm not sure. I have very little extra room. [4]
  g. In a garden/tool shed outside. [9]
  h. In my garage (protected from car exhaust and other potential
     damage). [7]
  i. In an observatory. [11]

Now add up your results and compare them to Table 3.2.

Table 3.2 **The Results Are In . . .**

If your score is between . . .	Then a good telescope for you might be . . .
17 and 30	Binoculars (7 × 35 or 10 × 50 for city dwellers, 7 × 50 or 10 × 70 for country residents)
25 and 45	Achromatic Refractor (3″ aperture on a *sturdy* alt-azimuth mount)
35 and 55	Newtonian Reflector (4″ to 8″ aperture on an alt-azimuth or Dobsonian mount)
45 and 65	Newtonian Reflector (6″ to 8″ aperture on an equatorial mount) -OR- Achromatic Refractor (3″ to 4″ aperture on alt-azimuth or equatorial mount)
55 and 75	Newtonian Reflector (12″ to 14″ aperture on a Dobsonian mount) -OR- Schmidt-Cassegrain Catadioptric (8″ aperture)—for astronomers who own small cars and must travel to dark skies
65 and 85	Schmidt-Cassegrain Catadioptric (8″ to 11″ aperture)—for astronomers who own small cars and must travel to dark skies -OR- Newtonian Reflector (10″ or larger aperture on an equatorial mount)
75 and 90	Cassegrain Reflector (8″ or larger aperture) -OR- Apochromatic Refractor (4″ or larger aperture) -OR- Newtonian Reflector (16″ or larger aperture on a Dobsonian or equatorial mount) -OR- Maksutov (3.5″ or larger aperture)

The results of this test are based on buying a new, complete telescope from a retail outlet and should be used *only as a guide*. The range of choices for each score is purposely broad to give the reader the greatest selection. For instance, if your score was 52, then you can select from either the 45 to 65 or the 55 to 75 score ranges. These indicate that good telescopes for you to consider include a 3- to 4-inch achromatic refractor or a 6- to 8-inch Newtonian reflector on an equatorial mount, a 12- to 14-inch Newtonian on a Dobsonian mount, or an 8-inch SCT. You must then look at your particular situation to see which is best for you based on what you have read up to now. If astrophotography is an interest, then a good choice would be an 8-inch SCT. If your primary interest was in observing faint deep-sky objects, then I would suggest the Dobsonian, while someone interested in viewing the planets would do better with the refractor. If money (or, rather, lack thereof) is your strongest concern, then choose either the 6- or 8-inch Newtonian. Of course, you might also consider selecting from a lower-score category if one of those suggestions best fits your needs.

Some readers may find the final result inconsistent with their responses. For instance, if you answer that you live and observe in the country, you already own a Schmidt-Cassegrain telescope, and you want a telescope to photograph faint deep-sky objects but are willing to spend $100 or less, then your total score would correspond to, perhaps, a Newtonian reflector. That would be the right answer, except that the price range is inconsistent with the answers, because such an instrument would cost anywhere from $600 on up. Sorry, that's just simple economics, but there are always alternatives.

Just what are your alternatives? Who makes the best telescopes for the money? For a survey of today's telescope marketplace, as well as a review of some of yesteryear's best amateur instruments, take a look at Chapter 4.

4

Attention, Shoppers!

It is time to lay all the cards on the table. From the discussion in Chapter 3, you ought to have a pretty good idea of the kind of telescope you want. But we have only just begun. There is an entire universe of brands and models from which to choose. Which is best, and where can you buy a good telescope? These are not simple questions to answer.

Let's first consider where you should *not* go to buy a telescope. This is an easy one! Do not buy a telescope from a department store, toy store, hobby shop, or any other mass-market retail outlet (yes, this includes those 24-hour consumer television channels) that advertise a "600-power × 60-mm (2.4-inch) telescope." These instruments are low quality and should be avoided! Though their seemingly hefty prices (usually between $100 and $300) might give the uninitiated consumer a false sense of security, their mediocre optics, flimsy mounts, and poor eyepieces leave much to be desired.

Where should you go to buy a telescope? To help shed a little light on this all-consuming question, let us take a look at the current offerings of the more popular and reputable telescope manufacturers from around the world, all of · which are a big cut above those department-store brands; you can buy from them with confidence. They are organized by telescope type, with manufacturers listed in alphabetical order. I have chosen to omit the department-store models because they are really more accurately classified as toys than as scientific instruments.

In the course of writing this book, I have looked through more than my fair share of telescopes. But it takes more than one person to put together an accurate overview of today's telescope marketplace. To compile this chapter (as well as the chapters to come on eyepieces and accessories), I solicited the help and opinions of amateur astronomers everywhere. The question I put to them was simple: Are you happy with your telescope? Notices were placed in both *Sky & Telescope* and *Astronomy* magazines, on computer bulletin boards, and at a number of conventions around the country. I found the results both interesting and enlightening and have incorporated many of those comments and opinions into this discussion. Read on.

Binoculars

Adlerblick

Manufactured by Carton Optics, Adlerblick binoculars have quickly won favor with amateur astronomers for their sharp images and reasonable prices. Finest of all are Adlerblick MC binoculars, the "MC" standing for "multicoated." In addition to the superior optical coatings, both the 7×50 and 10×50 Adlerblick MC binoculars feature conventional center focus and prisms of BaK-4 glass for bright, clear images across their full fields. Although a tripod socket is built into the central hinge, Adlerblick binoculars are exceptionally light, making them easy to support by hand. In fact, at just 24 ounces, the Adlerblick MC 10×50 is one of the lightest $10 \times$ binoculars around.

Other Adlerblick binoculars include fully coated (but not multicoated) 7×42, 8×42, and 10×42 models. All are constructed with superior BaK-4 prisms for bright images. Weighing only 21 ounces each, these Adlerblicks carry on the tradition of their multicoated cousins of being true lightweights in the binocular field.

Bausch and Lomb

With more than 100 models in their lineup, no other company offers as wide a variety of binoculars as Bausch and Lomb. Some are sold under the Bausch and Lomb tradestyle, while others are available with the Bushnell label. With such diversity, there just is not enough room in this book to evaluate every model that Bausch and Lomb sells. Only those suitable for astronomy will be discussed.

Of all the binoculars wearing the Bausch and Lomb name, only some of the glasses in their Discoverer and Legacy lines (listed in order of ascending quality and price) are suited to an astronomer's needs. Discoverer glasses come with fully coated optics, while multicoatings are found on Legacy models. All are characterized by BaK-4 porro prisms, one-piece American-style barrels, and conventional center focusing. Discoverer and Legacy binoculars range in size from 7×35 to 20×80.

When Bausch and Lomb acquired Jason Optics in January 1992, it was decided that all Jason binoculars would be incorporated into the Bushnell product lineup. As such, both will be discussed here jointly. In terms of both price and features, Bushnell/Jason binoculars are, in general, a cut below their

B&L cousins. Most come with rapid-focus seesaw thumb levers and dimmer BK-7 prisms, two traits that lessen their attractiveness to stargazers.

In fact, only half a dozen Bushnell binoculars qualify as astronomy-ready. Three belong to the Eagle line, carried over from Jason. They range in size from 7×40 to 10×56, with all featuring fully coated optics and BaK-4 prisms. The other three are referred to as specialty binoculars in company literature, ranging in size from 7×35 to waterproofed 7×50s, available in either black or yellow rubberized armor coating. They also have fully coated optics, but with individually focused eyepieces.

Celestron International

Celestron, best known for Schmidt-Cassegrain telescopes, currently offers four lines of binoculars suitable for astronomical viewing: the Pro, Ultima, ED, and Giant. Both the Pro and Ultima series feature center focus, built-in tripod sockets, fully multicoated optics, and BaK-4 prisms. Images are both bright and sharp, with edge definition a little sharper in the Ultima models. Many of the Pros and Ultimas enjoy long eye relief in excess of 20-mm, which is of special interest to observers who must wear eyeglasses. (Consult Chapter 5 for a definition of *eye relief*.)

A cut above the Pros and Ultimas are Celestron's ED (extra-low dispersion) binoculars. Available in 6.5×44 and 9.5×44 versions, the ED glasses make use of special glass in the objectives to virtually eliminate chromatic aberration. They also feature center focus, built-in tripod sockets, fully multicoated optics, and BaK-4 prisms.

Celestron Giant binoculars come in 11×80, 15×80, and 20×80 varieties. All include center focus, fully multicoated optics, and BaK-4 prisms, producing sharp, clear images that compare favorably to those of other glasses of similar aperture and magnification, such as Orion, Parks, and Unitron. These binoculars are too large and heavy to hold by hand, making an external support a must. Recognizing this need, Celestron includes a right-angle tripod adapter with each pair. All in all, while they are not the best money can buy, Celestron Giants represent a very good value.

Edmund Scientific

Edmund sells several different binoculars that are sure to attract the attention of stargazers. In each case, they are nearly twins to some of the binoculars sold by other outlets. Let's review the offerings.

The least expensive and most popular model in Edmund's catalog is a pair of 7×50s. They feature fully coated optics and BK-7 (Crown glass) prisms and are similar to Orion's 7×50 Observer II binoculars, which are reviewed later in this chapter. Are there any noteworthy differences? Only two: price and warranty. Orion's version sells for about 15% less and comes with a two-year warranty versus Edmund's one-year warranty.

Edmund also markets 25×100 "giant-giants." Weighing nearly 9 pounds and measuring over a foot long, these glasses must be mounted on a sturdy tripod before they can be used. Thanks to their high magnification, wide 2.6° field of view, fully coated optics, and BK-7 prisms, these glasses produce wondrous views of the Moon, deep-sky vistas, and even the planets. For the serious binocularist, these instruments are a dream come true. Similar binoculars, but with preferred BaK-4 prisms, are also offered by Parks and Orion.

Fujinon

A name that has come to be known among photographers for excellent film and camera products over the years is now synonymous with the finest in binoculars as well. Fujinon binoculars are famous for sharp, clear images that seemingly snap into focus. They suffer from little or no astigmatism, chromatic aberration, or other optical faults that plague glasses of lesser design. Binoculars just don't come much better.

Although not all Fujinon binoculars are designed for astronomical viewing, their Polaris series (also known as the FMT-SX series; see Figure 4.1) is made with the stargazer in mind. All models, ranging between 6×30 to 16×70, offer fully multicoated optics, BaK-4 prisms, tripod sockets, rubber eyecups, and individually focused eyepieces. The waterproof, sealed housings are purged of air and filled with dry nitrogen to minimize the chances for internal fogging of the lenses in high-humidity environments. In addition, the barrels are designed to permit adjustment of the tube assemblies, should the need ever

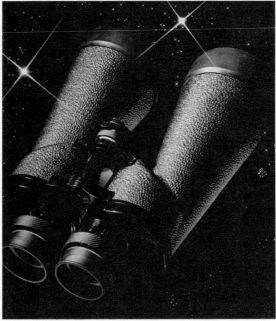

Figure 4.1 *Fujinon 10×70 FMT-SX binoculars. Photograph by Eric Hilton.*

arise. (Most binoculars nowadays are glued and pressed together, making adjustment impossible.) With all these niceties built right in, it should come as no surprise that Fujinon Polaris binoculars are expensive. But it is money well spent.

Fujinon also makes "giant-giant" binoculars called High Power Fixed-Mount Binoculars. Featuring the same quality construction as the Polaris line, Fujinon's Fixed-Mount binoculars are available in either 25×150 or 40×150 versions, the latter being the world's largest production binoculars. Optically, the Fixed-Mount glasses are identical to binoculars used by the U.S. Navy and Coast Guard; the only difference is their lighter weight. In this case, *lighter weight* is a relative term, as Fixed-Mount glasses tilt the scales at nearly 41 pounds! As such, though it is possible to support them on heavy-duty transportable mountings, Fujinon recommends that these binoculars be installed on a permanent mounting.

Minolta

Though best known for its sophisticated cameras, Minolta also manufactures several lines of binoculars. Most interesting from a technological point of view are Minolta's 8×22 and 10×25 autofocus binoculars, the first of their kind. Although certainly innovative, these glasses fail our astronomical litmus test because of their small exit pupils.

Stargazers will find some of the glasses in Minolta's Weathermatic and Standard families more attractive. Specifically, the Weathermatic 7×50 as well as the Standard 7×35, 7×50, 8×40, and 10×50 binoculars all have fully coated optics, BaK-4 porro prisms, rubber eyecups, and built-in tripod sockets. In addition, the 7×35, 8×40, and 10×50 glasses all offer extrawide fields of view of 11°, 9.5°, and 7.8°, respectively. Edge sharpness in all three is quite good, although some coma is evident. The other glasses have more typical fields, measuring 7.1° for the Weathermatic 7×50 and a wider 7.8° for the Standard 7×50. Moderately priced, Minolta binoculars represent some of the best buys in binoculars today.

Miyauchi

Not exactly a brand name that rolls off the tongue easily, Miyauchi nonetheless manufactures binoculars that are sure to make mouths water. The Miyauchi BJ80iA and BJ100iA are high-performance 20×80 and 20×100 instruments that rank among the most impressive binoculars made. Their long laundry lists of standard features include fully multicoated optics, BaK-4 prisms, semi-apochromatic objective lenses, and beautifully executed one-piece aluminum die-cast bodies. Each binocular barrel is evacuated and filled with dry nitrogen to minimize internal fogging. A pair of 27-mm reverse Kellner eyepieces (reminiscent of Edmund RKE eyepieces—see Chapter 5) provide fields of view covering 2.5° with long eye relief, an especially important feature for eyeglass wearers. Each eyepiece is tilted at a 45° angle for easier viewing. Both focus individually, with click-stop detents for precise setting. Optional accessories include screw-on dewcaps, a 3 × 12 finder, and a waterproof carrying case. The dewcaps can be made at home, and the finder can be done without, but it is a good idea to get the case.

Miyauchi binoculars, imported into the United States from Japan by Land, Sea, and Air (the 20 × 100s are also sold by Swift), are among the finest giant binoculars on the market today.

Nikon

Like Fujinon, the name Nikon may be more familiar to photographers than astronomers, but all that is changing fast. Nikon produces several lines of binoculars, including some of the finest for astronomical study. Best of the bunch are two models from Nikon's E, or Criterion, series. For handheld use, Nikon's 7 × 50 IF SP Prostars have almost all the right stuff. Included are fully multi-coated objectives made of low-dispersion glass to virtually eliminate coma as well as chromatic and spherical aberrations. Like the Fujinons, the Nikon Prostars are sealed and waterproofed, with the barrels purged of air and filled with dry nitrogen to prevent internal moisture from condensing. Surprisingly absent from these exceptional glasses is a built-in tripod socket.

The Nikon 10 × 70 IF SP Astroluxes are also missing a tripod socket, but they are otherwise remarkable giant binoculars. Sharing all the features of their smaller Prostar cousins, Astroluxe glasses have a 5.1° field of view, nearly identical to their Fujinon rivals. Which is better, Fujinon or Nikon? The answer may come down to simply a case of which can you get for the best price. Performance-wise, you will not be disappointed with either.

Orion Telescope Center

Few astronomy-related companies are into binoculars in as big a way as Orion. Besides offering models from Celestron and Fujinon, Orion also features a wide selection of models marketed under its own name. In the popular 7× to 8× range, Orion has four different grades of binoculars. Least expensive are their Observer II models, which feature fully coated optics, center focus, and fields of view between 7° and 8°. Being inexpensive models, it is not surprising that their prisms are made of BK-7 glass. You'll recall from last chapter's discussion that BK-7 prisms reduce contrast and brightness. Happily, the Observers do come with built-in tripod sockets.

Orion's Explorer series is the next step up. These glasses feature both fully coated optics as well as BaK-4 prisms. They produce bright, sharp images with good contrast. Unfortunately, the Explorer series do not have built-in tripod sockets, making external brackets necessary to attach them to photographic tripods.

Orion also markets 8 × 56 Mini-Giant binoculars, offering fully multi-coated optics and BaK-4 prisms for bright, well-defined views of the night sky.

The built-in tripod socket permits easy attachment of the glasses to any conventional tripod by using a 90° tripod adapter, available optionally. Views through the Mini-Giants compare favorably with glasses costing much more, and they are therefore highly recommended.

In the giant-binocular class, Orion sells 10×70s, 11×80s, 16×80s, 20×80s, and mammoth 14×100s. All feature fully multicoated optics and BaK-4 prisms for outstanding views of wide-field deep-sky objects such as the Andromeda Galaxy, the Pleiades, and the Double Cluster. Adapters for attaching the glasses to camera tripods also come standard. All have conventional center focusing, except the 14×100s, which come with individually focused eyepieces. As Orion points out, individual eyepiece focus eliminates flexing problems experienced by center focus glasses of this size.

Parks Optical

Parks, famous for its high-quality reflectors, also has an impressive assortment of binoculars from which to choose. The models of greatest interest to amateur astronomers are the Parks 80-mm giant binoculars. Three models, 11×80, 15×80, and 20×80, all sport fully multicoated optics and BPG-2 prisms. Parks tells us that porro prisms made of BPG-2 glass have better light transmission characteristics than traditional BaK-4 prisms, but in reality any improvement is difficult to detect.

Parks also touts the binoculars' rubberized armor plating. While rubber armor does protect binoculars from the inevitable bumps, it also adds a measurable amount of weight. To me the best way to protect binoculars is to keep them stored safely in their cases unless they are in use. Sales literature also claims that rubber armor provides "a good gripping surface." Handhold a pair of 80-mm binoculars? Have you ever tried to hold steady a pair of bricks at arm's length? It isn't easy, but that's just what supporting these monster glasses by hand would be like. Giant binoculars must be mounted on some sort of external support. Some commercially sold tripods and binocular mounts are highlighted in Chapter 6, while a build-at-home unit is detailed in Chapter 7.

Swift

Established in 1926, Swift offers one of the world's largest and most varied selections of binoculars. Most are designed for earthly pursuits such as birding and yachting, although a few are also appropriate for more ethereal viewing.

Looking first at binoculars that can be held by hand, many Swift models utilize prisms of BK-7 glass, making them somewhat less desirable from an astronomical vantage point. Standing above the crowd, however, are their 7×42, 8×42, and 10×42 Ultra Lite glasses. All three offer fully multicoated optics and BaK-4 porro prisms. Though the images they produce are quite sharp, the 7× and 8× glasses suffer from a certain degree of tunnel vision due to their small fields of view (6.5° and 6.3°, respectively). The 10× Ultra Lites, though more prone to dim images because of their comparatively small exit pupils, have a 6.5° field, considered quite wide for the magnification.

Swift also features 11×80 and 20×80 giant binoculars. Both have fully coated lenses (a plus) but use BK-7 prisms (a minus). Image quality is acceptable, but it does not pack the punch of similar binoculars from Celestron, Orion, or Unitron, all of which have BaK-4 prisms.

Finally, Swift offers a 20×100 giant: the Titan. It is virtually identical to the Miyauchi 20×100, with nitrogen-filled barrels, eyepieces tilted at 45° for ease of viewing, and fully multicoated optics.

Unitron

Although not extensively advertised in astronomy magazines, Unitron manufactures several binoculars that are useful for nighttime observations. Four compact porro-prism models, ranging from 7×42 to 10×42, make up their Pro binocular series. All include fully coated optics (multicoatings on the eyepieces), BaK-4 prisms, and built-in tripod adapters. The 6-mm exit pupil of the 7×42 glasses are of greatest interest to suburban and urban astronomers battling light pollution.

Also worth a look are Unitron's giant binoculars, ranging in size from 10×70 to 30×80. All come with fully coated optics, BaK-4 porro prisms, and built-in tripod adapters. I have owned and used a pair of Unitron 11×80 binoculars for years and am quite satisfied with them. Their performance is very good, comparing favorably with similarly priced models from Celestron, Orion, and Parks.

Zeiss (Carl)

Long known as one of the world's premier sources for fine optics, Zeiss continues the tradition by offering the finest (and most expensive) binoculars of all. Especially impressive from an astronomer's point of view are the Zeiss B Marine 7×50 binoculars. Fully multicoated optics, BaK-4 prisms, and individually focused eyepieces all combine to produce views that are unparalleled in brightness and sharpness across the entire 7.6° field of view. All that is missing to make these glasses *perfect* are longer eye relief (only 13 mm, too short for eyeglass wearers) and a built-in tripod socket.

The Zeiss B/GA T Dialyt binoculars are also worth noting. Only Zeiss could produce 7×42 and 8×56 roof-prism binoculars of this quality. Both include fully multicoated optics and conventional center focus but no tripod socket.

Refracting Telescopes

Though many small refractors continue to flood the astronomical telescope market, the discussion here is limited to only those with apertures of 3 inches and larger. As a former owner of a couple of 2.4-inch refractors, I found that I quickly outgrew those instruments' capabilities and hungered for more. Why? At the risk of possibly offending some readers, let me be brutally honest: 2.4-inch telescopes will produce acceptable views only of the Moon and maybe Venus, Jupiter's satellites, Saturn, and possibly a few of the brightest deep-sky objects. Most are supplied on shaky mountings that only add to an owner's frustration. If a 2.4-inch refractor is all your budget will permit, then I strongly urge you to buy a good pair of binoculars instead.

Astro-Physics

This is a name immediately recognizable to the connoisseur of fine refractors on rock-steady mounts. The performance of their Starfire telescopes (Figure 4.2) is unsurpassed by any other apochromat sold in North America, except perhaps Takahashi. And even then, the decision is close.

Currently, five triplet-objective refractors adorn the Astro-Physics line. Least expensive is the 105EDT, a 4.1-inch f/6 refractor measuring only 19 inches long, aptly named the Traveler. Like the larger Starfires, the Traveler's objective lens consists of two matching hard crown meniscus lenses sandwiching an element of Super ED glass, an expensive, premium glass whose refractive qualities are a lens designer's dream. Recall that chromatic aberration results from different wavelengths of light focusing at different distances away from a lens. Super ED glass can be used to minimize this dispersion effect, thereby eliminating, for all intents and purposes, false color. The result is an outstandingly sharp image that is also free of annoying spherical aberration and coma.

For those looking for a somewhat larger aperture and longer focal length, the 130EDT 5.1-inch f/8 Starfire should be given serious consideration. Like its smaller sibling, the 5.1-inch Starfire is outstanding both optically and mechanically. The rack-and-pinion focuser, for instance, moves the eyepiece in and out of focus smoothly, without any hint of binding or shift. All machining is done by Astro-Physics in their factory, ensuring top-notch quality control on all components. Others in the Astro-Physics line include the 155EDT 6.1-inch f/9, 155EDF 6.1-inch f/7, 180EDT 7.1-inch f/9, and 206EDF 8.1-inch f/8 Starfires. In addition to the Super ED objectives, the two EDF models feature field-flattener lenses, indicated by the "F" designation. All are outstanding instruments, with excellent optics and premium mechanics.

You may notice that I have not mentioned anything about Astro-Physics mountings. That is because none of the telescopes come with mountings; instead, you may select from their line of homegrown German equatorial mounts to suit your needs. All except the tripod-mounted 400 come on observatory-quality piers. (The 400 is designed for super portability, making it a perfect match for the Traveler.) The mounts don't just move, they flow across the sky thanks to their excellent construction of machined aluminum and stainless steel. The larger 600E, 800, and 1200 mounts come with highly accurate, built-

Figure 4.2 *Astro-Physics Starfire apochromatic refractor.*

in dual-axis drives that run on 12 volts DC for easy portable operation; a similar dual-axis drive is available optionally on the 400 mounting.

For either visual or photographic use, Astro-Physics apochromatic refractors are the finest of their kind made in America. Be forewarned that because of the high demand and limited production, delivery of Astro-Physics refractors may take several months, even a year. But their optimum performance has earned them a following of loyal and enthusiastic owners. Some buyers help talk themselves into paying the extremely high prices by considering Astro-Physics scopes an investment. Used ones will still command top prices 100 years from now.

Celestron International

Within the last decade, Celestron, best known for its Schmidt-Cassegrain telescopes, has branched out into other telescope markets as well. In the refractor category, Celestron features instruments made in Japan by Vixen that range in aperture from 2.4 to 4 inches. Especially noteworthy for amateurs who are in the market for a quality refractor is the Firstscope 80, a 3.1-inch scope available on a rugged alt-azimuth mount. The same basic telescope is also available on a Polaris German equatorial mount, though it is called the C80. Both come with 18-mm eyepieces and 45° star diagonals. Be aware that these diagonals prove less comfortable for night-sky viewing than the 90° diagonals supplied with most other refractors.

Two other excellent refractors from Celestron are the 4-inch f/9.8 C102 and the 4-inch f/8.8 C102F. Both appear nearly identical on the outside, with each well-supported by a Super Polaris German equatorial mount. The Super Polaris, or SP, mount is executed nicely and features a handy polar alignment scope built into the right ascension axis, though it is not as sturdy as Tele Vue's similar Equatorial Systems mount. (As this is being written, Celestron is beginning to substitute a new version of this mounting called the Great Polaris, or GP, mounting. The biggest improvement to the GP mounting is a superior latitude adjustment; otherwise, the SP and GP mounts are comparable.)

The biggest difference between the SP-C102 and the SP-C102F is in the optics. The objective lens of the C102 is a classic achromat, whereas the C102F is an apochromat with a fluorite element. As an owner of a C102, I can attest to its optical excellence with enthusiasm. Color correction is good for an f/10 achromat, with images that are both sharp and clear.

As good as the C102 is, the C102F is in an entirely different class. Though it costs considerably more than the C102, the higher price is offset by the instrument's exceptional quality. Contrast, false color suppression, and sharpness are all extraordinary. The images produced by the C102F are among the best of any 4-inch apochromat sold today. If you are in the market for a portable refractor capable of delivering tack-sharp views of the Moon, planets, and brighter deep-sky objects, then give the Celestron C102F strong consideration.

D&G Optical Company

D&G Optical is a small company that specializes in superb-quality achromatic refractors. Its line of complete refractors ranges in size from a 5-inch instrument (available at either f/10, f/12, or f/15) to a custom-made 15-inch f/12 monster. All optics are handmade and tested to exacting standards, producing the finest results. The 5- and 6-inch objectives receive a coating of magnesium fluoride as standard, whereas coatings on the 8-inch and larger objectives are available at extra cost and "at the customer's risk," according to the company. (Doesn't that give you a warm feeling?)

Unlike most refractors, whose objective lenses are mounted in nonadjustable cells, D&G objectives come with a *push-pull* cell, which allows the owner to align the lens's optical axis precisely, resulting in better images and a telescope that stays in collimation even with constant transport.

All D&G refractors come with a rack-and-pinion focusing mount and an 8×50 finderscope. The instruments may be purchased as either unmounted tube assemblies or with optional German-style equatorial mounts. The mountings are adequately strong for the task at hand and offer good rigidity with little flexure. For the adventurous amateur telescope maker, D&G also sells its objective lenses separately.

Recently, D&G added a 5.1-inch f/9.5 apochromatic refractor to their lineup. The D&G apochromat features a fully coated, air-spaced, two-element objective lens featuring Super ED glass for superior color correction. Images produced compare favorably with those of other refractors of similar size and type, though when outfitted with the suggested mounting and tripod (both sold separately), the instrument's cost is much higher than, say, the Meade 5.1-inch Apo.

Meade Instruments Corporation

To celebrate its twentieth anniversary in 1992, Meade introduced a line of new and exciting state-of-the-art apochromatic refractors (Figure 4.3). The Meade apochromats are among the most sophisticated refractors on the market today, outfitted with advanced computerized mountings and drives. Very impressive indeed, but how well do they work? For the most part, they perform quite well.

Like many other premium refractors today, Meade apochromats use ED glass in their objectives to minimize false color. The resulting images yield pinpoint star images and good field contrast, two characteristics of better refractors. When viewing brighter targets, however, some minor secondary spectrum becomes evident, as do some slight astigmatism and spherical aberration. This led one respondent to my survey to describe these as "semi-apochromatic" telescopes; perhaps they are not as good as some other multi-element apochromats such as Astro-Physics, Takahashi, or Tele Vue, but they're still better than traditional achromats at similar focal lengths.

As to the mechanical design of the telescopes and mountings, the supplied German equatorial mounts are very good, offering a good mix of stability, port-

Figure 4.3 *The Meade 152ED 6-inch f/9 apochromatic refractor. Photo courtesy of Meade Instruments Corporation.*

ability, and convenience. Both fast-speed slewing and slow-motion control are smooth, with little or no backlash evident. The only mechanical problem evident in the overall design is in the focusing mount, which is not as smooth as some of the competition's. The problem was especially evident in earlier production scopes but has since been lessened.

Meade also continues to offer a 3.1-inch f/11 achromatic refractor for the budget-minded hobbyist. Available as the Model 312 on an alt-azimuth mount or the Model 323 on a German equatorial mount, this instrument comes with a 6×30 finderscope and a poor-quality 25-mm, 1.25-inch diameter eyepiece. Both mountings are adequate for the task, but the alt-azimuth unit seems a bit steadier. The equatorially mounted version is nearly identical to Orion's Sky Explorer 80 (see the next section), although the latter comes with two eyepieces. Neither performs as well as Celestron's C80 variants.

Orion Telescope Center

Orion offers several small achromatic instruments ranging in aperture from 2.4 inches to 3.1 inches. The largest refractor in Orion's collection is the Sky Explorer 80, a 3.1-inch f/15 instrument featuring fully coated optics. The Sky Explorer 80 comes on a German equatorial mount equipped with manual slow-motion controls, though an AC-powered clock drive is available as an option. The mounting provides adequate support as long as winds are light. Also included are a 6×30 finder and two 1.25-inch Kellner eyepieces. The Sky Explorer 80 is a good choice for anyone looking for a small refractor. By shopping around, however, you might be able to get a better deal on a Celestron/Vixen 80-mm refractor. The latter comes with only one 1.25-inch eyepiece, but it has a sturdier mount (once again, a motorized clock drive is available as an option).

Pentax

Best known for its cameras, this company offers some exciting achromatic and apochromatic refractors for astronomical viewing. The Pentax-J80, highlighted by a 3.1-inch f/11 achromatic objective, is the least expensive of the Pentax scopes. It is set in a pair of rotating rings atop a sturdy alt-azimuth mounting. Pentax also offers three equatorially mounted f/11 achromatic refractors: the 2.6-inch Pentax-65, the 3.3-inch Pentax-85, and the 4-inch Pentax-100. The mountings are very stable, with polar-alignment scopes built right into the mounts supplied with the 3.3- and 4-inch instruments (a polar-alignment scope is available as an option on the 2.6-inch). Manual slow-motion controls are standard on all three mounts, while single-axis clock drives may be purchased optionally. Images produced by all of the achromatic Pentax scopes are clear and sharp, although some secondary spectrum is inevitable, especially around brighter targets.

Pentax produces several of the same telescopes in two-element apochromatic versions as well. The objective lenses consist of special ED glass to reduce aberrations common among the achromats. The results are impressive indeed, with little spherical or chromatic aberration evident. The views produced by the Pentax apochromats appear nearly identical to those through the Meade apos but not as color free as Astrophysics or the Tele Vue Genesis-SDF. Pentax refractors fall short in the standard-accessories department, as all are supplied with small 6×30 finderscopes and, worse yet, subdiameter (0.965-inch) eyepieces. The latter seriously impacts the versatility of these otherwise fine instruments. (Yes, adapters are available, but they can narrow the field of view.)

Takahashi

These apochromatic refractors have well-deserved reputations for their excellent performance. In fact, to many they represent the pinnacle of the refractor world . . . they just don't come any better. The only drawback is that this superior quality does not come cheap.

Takahashi apos may be divided into two product lines, the only difference being objective lens design and focal length. The FC refractors, characterized by doublet objectives, include six aperture/focal ratio combinations. These include the 2-inch f/8 FC-50, 2.4-inch f/8.3 FC-60, 3-inch f/8 FC-76, 4-inch f/8 or f/10 FC-100, and 5-inch f/8 FC-125. The more expensive FCT triplet objectives include the 3-inch f/6.4 FCT-76, 4-inch f/6.4 FCT-100, 5-inch f/5.6 FCT-125, and 6-inch f/7 FCT-150.

Astrophotographers will find the faster FCT refractors of interest, but visually, the difference between them and the FC series is negligible. Both yield color- and aberration-free images that can only be described as perfect. Planetary detail is unrivaled, with each Takahashi showing features that remain invisible in similar-aperture refractors, even those by Astro-Physics and Tele Vue. Given steady atmospheric conditions, the optical quality of Takahashi instruments routinely permits the use of magnifications in excess of $100\times$ per inch (see Chapter 5 for a discussion of magnification and its limits). All optical components carry a lifetime guarantee.

All FC and FCT refractors ride aboard unyielding German equatorial mounts that are among the best production systems in the world. An illuminated polar-alignment scope is built right into the right ascension axis on all Takahashi mounts. A single-axis motor drive is included on mounts supplied with the 3-inch, while dual-axis drives are standard on the larger scopes. All come with adjustable wooden legs, and observatory-style metal piers are available for 4- and 5-inch models.

As previously mentioned, Takahashi apochromatic refractors are the finest production instruments sold today, yet they are certainly not as popular as, say, either Astro-Physics or Tele Vue. Why? Could it be their prices? There can be no doubt, because even the 3-inch FC-76 sells for around $2,500, while the 6-inch FCT-150 retails for more than $27,000! But for those of us who can proclaim proudly that money is no object, Takahashi refractors may be the way to go.

Tele Vue, Inc.

Tele Vue manufactures four outstanding short-focus deep-sky refractors for the amateur. The images they produce are among the sharpest and clearest, with excellent aberration correction.

Least expensive of the Tele Vue refractors is the Pronto, a compact 2.8-inch (70-mm) f/6.8 instrument that measures a mere 18 inches from end to end. Though technically below the 3-inch threshold I set earlier in this section, the Pronto's sharp, wide-field performance warrants its mention. Included with the Pronto is a 2-inch focusing mount, 2-inch star diagonal (a 1.25-inch adapter is included), a 21-mm Plössl eyepiece (discussed with other eyepieces in Chapter 5), and a sliding mounting ring for attaching the instrument to a camera tripod (not included). No finder is provided, but Pronto's wide field of view at low power makes one unnecessary. Although designed for wide starfield views, the Pronto will also produce reasonable images of the planets when steady sky conditions permit the use of short-focal-length eyepieces.

Also from Tele Vue is the Renaissance, a 4-inch f/5.5 instrument. Outfitted with a distinctive brass tube, a throwback to the refractors of a century ago, the Renaissance comes with a four-element semi-apochromatic objective to render sharp images of wide starfields. Images of the planets are also good, though not as good as those available through true apochromats, such as Tele Vue's Genesis, Celestron's C102F, or those by Astro-Physics or Takahashi. Once again, a mounting ring is supplied, but you have to supply the tripod. The optional equatorial mount sold for the Renaissance is a variation of the mounting supplied with many less expensive instruments, but it does not do justice to this fine instrument. If you really want the Renaissance, spend the extra money to purchase either the Gibraltar or Equatorial Systems mounts noted below.

If glimpses of exceptional planetary detail are what you crave, then consider Tele Vue's flagship Genesis-SDF 4-inch f/5.4 apochromatic refractor. Although the short focal length keeps the optical tube assembly to a toss-it-over-your-shoulder length, the four-element special-dispersion-plus-fluorite (SDF) objective permits the use of a broad range of eyepiece focal lengths. The result is a compact instrument of outstanding optical quality that is just as useful for close-up planetary work as it is for rich-field deep-sky observing. Included with the Genesis-SDF is a 2-inch focusing mount and star diagonal, a 1.25-inch adapter, and a 17-mm Plössl eyepiece, but—surprisingly—no finderscope. (A one-power Starbeam aiming device is available optionally; it is described in Chapter 6.) The Genesis-SDF is available on either the Gibraltar alt-azimuth or the Equatorial Systems mounting, both sturdily supported atop wooden tripods.

Still available, and less expensive, is the standard Genesis apochromat. A 4-inch f/5 telescope, the Genesis features a four-element objective including an internal fluorite element. Wide starfields are sharp and aberration free, though planetary views are not equal to the SDF version. Both Genesis telescopes have many small, thoughtful features, such as a threaded lens cap, sliding dew shield, and blackened threads inside the star diagonal to reduce light scattering. Unquestionably, the Tele Vue Genesis and Genesis-SDF are two of the finest deep-sky apochromatic refractors for sale today.

Lastly, Tele Vue offers the Solaris, a dedicated solar telescope. The Solaris is designed to work with the T-Scanner hydrogen-alpha filter system by Daystar (sold separately); the combination yields sharp views of solar prominences, filaments, and other details invisible with white-light solar filters. For a complete discussion of the T-Scanner, see Chapter 6. If solar observing is your prime interest, then the Solaris is certainly worth considering.

Unitron, Inc.

Since its founding in 1952, Unitron has become widely respected for its excellent achromatic refractors. Chances are that if an amateur astronomer in the

1950s or 1960s wanted a high-quality refractor, he or she would have bought a Unitron. Although it has much more competition today than it did back then, Unitron still produces fine instruments that compare favorably with others in their size and price class.

A quick glance through this market survey reveals that most manufacturers produce achromatic refractors around f/10 to f/12. Although this lets the telescopes' tubes shrink in length, the shorter focal length makes the instruments more prone to secondary spectra. Unitron, on the other hand, has chosen to stay with the classic approach of focal ratios between f/14 and f/16. The instruments can be up to 50% longer, but they are less apt to suffer from this aberration, and so typically produce superior views of the planets and double stars. If tube length is of concern, there is always Unitron's Model 131C, 3-inch f/16 folded refractor. Two flat mirrors bounce the light around to cut the tube down to only 19 inches in length. Although aperture and focal ratio are the same as Unitron's conventional 3-inch models, the images yielded by the folded telescope will be slightly dimmer and with decreased contrast because of light lost and scattered by the mirrors.

Unitrons range in aperture from 2 to 5 inches. Each telescope features fully coated optics, a rack-and-pinion focusing mount, two or more 1.25-inch eyepieces (except for the 2-inch refractor), a small (too small) finderscope and, in some cases, a solar projection screen. Mountings include both alt-azimuth and German equatorial designs on wooden tripods. The equatorials are especially well crafted and sturdy and may be outfitted with optional motor drives. Another nice option for Unitron scopes is the Unihex rotary eyepiece turret, which holds up to six eyepieces at any one time. To change power, rotate the turret until the desired eyepiece is in position.

Unlike so many things in life that change simply for the sake of changing, Unitron refractors have remained basically the same since their introduction: well-designed instruments that yield good images.

Vixen Optical Industries

In addition to supplying Celestron with its refractors, Vixen also offers several fine refractors under its own name through Celestron dealers in the United States and Canada. Vixen refractors range in aperture from a too-small 40-mm to 4 inches. Vixen's Custom 80M 3.1-inch f/11 and Custom 90M 3.5-inch f/11 achromatic refractors have features virtually identical to Celestron's Firstscope 80. Celestron's version, however, features 1.25-inch eyepieces, whereas the Vixens are equipped only with substandard, inferior-quality 0.965-inch oculars.

Vixen's Polaris 102M 4-inch f/10 achromatic refractor has been marketed in the past as Celestron's SP-C102. As an owner of the C102, I can attest to its fine solar, lunar, and planetary performance. However, once again the Vixen version is outfitted with substandard 0.965-inch eyepieces, degrading an otherwise fine instrument.

Of the Vixen refractors, only the Vixen Polaris FL-90S comes with a 1.25-inch eyepiece holder. The FL-90S, a 3.5-inch f/9 apochromatic/fluorite refractor supplied on a sturdy Polaris German equatorial mount, is only offered for sale under the Vixen name. Images are virtually color free, with excellent correction for chromatic and spherical aberrations as well as coma. All in all, the FL-90S is an excellent telescope, though it can be hard to find here in North America.

Reflecting Telescopes

Celestron International

Celestron offers two small Newtonian reflectors that are also made in Japan by Vixen: the C4.5 (4.5-inch f/7.9) and the C6 (6-inch f/5). Both come on wooden tripods with German equatorial mounts that are the sturdiest in their class. Unfortunately, tripod height is fixed on the C4.5, making it less flexible for storage, transport, and use than the adjustable tripod supplied with the C6.

Image quality in the C4.5 is good, but some owners remark that image brightness is a bit off when compared with other scopes of similar aperture. Why? *Astronomy* magazine, in a review of small telescopes in their December 1991 issue, seems to have hit the nail on the head: "The C4.5 projects the focus point farther out from the side of the (telescope) tube . . . so the telescope can accommodate a camera. The C4.5 should therefore have a larger secondary mirror to intercept the entire cone of light coming from the main mirror. But it doesn't." Because the minor axis (the width) of the secondary mirror is less than the diameter of the cone of light reaching it from the primary, the telescope's effective aperture is reduced. Still, the C4.5 is a good first telescope for aspiring astronomers.

The same review in *Astronomy* cast doubt on the optical quality of the 6-inch f/5 C6. However, of those I have looked through, I have found images to be quite good for an instrument of such short focal length. Stars formed crisp points with little evidence of aberrations. One weak point in the C6 design, however, is its unusual focusing system. Instead of moving the eyepiece in and out, as is the convention, the eyepiece holder and diagonal mirror slide back and forth as a unit, thereby varying their common distance from the primary. The effect is the same, but collimation can be adversely affected if the assembly is not positioned perfectly on the tube or if it loosens up (as mechanical things have a habit of doing over time).

Coulter Optical, Inc.

When it comes to getting the largest aperture for the money, Coulter's Odyssey Dobsonian-style Newtonian reflectors rate as a best buy. No other manufacturer offers as large a telescope for so little money. For under $1,000, you can choose from either the 8-inch f/4.5 or f/7 Odyssey 8, the 10.1-inch f/4.5 Compact Odyssey, or the 13.1-inch f/4.5 Odyssey I. The flagship 17.5-inch f/4.5 Odyssey

II can be yours for less than $2,000. The new 8-inch f/7 is worth a second mention, as it represents an outstanding value for beginning astronomers. Its longer focal length will produce good views of both deep-sky objects and the planets, making it a better overall choice than the 8-inch f/4.5.

How does Coulter do it, and what are you getting for the money? The Odysseys are well known as no-frills, albeit serviceable, telescopes. The tubes are made of spiral-wound cardboard, and the Dobsonian mounts are made from pressboard (aka chipboard or particle board). Each Odyssey comes with a 27-mm Kellner eyepiece but no finder. That's it. Some will argue that the Odysseys, as delivered, are ill-equipped, but more creative hobbyists see these telescopes as great opportunities to customize their equipment exactly as they want. With the money saved, one can buy better eyepieces, a big finderscope, a set of counterweights (a must to counterbalance the telescope for the added weight of the new eyepieces and finder), and even a state-of-the-art focusing mount, if desired.

What about optical quality? I was surprised to learn during research for this book that only Coulter's 17.5-inch Odyssey uses mirrors made of Pyrex glass; the 8-, 10.1-, and 13.1-inch Coulter mirrors are made of plate glass, unheard-of in this day and age. Pyrex is much stronger than plate glass, thereby reducing the risk of breakage. It also has a lower coefficient of thermal expansion, meaning it is less affected by changes in temperature than plate glass. Nearly all other telescope manufacturers use mirrors of Pyrex in their reflectors. So why does Coulter use plate glass instead of Pyrex? Plate glass is cheaper. Readers should note, by the way, that of Coulter's two main competitors in the cheap-but-big telescope market, Orion Telescope Center's Deep-Sky Explorer telescopes also use mirrors made from plate glass, while the new Meade Starfinder Dobsonians feature mirrors of Pyrex.

Having said that, I have looked through many Odysseys that could meet or beat the image quality of any same-size Newtonian telescope regardless of manufacturer. I have also looked through a couple of lemons. Most Odysseys fall somewhere in between; that is, they are maybe not the best optics around but a very good deal for the price. Personally, my homemade 13.1-inch Newtonian uses Coulter optics. The verdict? Although the mirrors are not perfect (the primary is slightly astigmatic), they work well *when properly collimated.* That is the key. Collimation is critical, especially in the f/4.5 models (true for ALL short-focus reflectors, not just Coulters). Even a slight misalignment will cause the view to go to pieces. Collimation must be inspected before every observing session. Consult Chapter 8 for more details.

As for the mechanical aspects of the Coulter scopes, they are quite stable and vibration free (ahh, the beauty and simplicity of wood!). Of course, being Dobsonians, they are not readily usable for astrophotography, although they might possibly work for shots of the Sun and Moon. But for a visual tour of the universe, Odysseys offer exceptional value.

The sling-type mirror mount used in older Odysseys (that is, the originals with blue tubes) required the installation of the primary mirror before each use and its removal afterwards. I have heard more than a few horror stories of mirrors being accidentally dropped when being placed in or taken out of the telescope in the dark of night. When Coulter redesigned its telescopes a few years ago (changing from blue to a red-and-black motif), they modified the mirror mount to hold the primary permanently in place. That's a big plus, but the downside is that the mirror mount currently supplied is not very well designed. If you are going to buy an Odyssey, consider replacing the mirror mount with a better-quality unit.

The blue-to-red design shift also saw a change in focusing mounts. Rather than a rack-and-pinion system, today's Odysseys come with a PVC pipe slip-fit mount. The eyepiece is slid up and down to achieve focus, and then a ring is tightened to hold it in place. This is crude and not very useful, especially at higher powers where focusing is critical. My advice is to replace it with a better unit from another supplier.

When it comes to delivery, Coulter gets into a bit of hot water. Just as Odysseys are famous for their value, so too are they famous (or infamous) for their long delivery times, which typically range from two to six months; in the past delivery often took a year or more for the popular 13.1- and 17.5-inchers. This has left many amateurs with a bad taste in their mouths for the company. Although Odysseys are usually ordered directly from the factory, a few of the distributors may have the telescopes in stock. If you are considering purchasing an Odyssey, then these outlets are certainly worth investigating. Consult Appendix A for selected distributors. Also consider the Meade Starfinder and the Orion Deep-Space Explorer series of Newtonians, reviewed later in this chapter.

Edmund Scientific Company

Once one of the most often-mentioned sources for telescopes and accessories, Edmund's share of the astronomical marketplace has dwindled over the last quarter-century. Many of its once-popular products are no longer sold, having been replaced by a new, limited assortment of astronomical gear. The most popular of all Edmund telescopes around today has to be the small-in-size-but-not-price 4.25-inch f/4 Astroscan 2001 rich-field Newtonian. The Astroscan is immediately recognizable by its unique design, which resembles a bowling ball with a cylinder growing out of one side. The primary mirror is held inside the 10-inch ball, opposite the tube extension, which supports the diagonal mirror and eyepiece holder. The entire telescope weighs about 13 pounds and is supported in a three-point tabletop base that may also be attached to a camera tripod.

The primary mirror, advertised as ⅛ wave, yields good star images when used with the 28-mm RKE eyepiece provided. However, as magnification increases, the image quality is seen to degrade. This should come as no great shock, because small RFT Newtonians are really not suitable for high-power applications. While the rings of Saturn and the bands of Jupiter are visible, their clarity is inferior to same-aperture NFT Newtonians.

To prevent dust infiltration, Edmund seals the front of the Astroscan's tube with a clear optical window that also holds the secondary mirror in place. This is a good idea on paper, but I have seen some Astroscan mirrors contaminated with dirt and, worse yet, mildew as a result of improper storing, which is the owner's fault, not the telescope's. Although most Newtonians are easy to disassemble and clean, the Astroscan is cemented shut, making it necessary to return it to the factory for any service.

Though a finder in the classic sense is not supplied with the Astroscan, it does come with a one-power aiming device that permits easy aiming of the telescope. This little finder is fine for the job because the 28-mm eyepiece that comes with the Astroscan produces a 3° field of view. Other accessories offered include a tripod bracket, tote bag, solar viewing screen, and camera adapter. The optional tripod bracket is a must if you want to try short-exposure photography through the Astroscan, because the weight of a camera will throw off the telescope's balance in the standard cradle.

Jim's Mobile Industries (JMI)

As a schooled mechanical engineer, I appreciate a product that emphasizes excellence in design and workmanship. So when I first laid eyes on JMI's Next Generation Telescopes (NGT, for short), my heart skipped a beat! Of all the giant Newtonians on the market today, no other has the innovation or craftsmanship of the NGT series (Figure 4.4).

Three models comprise the NGT line of telescopes: a 16-inch f/5, an 18-inch f/4.5, and a monster 25-inch f/5. Sibling resemblance is unmistakable, with each instrument sporting an open-truss tube mounted on a sturdy split-ring equatorial mount similar in design to the 200-inch Hale telescope atop Mount Palomar. The telescopes and mounts disassemble for transport to and from remote observing sites. Both the 16- and the 18-inchers can squeeze into the back of a subcompact station wagon, whereas the 25-inch requires a little more car to cart it around. All components are assembled using large knobs, making extra tools unnecessary. Unfortunately, the black knobs are not held captive to the telescopes, making it possible to lose them at night (see the note in the section covering Obsession telescopes.)

The image quality of all of the NGTs is consistently excellent, thanks to their fine optics. Of those I have looked through, all sported mirrors with resolution limited only by seeing conditions, not by mirror imperfections. For this, a pat on the back must go to Galaxy Optics, manufacturer of the mirrors used in all JMI scopes.

The NGTs are adorned with all sorts of neat little gadgets. Features such as no-tools-required assembly, a rotatable nose assembly to permit comfortable viewing of any part of the sky, and built-in lifting handles are standard on all models. In addition, the 18- and 25-inch scopes come with dual-axis clock drives with integral quartz drive correctors and electric eyepiece focusers. All of these latter items are available as extra-cost options for the 16-inch.

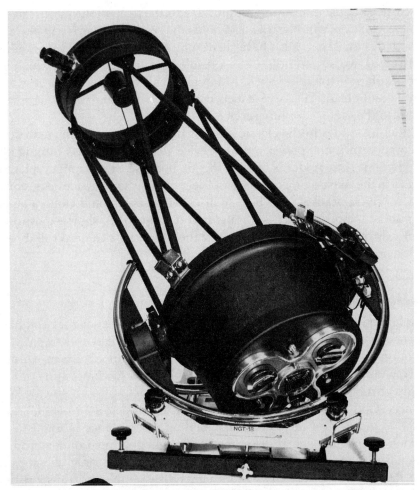

Figure 4.4   *The NGT-18, a sophisticated 18-inch f/4.5 Newtonian reflector on a horseshoe-style equatorial mount. Photo courtesy of Jim's Mobile Industries (JMI).*

The most sophisticated innovation of all has to be the NGC-miniMAX digital setting circles, featuring a built-in computerized catalogue of 1,950 deep-sky objects. An upgrade to this, the NGC-MAX, holds data on more than 8,100 sky objects and is available for all NGTs. Though I am *not* a strong proponent of computerized telescopes (see my sermon in Chapter 6), both MAXs greatly ease polar-aligning the NGT telescopes. Without them, polar alignment of the NGT can prove daunting.

Two other options offered for NGTs are almost mandatory if you observe anywhere near light-polluted or damp environs. One is a black sleeve to prevent stray lighting or cross winds from infiltrating the open tube and hampering the view. When wrapped around the telescope's skeletal structure, the sleeve prevents light and wind from crossing the instrument's optical path. An added benefit is that it slows dewing of the optics (a problem for all open-truss telescopes, not just the NGTs).

The other item that is a must is a tube extension that sticks out beyond the telescope's front. The extension serves two purposes. First, it helps prevent fogging of the diagonal mirror. Second, because the NGT's eyepiece and diagonal are so close to the end of the tube, the extension prevents stray light from scattering onto the diagonal and washing out the scene.

JMI's NGT telescopes are the finest portable equatorially mounted Newtonians available to today's amateur astronomer. They just about have it all, including, unfortunately, very high prices. But for those who can afford the expense and plan on observing from either a dark suburban or a darker rural site, JMI telescopes are without peer.

The newest addition to JMI's growing line of telescopes is the NTT-25, short for New Technology Telescope. The NTT-25 is a 25-inch f/5 folded Newtonian, the only telescope of its kind on the market. By placing a second, optically flat mirror between the primary and the diagonal mirrors, the instrument's overall length is cut dramatically. Now, instead of having to climb to towering heights to look through the eyepiece (as is the case with all 25-inch unfolded Newtonians), the eyepiece is never more than 6 feet off the ground and so requires only a stool at most. The NTT comes on a state-of-the-art clock-driven alt-azimuth mounting. Although the basic NTT-25 cannot be used for long-exposure photography due to field rotation, the deluxe NTT-25C incorporates something called a *field-rotation drive* that actually turns the eyepiece holder to eliminate the problem.

The folded Newtonian is a design not without faults, however. The biggest problem is the huge flat secondary mirror at the front of the tube required to redirect the light back toward the diagonal. In the case of the NTT-25, it measures 9 inches across—larger than some primaries! This translates to a central obstruction equal to 36% of the primary's diameter (13% by area), much greater than any other Newtonian. Not surprisingly, the large central obstruction dramatically reduces image clarity and contrast when compared to a conventional Newtonian. The net result is a telescope with excellent optics delivering only a lukewarm performance. As a result, it cannot be recommended.

Jupiter Telescope Company

A telescope company on Saturn Street in Jupiter, Florida? Sounds perfect! Andy Johnson, an amateur astronomer and owner of Jupiter Telescope Company, has an interesting philosophy when it comes to selling telescopes. Rather than offer a number of different, possibly mediocre instruments, he has chosen to concentrate all his efforts on perfecting one telescope design available in three apertures. Those telescopes are the Juno-12.5 (12.5-inch f/4.5), Juno-15 (15-inch f/4.5), and Juno-18 (18-inch f/4.5) Newtonians.

At first glance, the Junos look like standard Dobsonian-style reflectors. The Juno-12.5 features a flawless octagonal wooden tube that unclamps for easier storage—a nice touch. The Juno-15 and Juno-18 come with a user-friendly open-truss tube. The truss tubes are held in clamps using large knobs, rather than with bolts passing through the tubes themselves. The result is a superior system that allows the telescope to be assembled and disassembled without fumbling in the dark with loose hardware or a wrench.

Optically, the Junos are about as good as short-focus Newtonians get, thanks to the excellent primary and secondary mirrors. Tack-sharp stars even at high magnifications are a common characteristic of all Juno instruments.

So far, so good, but the Junos aren't done yet. It's what's underneath the telescopes that makes them even more special. All Junos come mounted on a special d'Autume equatorial table, as it is called, which allows the telescope to track the stars like a conventional equatorial mount while keeping the overall dimensions of the telescope to approximately the same as a conventional Dobsonian of similar size. The equatorial table makes tracking sky objects possible with the Junos, while it is not practical with a traditional Dobsonian. Drawbacks to the design are an increase in overall weight and the fact that different tables are required for different latitudes. Consumers may choose one table for their location when ordering a Juno and purchase others later at additional cost, if desired. Long-exposure photography is possible with the Junos but only after adding some counterweights to offset the camera's weight. (Exact balancing is critical due to the lack of an altitude-axis lock.)

Owners rave about Juno instruments, recounting tales of optical and mechanical performance that meets or beats the competition. Although the equatorial tables increase their prices over conventional Dobsonians, Juno telescopes are highly recommended.

Meade Instruments Corporation

At the same time they introduced their new apochromatic refractors, Meade also unveiled a revamped line of economical Newtonian reflectors called Starfinders. Consumers can select between Starfinders on clock-driven German equatorial mounts or economical Dobsonian-style altazimuth mounts.

Starfinders come in five apertures: 6-inch f/8, 8-inch f/6, 10-inch f/4.5, 12.5-inch f/4.8, and 16-inch f/4.5, with all optical components made from fine-annealed Pyrex glass. Optically, Starfinders perform quite well, and represent some of the best values on the market today. Images are sharp and clear even at relatively high magnification. Unfortunately, the supplied 25-mm 1.25-inch eyepiece offers mediocre performance, at best.

At first glance, the Starfinder optical tube assemblies appear identical regardless of mounting. All have tubes made of spiral-wound cardboard tubes and share the same 1.25-inch rack-and-pinion focusing mounts (except the 16-inch equatorial model, which has a 2-inch focuser), but there are subtle differences beneath the surface. The primary mirror mounts are one example. The equatorial telescopes use mirror mounts that are superior to those used in the cheaper Dobsonians, making adjustment easier. Another instance is that the equatorial Starfinders come with finderscopes (the 6-, 8-, and 10-inchers feature 6×30 finders, and the 16-inch comes with an 8×50), while the Starfinder Dobsonians are sold finderless. Both 6×30 and 8×50 finderscopes are sold separately by Meade as well as from a number of other sources, if desired. Readers should note that the Dobsonians' tubes come with predrilled holes to accommodate Meade's finder mounts.

The Starfinder Dobsonian line use Teflon-on-Nylon bearing surfaces for smooth motions in both altitude and azimuth. The mountings themselves are made of laminated pressboard, nearly identical in appearance to Orion's Deep-Space Explorer line.

Of the equatorially-mounted Starfinders, the 6-, 8-, and 10-inch models are each held by a pair of flexible straps on German equatorial mounts atop 4-inch diameter piers. The straps make it possible to rotate the tube for easier access to the eyepiece. Each mount comes outfitted with an AC-powered spur-gear drive, passable for visual use as well as some simple astrophotography. Although the mount is adequate for the 6-inch, it begins to teeter under the weight of the 8- and 10-inchers. The 12.5-inch Starfinder is not available on an equatorial mount at this time. (By the by, don't be fooled by what look like tapered shafts on the Starfinder mounts; the shafts are straight. Recall from an earlier discussion that tapered shafts are more desirable.)

I singled out the 16-inch equatorial Starfinder, since it is a little different from its smaller siblings. Included with the package is a 2-inch focusing mount (an adapter for 1.25-inch eyepieces is included) and an inexpensive 25-mm eyepiece. Optical quality is very good, the equal to many other more expensive telescopes, although Meade's tall eyepiece holder requires a large (4-inch) secondary mirror. The 16-inch comes on a German equatorial mount that rides low to the ground. Though it is more substantial than the mounts supplied with the 6-, 8-, and 10-inch Starfinders above, it proves wobbly under the weight of the 16-inch. Another drawback to the Starfinder 16's mounting is that the tube is attached directly to the mount's cradle, rather than with straps like the others. Tube rotation, therefore, is impossible, making it difficult to get to the eyepiece at times. This will also cause undue stress to the tube over time, possibly resulting in tube sag or perhaps greater damage.

Summing everything up, Starfinders are very good choices for amateurs who want well-made instruments without spending a fortune. The Dobsonians are superior to both Coulter's and Orion's big-but-cheap reflectors in just about every way. (For a further comparison of the Starfinder Dobsonians to Orion's Deep-Space Explorers, see the discussion under Orion.) If finances permit, however, consider the Starfinder equatorials over their Dobsonian brethren. Congratulations to Meade for making telescopes that everyone can use.

One other reflector belongs to the Meade stable of Newtonians: an imported 4.5-inch f/8 instrument that is just about identical to Orion's Space-Probe, which is discussed later in this chapter. The Meade model 4420 comes on a less-than-adequate equatorial mount. Happily, the telescope accepts standard 1.25-inch diameter eyepieces but is still plagued by a tiny 5×24 finder. If this is the aperture telescope you want, then consider the Celestron C4.5 over the Meade 4420.

Obsession Telescopes

Obsession telescopes (Figure 4.5) are considered by some observers as the finest large-aperture alt-azimuth Newtonians sold today. They are famous for their sharp optics, clever design, fine workmanship, and ease of assembly and use. All add up to a winning combination.

Six models adorn the Obsession telescope line: 18-, 20-, 25-, 30-, 32-, and 36-inchers. Company owner Dave Kriege chose to use primary and secondary mirrors manufactured by Galaxy Optics, famous for producing excellent images. The secondary mirrors feature enhanced aluminizing to increase their reflectivity, while the primaries come with standard aluminizing.

The overall design of the Obsessions makes them some of the most user-friendly large-aperture light buckets around. Their open-truss tube design allows the scopes to break down for easy transport to and from dark-sky sites (keep in mind, we're still talking about BIG telescopes that are much more unwieldy than smaller instruments). Once at the site, the 18- and 20-inch Obsessions can be set up by one person in about 15 minutes without any tools; the bigger scopes really require two people but are still quick to assemble. Their no-tool design is especially appreciated by those of us who know what it is like to drive to a remote site, only to find out that the telescope cannot be set up without a certain tool inadvertently left at home. Even nicer is the fact that all

Figure 4.5 *The Obsession-20, an outstanding 20-inch f/5 Newtonian reflector on a state-of-the-art Dobsonian-style alt-azimuth mounting. Photo courtesy of Obsession Telescopes.*

hardware remains attached to the telescope, so it is impossible to misplace anything. (Ever drop a wing nut in grass at night? I'm convinced that grass eats them.) The only way to lose Obsession hardware is to lose the telescope.

Another plus is the Obsessions' low-profile Crayford focusing mount. A Crayford focusing mount uses rollers instead of a rack-and-pinion gear to move the focusing tube in and out. With the eyepiece held closer to the tube, a smaller secondary mirror can be used to keep the size of the diagonal (and the central obstruction) to a minimum.

Many other conveniences adorn the Obsessions. For instance, a small 12-volt DC pancake fan is built into the primary mirror mount to help the optics reach thermal equilibrium with the outside air more quickly. Another advantage is that each instrument comes with a pair of long, removable handles attached to a set of rubber wheels. The handles quickly attach to either side of the telescope's rocker box, letting the observer wheel the scope around like a wheelbarrow—a good thing too, as the 20-incher's rocker and mirror boxes tip the scales at about 150 pounds. Of course, these handles do not solve the problem of getting the scope in and out of a car. If you own a van, hatchback, or station wagon with a low rear tailgate, the scope may be rolled out using a makeshift ramp, but if it has to be lifted, two people will be required.

It should come as no surprise, given the aperture of these monsters, that their eyepieces ride high off the ground. When aimed toward the zenith, the eyepiece of the 20-inch is at an altitude of about 95 inches. Unless you are as tall as a basketball player, you will probably have to climb a stepladder to enjoy the view. A tall ladder is definitely in order to look through the bigger Obsessions; the eyepiece of the 36-inch scope (when aimed at the zenith) is a towering 15 feet off the ground. What might come as a surprise is that Obsession scopes do not come with any eyepieces or a finderscope. Instead, each comes outfitted with a one-power Telrad aiming device (see Chapter 6). Like all open tubes, a cloth shroud is recommended when the telescope is used in light-polluted locations; Obsession offers optional custom-fit black shrouds for all of their instruments. (John Dobson himself recommends using *not* a black cloth shroud but instead a "space blanket," which he feels better prevents extraneous heat—notably the observer's body heat—from fouling the telescope's view.)

Nothing is small about the Obsession scopes. Besides large apertures, considerable weight, and substantial girth, their cost is quite heavy. Yes, they are expensive, but after all, they are a lot of telescope—and the standard by which all other Dobsonian-style Newtonians are judged! For the observer who enjoys looking at the beauty and intrigue of the universe and has ready access to a dark sky, using Obsession telescopes really can become obsessive.

Optical Guidance Systems (OGS)

For the rich-and-famous die-hard amateur astronomer, Optical Guidance Systems offers some impressive telescopes that are worth a look. OGS specializes in medium-to-large–aperture Newtonians and Cassegrains that rank among the finest instruments in their respective classes.

Looking at its Newtonian line first, OGS offers several different instruments ranging in size from an 8-inch f/6 to a huge 24-inch f/4. Each may be purchased either as a tube assembly only or complete with a pier-supported heavy-duty German equatorial mount. In either case, each telescope comes with a 2-inch focusing mount, straight-through 8 × 50 finderscope, and a choice of one high-quality eyepiece. All are useful accessories, though for the prices charged I think at least two eyepieces could have been included.

Optical Guidance Systems is one of the few companies in the amateur market to offer both Classical and Ritchey-Chretien Cassegrain reflectors. Models range from 10 to 32 inches in aperture. The Classical design is available with focal ratios around f/15, whereas the Ritchey-Chretien variants are around f/9. Perhaps the most important benefit of the Ritchey-Chretien (RC) design is its complete freedom from coma. This is especially noteworthy if the instrument will be used for astrophotography.

The smallest OGS Cassegrain is a 10-inch f/8.5 RC on a modified Tele Vue equatorial mounting. Quite candidly, this is too heavy a telescope to place on such a light mount, being better supported on a mount such as the Celestron/Losmandy G11 or OGS's own HP50 equatorial. At the opposite end of the price and size scale is its 32-inch f/12 Ritchey-Chretien complete with a computerized mount.

The strongest selling point of Optical Guidance System's telescopes is their superb optics, which are supplied by Star Instruments of Flagstaff, Arizona. Unlike some manufacturers, who make all sorts of optimistic claims about optical quality, OGS simply states that its telescopes are guaranteed to have a *final* wavefront of ¼ wave or better. Further, they certify this claim by testing the optics with an interferometer, thus eliminating any guesswork.

When assembled, the instruments are rock-steady but also quite heavy. Their 10-inch f/5 Newtonian on the HP50 mount weighs in at about 130 pounds, while a 10-inch f/8.9 Ritchey-Chretien on the HP50 mount tips the scales at 120 pounds. In both cases, the mounting contributes about three-quarters of the weight. Of course, the telescopes and mountings may be broken down into several pieces to facilitate their transport, but still it is best if these instruments can be stored right at the observing site.

Orion Telescope Center

Orion recently added a line-up of four Dobsonian-style Newtonian reflectors to its inventory. The Deep-Space Explorer series comes in four sizes: a 6-inch f/8, 8-inch f/6, 10-inch f/5.6, and 12.5-inch f4.8. All look suspiciously like telescopes offered a couple of years ago by a small company called Pirate Instruments of Temecula, California, but no one is offering an explanation.

Like their chief competition, Coulter's Odyssey telescopes and Meade's Starfinders, Orion's Deep-Space Explorer reflectors minimize on frills and standard accessories to keep prices low. All of the telescope tubes are made of spiral-wound cardboard, while the Dobsonian mounts are made from laminated pressboard. This yields a sturdy mounting with excellent vibration dampening.

Although the Orion Deep-Space Explorers all represent excellent buys, they fall a little short of the similarly priced Meade Starfinder Dobsonians. One of the biggest pluses to the Meades is their rack-and-pinion eyepiece-focusing mount, which is much easier to use than Orion's helical-style focuser. Another point in favor of Meade is that their optics are made of Pyrex, a preferred material to Orion's plate glass mirrors. Starfinders also include four-vane spider mounts to hold their secondary mirrors in place, which are sturdier to the single-vane diagonal mirror holder on the Orion 6- and 8-inch models. (This problem does not hamper Orion's 10- and 12.5-inch models, both of which come with four-vane spider mounts.) A big plus for the Deep-Space Explorers are that they come with a 25-mm Plössl eyepiece that is far superior to Meade's standard eyepiece.

Let us not forget Coulter's Odyssey telescopes either. Coulter's biggest advantages are their lower prices (about $100 to $200 cheaper), somewhat sturdier mounts that feature larger-diameter altitude bearings on the bigger models (Orion and Meade use hockey-puck size bearings for all instruments), and handles to help in lifting. Coulter's drawbacks include poorly designed mirror mounts and eyepiece holders, and a poor standard eyepiece. The biggest failing of the Coulter telescopes, however, can be their long delivery times—in the past, sometimes as long as a year or more.

One other reflector bears the Orion name on its side. The Orion Space-Probe, a 4.5-inch f/8 Newtonian, is essentially the same instrument sold by many department and camera stores under a variety of different brand names. To me, there is a lot more bad than good with those instruments, which is why you will not find any reviewed in this book. But the SpaceProbe is a different story, because Orion makes a number of intelligent improvements to the telescope before it is sold. For instance, the SpaceProbe accepts standard 1.25-inch eyepieces, rather than the subsized 0.965-inch eyepieces used by most instruments of this size class. This gives the owner a much wider and better selection of eyepieces from which to choose. (More on this in the next chapter). Another plus to the SpaceProbe is its finderscope, a straight-through 6×30. Although this is still too small, it is a big improvement over the tiny 5×24 finder supplied with most other 4.5-inch reflectors. On the down side, however, the German equatorial mount supplied with the SpaceProbe is the same as those supplied with the department-store versions of this imported scope. Its weak equatorial head and wobbly wooden legs are just not up to the task. The SpaceProbe retails for a little less than the Celestron C4.5 but lacks the Celestron's solid mounting.

Parallax Instruments

One telescope manufacturer to break the long-focal-length–Newtonian taboo is Parallax Instruments. Parallax, a little-known name among amateurs, has six long-focus Newtonian tube assemblies from which to choose, ranging in aperture from a 4.25-inch f/10 reflector to a 14.5-inch f/7 monster. Just to appease fans of short-focus Newtonians, Parallax also manufactures telescopes down to f/5 across the same aperture range.

Each instrument sold by Parallax boasts a diffraction-limited primary mirror hand-figured by well-known amateur telescope maker Richard Fagin and coated with enhanced aluminum. In addition, all come with low-profile eyepiece focusing mounts, aluminum tubes, and 8 × 50 finders (4.25-inch models are supplied with 6 × 30 finders). The only extras needed are a couple of eyepieces and an adequate mounting.

If the telescope is going to be used for informal visual observations only, then a Dobsonian-style alt-azimuth mounting is one of the simplest and sturdiest ways to get a Parallax instrument up and running. If, however, the telescope will be used for extended planetary studies or photography, then an equatorial mount is a must. For their 6-inch and larger models, Parallax sells the GEM 150, a massive German equatorial mount. These come with worm-gear clock drives, steel piers, wooden storage cases, and optional declination axis drives.

Parks Optical

When it comes to manufacturing the perfect Newtonian reflector, Parks Optical comes mighty close. Both optically and mechanically, Parks Newtonians are among the best in the business, though their cost and weight may be a bit heavy for many hobbyists to bear. They come with apertures ranging from 6 to 16 inches and in no fewer than five different versions, or systems, as Parks describes them. All optical assemblies feature fiberglass tubes held in rotating rings (an important convenience feature) atop German equatorial mounts.

Options vary from system to system. The least expensive 6- and 8-inch Precision and Astrolight Newtonians feature only lone 25-mm Kellner eyepieces and right-angle 6 × 30 finders (although setting circles and clock drives are available separately). At the opposite end of the cost and weight scales are the massive 8-, 10-, and 12.5-inch Superior telescopes, as well as 12.5- and 16-inch Observatory instruments. Although outstanding for their quality, these heavy scopes are only recommended for amateurs who do not need to travel to dark-sky sites or who enjoy weightlifting. For instance, the 8-inch f/6 Superior totals 183 pounds, with the mount alone contributing 125 pounds. At the opposite end, the 16-inch Observatory model weighs in at 755 pounds! No one ever accused Parks telescopes of being lightweights in any sense of the word.

If wide-field photography of deep-sky objects is your forte, then the Parks Nitelight series is especially noteworthy. Nitelights range from 6- to 16-inches in aperture, all with an amazing f/3.5 focal ratio. This makes them the fastest telescopes sold, greatly reducing the exposure times required to record faint sky objects. The 6- and 8-inch models are available on undriven Super Polaris mounts, while another version of the 8-inch as well as the 10-, 12.5-, and 16-inchers are mounted on sturdy though heavy, clock-driven German equatorial mounts. All come with 8×50 finders and one or two eyepieces, depending on the model.

Finally, fans of the Cassegrain telescope hail the unique H.I.T. series of 6- to 16-inch telescopes by Parks. These might be more accurately called Jekyll and Hyde telescopes because they can switch back and forth between being a long-focus Cassegrain and an f/3.5 RFT Newtonian, thereby offering the best of both worlds. To permit this versatility, H.I.T. telescopes come with two interchangeable secondary mirrors. Simply remove one, insert the other, check collimation, and the telescope's alter ego is ready to go. H.I.T. scopes are available on a variety of German equatorial mounts with or without clock drives. The 6-inch comes with a too-small 6×30 finderscope, while the others feature 8×50 finders.

Safari Telescopes

Nowadays, it is difficult for a small company to stand out in the jungle of large-aperture Newtonians. Most of these open-tube-on-Dobsonian-mount instruments end up looking like clones of one another. Owner Allan Green wanted to do something a little different with his telescopes while still keeping the sizes and prices manageable. The result is the Safari 180, an 18-inch f/4.5 Newtonian. Like many of the other instruments in its size class, the Safari 180 uses a modified Seurrier truss to support the optical components . . . nothing new about that. But what is different is that unlike the others, the Safari 180 is fabricated of reinforced fiberglass and aluminum for a unique look. Thermal properties are also very good, allowing for quick cool-down of the optics when the instrument is brought out on cold nights. Perhaps nicest of all is the fact that the Safari 180 is one of the lightest telescopes of its size, making transport to dark-sky sites easy. The instrument may be set up and collimated in a matter of 10 to 15 minutes without using any tools. However, the design is still inferior to others that use captive hardware that remains attached to the telescope even between setups.

The Safari 180 features excellent primary and secondary mirrors from Galaxy optics. Images are both clear and distinct, with minimal aberrations. Also included with the Safari 180 is a low-profile rack-and-pinion eyepiece focusing mount and a Cheshire collimation eyepiece.

Drawbacks to the Safari? There are a couple. First, Safaris are not supplied with either finderscopes or eyepieces. Also, while the Safari's mounting moves smoothly, it is not quite as stable as, say, Obsession, Starsplitter, or Tectron telescopes. Finally, there is no provision for attaching wheels to move the instrument around once it is set up. The only way to move the instrument is to tear it down, relocate it to where you want, and then set it up all over again. Still, with a price that is hundreds of dollars less than the competition, the Safari 180 represents an excellent buy in large-aperture telescopes.

Sky Designs

From Colleyville, Texas, comes a trio of large-but-portable Newtonian/Dobsonian telescopes by Sky Designs. Sky Designs instruments—a 14.5-inch f/4.5, 18-inch f/4.5, and 20-inch f/4—all follow the popular open-tube theme. Connecting the upper secondary-mirror cage to the lower primary-mirror box are eight aluminum poles, all secured in place by hex nuts and bolts. Unfortunately, as supplied, the Sky Designs system requires using a small wrench to tighten everything up—an inconvenience. I would suggest that owners substitute either threaded knobs, or at least wing nuts, to make the job a little easier.

Fit and finish to both the wooden primary-mirror box and the secondary-mirror cage are excellent, with birch veneer plywood used throughout. All primary mirrors in Sky Designs instruments are made by the company rather than by a third-party supplier. Each is tested both on the bench as well as in the final optical tube assembly to ensure quality.

The biggest drawback to Sky Designs telescopes is that the secondary mirror is glued directly onto a wooden dowel. Adjustment is made by turning two sets of opposing screws into the dowel, more difficult than with conventional spider mounts. Like the Safari, Sky Designs does not include a provision for attaching wheels to move the instrument around once it is set up. Once again, the only way to move the instrument is to tear it down, relocate it, and then set it up all over again.

Star-Liner Company

For more than three decades, the name Star-Liner has been associated with elegant, albeit costly, Newtonian and Cassegrain reflectors. Currently, its telescopes are available in several different product lines. The Star-Liner Econo-Line features a pair of Newtonians: a 6-inch f/8 and an 8-inch f/7. Both come mounted on heavy-duty German equatorial mounts. Econo-Line features include a rotating fiberglass tube, a much-too-small 6×24 finder, and two eyepieces—surprisingly spartan for telescopes that cost in excess of $1,200. Only the Deluxe Econo-Line instruments come with a single-axis AC clock drive; the Standard mounting is undriven. Neither a manual nor an electric control for the declination axis is available.

Next up is the Quality-Line echelon of 6- to 14.25-inch Newtonians. Like the Econo-Line reflectors, the Quality-Line 6- and 8-inch instruments come in both standard (with no clock drive) and deluxe (with clock drive) models. The 10-, 12.5-, and 14.25-inchers all feature AC-powered single-axis clock drives. In addition, a choice of either manual or electric declination axis control is available at extra cost. Included with the 6- and 8-inch models are two eyepieces and 7× finderscopes; the others feature 10×50 finders and three eyepieces. The 6- and 8-inch Quality-Line scopes are light enough to transport from home to a dark-sky site, while the 10-, 12.5-, and 14.25-inch models are described as "transportable." In telescope lingo, that means *heavy*: the 14.25-incher weighs in at 290 pounds (of course, it can be broken down into several subassemblies for transport).

Star-Liner is one of the few companies to offer a line of Cassegrain telescopes. Ranging in size from 8 to 24 inches, Star-Liner Cassegrains are available in two optical configurations: a straight Cassegrain or a convertible Cassegrain-Newtonian (like the Parks H.I.T. series, this requires the replacement of the secondary mirror.) This latter approach allows the best of both worlds: a fast f/3.8 Newtonian for wide-angle deep-sky observing and a long-focus f/12 Cassegrain for planetary scrutiny. Once again, the 6- and 8-inch telescopes are portable; the 10-, 12.5-, and 14.25-inchers are "transportable," while the larger apertures are "observatory" class. These are serious telescopes carrying serious prices, though they are less expensive than comparable instruments by Parks.

Starsplitter Telescopes

Starsplitter telescopes, a new name on the market, are typical of the low-profile Dobsonian-style instruments that are quite popular among serious deep-sky observers today. Available in seven aperture sizes from 12.5 through 30 inches, all feature primary and secondary mirrors by Galaxy Optics. (As previously noted, Galaxy mirrors are considered among the best available.)

Starsplitter telescopes are designed around open-truss tubes that allow for comparatively easy storage. Their no-tool design uses captive hardware, making separate tools, such as wrenches and screwdrivers, unnecessary for setting up and tearing down the instrument, an especially nice feature when taking the telescope to remote observing sites. The mounting and baseboard are made of fine-quality wood and feature smooth Teflon®-on-Formica® bearing surfaces. In addition, each comes with a low-profile focusing mount, a Telrad aiming device, and wheelbarrow handles to help move the instrument without having to disassemble it first.

In fact, I was so impressed with Starsplitter telescopes when I first saw them in 1993 that I now own one of their 18-inchers. It is a joy to use. Optical and mechanical quality and workmanship are equal to the finest instruments on the market. And because Starsplitters are made on an order-by-order basis, each may be customized to a person's needs. In many ways, Starsplitters are clones of the better-known Obsession instruments, but with prices hundreds of dollars less. They come highly recommended.

Tectron Telescopes

Like the Obsession and Starsplitter instruments mentioned above, these have redefined the meaning of "large-aperture light-bucket telescopes." To sum them up in one word, Tectron telescopes are superb. Like the others, they feature open-truss optical tube assemblies set on finely crafted Dobsonian-style alt-azimuth mounts made of oak plywood. Assembly of the 15- and 20-inchers can be performed by one person, while the 25- and 30-inchers really require two.

Once erected, the telescopes are wonderfully stable and a joy to use and steer across the sky, even at high power. Disassembly and storage are also easy, as the secondary-mirror cage is designed to nest inside the primary mirror box; only the six truss tubes remain outside the box. The Tectron 30, for instance, folds down into a 38″ × 39″ × 33″ cube, a great feature for us telescope-rich-but-storage-space-poor hobbyists.

The optics used in Tectron telescopes are excellent. I can recall a view of the Orion Nebula I had through a Tectron 25-inch a few years ago. The sight bordered on a religious experience. Even planetary images are sharp and clear at full aperture (a quality for which many fast Newtonians are not known). Tectron's secondary mirrors have enhanced-aluminum coatings to increase their reflectivity; the primaries come with standard aluminizing.

The sundry items on Tectron scopes, such as the low-profile 2-inch focuser and 18-point mirror cell, are all made by Tectron in its own shop, ensuring quality control. All features of Tectron telescopes come together to produce excellent instruments that are well regarded by their owners. Two items missing from Tectron telescopes are a finder and an eyepiece; all come with only a one-power Telrad aiming device (again, like Obsession and Starsplitter).

Which is better: Tectron, Starsplitter, or Obsession? Optically and mechanically, they are all excellent. As I see it, there are two shortcomings to Tectron telescopes. One surfaces when moving them around once set up. The 20-inch and larger Tectrons offer snap-on casters as an option, but they do not work as well as the wheelbarrow approach, especially on irregular terrain. Also, Tectron's trusses are assembled using large knurled knobs. Although they do not require any tools to be tightened, they can get misplaced easily at night. Though Obsessions are a little more convenient to assemble and move around, Tectrons are a better buy.

Torus Optical

Located in Iowa City, Iowa, this is a small company in the business of making big telescopes. Following the lead set by Obsession and Tectron, Torus offers for sale large-aperture open-tube Newtonians set on furniture-grade wooden mounts of birch and mahogany veneer plywood. All woodworking is done by a professional boat builder who obviously takes great pride in workmanship. Torus owner James Mulherin, a physicist, oversees the grinding of the mirrors right on the premises, promising diffraction-limited results. The mirrors are then sent to Clausing, Inc., to be coated with their Beral coating, which is a variation on aluminizing that is not as reflective as so-called enhanced aluminizing.

Torus telescopes are available in 12.5- to 24-inch apertures, all between f/5 and f/5.6, although focal ratios to customer specifications are also available. All instruments are outfitted with a well-made, low-profile focusing mount and a Telrad aiming device.

Assembly, disassembly, and collimation of Torus telescopes requires the use of a ball-driver tool—not quite as nice as some no-tool designs, but at least the tool is supplied with each instrument. Two important features missing from Torus scopes are add-on wheels and handles for moving the instrument around after setup. These are sorely needed, especially with the big-aperture models, but thankfully the telescopes are light for their size; indeed, they are among the lightest available.

Vixen Optical Industries

While most of Vixen's telescopes are sold in North America under the name Celestron, a few are offered under Vixen's own label. Of these, only a pair of 4.5-inch f/8 instruments highlight this outfit's line of Newtonian reflectors. Both feature the same optical tube assembly as Celestron's C4.5 model but are inferior to the C4.5 mechanically. The Vixen R-114 comes on a wobbly alt-azimuth mounting, while the D-114E features a department-store–esque German equatorial mount. Both are supplied with aluminum tripods. Another drawback to the Vixen twins is that they can only use subdiameter (0.965-inch) eyepieces. By contrast, both the C4.5 and Orion's SpaceProbe use higher-quality 1.25-inch oculars.

Catadioptric Telescopes

Celestron International

Celestron is world renowned as the first company to introduce the popular 8-inch Schmidt-Cassegrain telescope back in 1970. While only one basic model—the Celestron 8—was sold back then, there are now many variations of the original design from which to choose. They range from bare-bones telescopes to extravagant, computerized instruments. Celestron tells us, however, that all of its 8-inch SCTs share similar optics with the same optical quality, regardless of model and price. All except the most basic C8 feature Starbright enhanced coatings for improved light transmission and image contrast. To my eye, these coatings are the difference between fair and good views. They are strongly recommended. Readers should note, however, that for years these special coatings were considered optional. If you are looking to buy a used Celestron 8, double-check to make sure that it has enhanced coatings; otherwise, reject it and continue your search.

It is important to note up front that both Celestron and Meade (to be discussed later) Schmidt-Cassegrain telescopes suffer from something called *mirror shift*. To focus the image, both manufacturers chose to move the primary mirror back and forth rather than the eyepiece, which is more common. Unfortunately, as the mirror slides in its track, it tends to shift, causing images to jump and impeding exact focus. The current telescopes produced by both companies have less mirror shift than earlier models, but it is still evident.

Celestron offers several variations of the C8, with equipment levels varying dramatically from one model to the next. At the low end of the scale are the Classic 8 and the Great Polaris-Celestron 8, or GP-C8. Both come with so-so 25-mm Kellner eyepieces and small 6×30 finderscopes. Owners quickly discover they need to buy new finderscopes (an 8×50 would be nice) and a couple of better-quality eyepieces to make the packages complete.

The GP-C8, the next generation of the popular Super Polaris-C8 (SP-C8), is mounted on a tripod-supported Great Polaris German equatorial mount. The GP features an improved latitude adjustment system but is otherwise comparable to the older SP mounting. A motor drive is not supplied as standard, although an AC-powered drive is available as an option. The GP mount is surprisingly steady and proves quite adequate for supporting a telescope of this size.

The Classic 8 comes on a less-steady fork mount outfitted with an AC spur-gear clock drive but without a much-needed tripod (for that you will have to pay extra). While a spur-gear drive is fine for the visual observer who just wants to keep a target *somewhere* in the field of view for an extended period, it is not as accurate as a worm-gear drive. As a result, long-exposure astrophotography through the telescope is quite tedious, especially without a dual-axis drive corrector (available from a number of different manufacturers—see Chapter 6).

Next up in the price range is the C8+, available both with and without a built-in computer. The C8+ is a new and somewhat simplified version of the older Powerstar 8. Each C8+ comes equipped with a self-contained clock drive that is powered by a single 9-volt alkaline battery—a great feature for observers who spend most of their time under the stars far from civilization. (Once again, this is a recent feature for Celestron. Older C8s, such as the Super C8 and the Super C8+, require external AC/DC drive correctors to run their clock drives off portable batteries.)

The C8+ Computerized features a built-in Advanced Astro Master with a resident memory capable of locating over 8,000 sky objects once the mounting is polar aligned. Impressive? Yes. Desirable? I remain unconvinced. It takes more than just a fancy telescope to be an amateur astronomer; for that, you must know the sky. Sure, a computer-controlled telescope will show more objects in a given time period than an observer could possibly find by eye, but where is the challenge in that? To my way of thinking, half of the fun observing the night sky is in the thrill of the hunt. Is there a place in the amateur world for computerized telescopes? Yes, if you are involved in an advanced observing program, such as estimating the brightness of variable stars or searching for extragalactic supernovae. But if you just enjoy looking around, I feel you would do better with a simpler instrument and find your own way around the sky. 'Nuff said (at least for now).

At the high end of the price scale is Celestron's flagship Ultima 8 (Figure 4.6), which features a sophisticated, self-contained clock drive that also runs on a single 9-volt alkaline battery and a beefed-up fork mounting for added rigidity. This latter innovation will be of special interest to all who aspire to take long-exposure photographs through the telescope.

Figure 4.6 *The Ultima 8, Celestron's flagship 8-inch Schmidt-Cassegrain instrument. Photo courtesy of Celestron International.*

The Ultima's clock drive uses oversized worm-gear systems and has a built-in Periodic Error Control, or PEC, circuit for greater tracking accuracy. Theoretically, a worm-gear clock drive should track the stars perfectly if it is constructed and aligned accurately, but this is not the case in practice. No matter how well-machined a clock drive's gear system is or how well-aligned an equatorial mount is to the celestial pole, the drive mechanism is bound to experience slight tracking errors that are inherent in its very nature. These errors occur with precise regularity, usually keeping time with the rotation of the drive's worm. The PEC eliminates the need for the telescope user to correct continually for these periodic wobbles. After the observer initializes the PEC's memory circuit by switching to the record mode and guiding the telescope normally with the hand control (typically a five- to ten-minute process), the circuit plays back the corrections to compensate automatically for any worm-gear periodicity. Be aware, however, that the PEC will *not* perform its function without being retaught every time that the clock drive is switched on. Neither will the PEC make up for sloppy polar alignment. (Again, see the discussion in the section on Meade.)

Celestron recently reintroduced the C5, a 5-inch f/10 SCT that comes in three different versions: the C5 Spotting Scope, the Classic C5, and the C5+. Because all C5s feature identical optical components enhanced with Starbright coatings, the only real difference between the models is in their mountings.

The spotting scope is sold unmounted, making it necessary to attach it to a camera tripod before use. This makes it better suited for terrestrial rather than astronomical viewing, although it certainly can be aimed skyward if desired. Both the Classic C5 and the C5+ come mounted on a single-arm fork mount featuring a built-in clock drive. The Classic's drive runs only on 110-volt AC power and so must be mated with a portable AC/DC power supply for powered operation in the field. The C5+ comes with a built-in drive powered from a single 9-volt battery. Celestron claims a single battery will last an estimated 50 hours. The optical tube assemblies for both the Classic and the C5+ are mounted on dovetail plates that permit easy removal for transport, a nice touch for jet-setting astronomers. Most others, however, may find the small aperture rather limiting, making it worthwhile in the long run to spend an extra few hundred dollars to get an 8-inch SCT instead.

For those who want a larger SCT, Celestron makes the Ultima 11-PEC, CG-11, and the Celestron 14. The Ultima 11-PEC 11-inch f/10 SCT is basically an overgrown version of the Ultima 8, sharing all of its strengths (portability, extensive choice of accessories, etc.) and weaknesses (mirror shift, low contrast, etc.). Housed in the base of the Ultima 11 is an accurate dual-axis drive system designed by Edward Byers Company and featuring a 7.54-inch worm gear and matching stainless-steel worm. If you are purchasing an Ultima 11, make sure you get the newly redesigned mount that incorporates improved fork tines, wedge, and tripod. Its performance is far superior to older models. In addition, the instrument comes with an 8×50 finder, a 2-inch star diagonal, 18- and 30-mm Plössl eyepieces, sliding counterweights to make balancing cameras and other accessories easier, and vibration suppression pads as standard equipment.

If astrophotography is your passion, then Celestron's new CG-11 will be of great interest. While the optical assembly and accessories are the same as the Ultima 11-PEC, the CG-11 comes mounted on an absolutely rock-steady German equatorial mounting made by Scott Losmandy's Hollywood General Machining. Telescope and mounting slide together by means of a 17-inch-long dovetail bar that spans the length of the optical tube assembly. This freedom lets the user set telescope balance precisely—a big plus when adding cameras or other accessories.

The G-11 mount is one of the few commercial units that successfully combines portability with stability in one neat parcel. The dual-axis drive system marries two pairs of 5.625-inch worm gears and matching stainless-steel worm gears to a crystal-controlled circuit that can run on either 115 volts AC or 12 volts DC. A PEC circuit is also included to help the user correct for any periodic gear error in right ascension, while the declination drive features a unique Time Variable Constant (TVC) setting to eliminate backlash. One of the neatest features of all is the mount's unusual polar-alignment telescope. Through the illuminated reticle, the observer sees patterns of the Big Dipper and Cassiopeia; when the patterns line up with the stars in the sky, the mounting is aligned. All of these features, and more, combine to make the Celestron CG-11 the most versatile large-aperture SCT available today.

The Celestron 14 is truly an observatory-class instrument that can also be made transportable with two or more people. Like the smaller C8+ and Ultima series, the C14 comes with motorized slow-motion controls on both the right ascension and declination axes (a must for astrophotography) but does not feature the PEC circuit. And for the rich and famous, Celestron offers the computer-controlled Compustar 14. In its memory are data on just about every sky object visible in a 14-inch telescope. How much does the Compustar 14 cost? If you have to ask, you can't afford it!

So which brand is better: Celestron or Meade? That question has been pondered by amateurs now for about two decades. Though both companies have been criticized openly in the past for poor quality, they seem to have cleared up most of the deficiencies. As it is now, both produce about equal-quality optics, with a slight nod going to Celestron. (Though most observers agree that Schmidt-Cassegrains universally produce inferior images to high-quality refractors and Newtonians.) As to the mountings, Meade's new 8-inch LX200 series wins out over the Celestron 8 variants, but for larger mountings, the CG-11 cannot be beat. How will Celestron answer the challenge of the Meade LX200? Stay tuned—the bell for round two is about to sound, because Celestron has just introduced a new line of computerized telescopes that take direct aim at the Meade LX200.

Ceravolo Optical Systems

Peter Ceravolo, formerly on the editorial staff of the defunct *Telescope Making* magazine, is currently producing a limited number of outstanding 5.7-inch f/6 Maksutov-Newtonian telescopes. These fine instruments feature all-aluminum construction and customized low-profile helical focusing mounts for both 1.25- and 2-inch eyepieces. The HD145 Mak-Newt is unhindered by coma, astigmatism, and other aberrations that plague lesser instruments. Contrast is also very good, thanks to a small 1-inch secondary mirror. (As mentioned previously, larger secondary mirrors lessen image contrast.) When mated with a Tele Vue 22-mm Panoptic eyepiece, for instance, the HD145's view encompasses 1.8°, nearly four times the diameter of the full moon, with pinpoint star images seen across the entire field. At higher powers, fine planetary detail comes through clearly, rivaling that seen with apochromatic refractors of similar aperture and focal ratio but costing double or triple the price. Few telescopes are able to compete with such impressive performance.

The HD145 is not available from stock but rather is constructed only on a per order basis. As a result, delivery times range between three and six months. The telescope as delivered does not include an eyepiece, finderscope, or mounting. Mounting rings to fit the instrument to either Tele Vue or Celestron/Vixen equatorials are available. For those looking for the performance of a short-focus apochromatic refractor but with a slightly different spin, the Ceravolo HD145 Mak-Newt may be for you.

INTES

Many eyebrows were raised when advertisements for this Russian-made 6-inch f/10 Maksutov first began to appear in both *Sky & Telescope* and *Astronomy* magazines a few years ago. The price of under $2,000 for both the optical tube assembly and clock-driven fork mount made it especially attractive. But the question remained: How well does the INTES perform?

Mechanically, the INTES Maksutov operates quite well. Both the primary and secondary mirrors are held fixed in the tube, thereby eliminating the problem of mirror shift experienced by Meade and Celestron SCTs. However, the helical focuser supplied with the telescope seems weak and not as well designed as it ought to be.

The INTES fork mount is also a little wobbly (as are just about all commercial fork mounts), but it and the single-axis clock drive built into its base are adequate to the task. The worm-gear drive may be powered by either 115 volts AC or an external 12-volt DC battery, making it handy for field operations. Only a manual declination slow-motion control is included with the instrument.

In a side-by-side test between the INTES and a Celestron 8, the INTES images were not up to par. Though they appeared sharp, the images lacked contrast and pizzazz, even though the optics are proclaimed as being multi-coated and hand-corrected by a Russian optician.

The INTES comes with a narrow-field 10×30 finderscope as well as a 2.4-inch Maksutov guidescope mounted piggyback on the tube. Also included are a wooden carrying case, mirror-type star diagonal, and a camera adapter. Although a tripod is not included, one is available separately.

Meade Instruments Corporation

Meade did it again with the introduction of its new LX200 series of 8- and 10-inch Schmidt-Cassegrain telescopes, both available with either f/6.3 or f/10 focal ratios. These telescopes are state of the art, with the most sophisticated mountings ever packaged for the amateur astronomer.

The LX200 series are mounted on computer-controlled fork mounts that automatically slew the telescope across the sky from object to object simply by entering the desired target into the hand-held controller. Best of all, the telescopes do not have to be polar-aligned for the 12-volt DC clock drive to locate or track an object. To calibrate the system, set the correct time and date, then simply aim the telescope at one of the reference stars listed in the instruction manual. That's all there is to it; the instrument is now set to find and track any of the 747 objects (expandable to 8,000 objects) stored in its memory. (Uh-oh, another computerized telescope! See my earlier remarks concerning Celestron's C8+ Computerized.) This allows the telescope to be set up in an alt-azimuth configuration, proving much steadier than when the fork mount is tilted by an equatorial wedge. In the alt-azimuth mode, the LX200s are the sturdiest fork-mounted SCTs ever. (Of course, in this configuration, they are more like Dobsonian alt-azimuth mounts.) The only shortcoming is that for long-exposure photography, alt-azimuth–induced field rotation can only be eliminated by using an equatorial wedge. Although Meade does not supply a wedge with any of the LX200s, one may be purchased separately.

Like Celestron's Ultima 8, the LX200's Permanent Periodic Error Correction circuit allows the user to compensate for minor periodic worm-gear inaccuracies. However, the Celestron must be reprogrammed each time the drive is switched on; the Meade PPEC remembers the steps needed to compensate for the inaccuracies forever.

Meade also offers the less expensive LX100 series of 8- and 10-inch f/6.3 and f/10 Schmidt-Cassegrains. Though mounted on the same fork mounts as the LX200s, the LX100s do not feature the same sophisticated electronics. As such, they must be polar-aligned before the instruments will track the sky. (Oh well, back to reality.) What they do offer, however, is an accurate 12-volt DC Smart Drive tracking system with the same built-in PPEC circuit as the LX200.

The LX100 and LX200 series have identical optical-tube assemblies featuring fully "super multi-coated" optics. As with Celestron's Starbright coatings, these are a must! Meade's light transmission and image contrast are good for Schmidt-Cassegrains, though I still feel that Celestron has a slight edge. (But neither can compare to a good refractor or Newtonian.)

As to choosing between focal ratios, I would almost invariably advise against the f/6.3. The reason is simply a matter of image contrast. Because of the large central obstructions from their secondary mirrors, all SCTs suffer from lower image contrast than refractors and most Newtonian reflectors. Therefore, it makes sense to make the secondary mirror as small as possible. To achieve the faster focal ratio, however, the Meade f/6.3 scopes must use a larger secondary mirror than the equivalent f/10s, thereby decreasing contrast. If focal ratio speed is of concern to you (attention photographers), consider purchasing an f/10 SCT and a Celestron Reducer/Corrector attachment that cuts the focal ratio to f/6.3 (see its review in Chapter 6).

With each LX100 and LX200, Meade includes a variable-height metal field tripod, a foam-fitted carrying case for the telescope, a 26-mm Super Plössl eyepiece, and a straight-through 8×50 finder (6×30 with the 8-inch LX100).

Though not heavily advertised, Meade also offers its 8- and 10-inch f/6.3 and f/10 Schmidt-Cassegrains on the LXD600 German equatorial mount, the same mounting as its 4- and 5-inch apochromatic refractors. Although these are recommended over the base 2080 and 2120 models, Meade should have used the larger, sturdier LXD700 mount from its 6- and 7-inch apos instead. (Future plans call for 12- and 14-inch Schmidt-Cassegrain telescopes on the LXD700 mount.)

Least expensive of the Meade 8- and 10-inch SCTs are the 2080 8-inch f/10 and the 2120 10-inch f/10. Both are throwbacks to earlier times at Meade, with weaker fork mounts, 115-volt AC motor drives, and manual-only slow-motion controls. In addition, the 2080A and 2120A do not come with the all-important optical coatings on the corrector plates, and as such they should be avoided.

The 2080B and 2120B do feature these coatings, producing somewhat better images.

Lastly, Meade has the 2045D, a 4-inch f/10 mini-SCT. The 2045D comes with a super multicoated corrector plate, built-in 12-volt DC clock drive, and three-legged tabletop support. Optional accessories include an equatorial wedge and field tripod and a carrying case. It is a nice, portable instrument, but the 2045D is quickly outgrown by many owners who yearn for more than the small 4-inch aperture can deliver.

Questar Corporation

The most popular Maksutov-Cassegrain models are made by Questar, which sells them in three different apertures: 3.5-, 7-, and 12-inch. Due to the already hefty price, most amateurs choose the 3.5-inch—*big* Maks just cost too much. For instance, the Standard 3.5-inch Maksutov from Questar retails for about twice the price of a Meade or Celestron 8-inch Schmidt-Cass!

Questar telescopes (Figure 4.7) come outfitted with many little niceties that add to the user's pleasure. One of the best is actually very simple: a built-in telescoping dew cap that effectively combats fogging of the corrector. Another welcome plus is a screw-on solar filter that allows safe viewing of our nearest star. No other telescope comes with one as a standard accessory. And the quality of assembly is without parallel.

Figure 4.7 *A piece of art plus a telescope all in one: the Questar 3.5-inch Maksutov-Cassegrain telescope. Photo courtesy of Questar Corporation.*

As nice as Questars are, a couple of idiosyncrasies plague the 3.5- and 7-inch models. For one, only Questar's own custom-made Brandon eyepieces fit into their eyepiece holders. Owners should note that Tele Vue now offers an adapter that permits use of standard 1.25-inch eyepieces (they hope theirs) in Questars. The other weakness in my opinion is actually looked upon as a plus by many people. Instead of equipping the telescope with a separate finder, Questar built one right into the instrument. By flipping a lever, the observer can switch back and forth between finder and main telescope without ever leaving the instrument's eyepiece. Though the design is clever, in practice it takes a lot of getting used to before the telescope can be accurately aimed toward a target. Many people believe this to be a great convenience, so I guess it is largely a matter of personal preference.

Naturally, the small size of the Questar 3.5 dramatically limits what can be seen, but of what is visible, the images are exquisite. This appeals to the many amateurs who prefer image quality over aperture quantity, and there is certainly something to be said for that philosophy. Think of it in terms of a fine painting versus a snapshot photograph. For most people, the snapshot adequately shows the scene. Although it may miss some of the fainter details, the painting reaches deeper to touch the soul of those who appreciate such things. After all, a Questar is as much a work of art as it is a scientific instrument.

Takahashi

Best known in North America for its refractors, Takahashi also manufactures a 9-inch f/12 Schmidt-Cassegrain telescope that is considered by most to be optically superior to all others of the genre. The Takahashi TSC-225 comes standard with an oversized 9.3-inch primary mirror that is hand-figured to deliver a 1/8-wave final wavefront at the eyepiece. Both it and the secondary mirror have multilayered, enhanced aluminum coatings to deliver the brightest, clearest images possible in a scope of this type. (Image contrast is better than in either Meade or Celestron telescopes due to superior baffling and the secondary mirror's comparatively small 28% obstruction of the primary's diameter—less than the competition's SCTs.) Like Celestron and Meade SCTs, the TSC-225 focuses light by moving the primary mirror back and forth. As a result, mirror shift seems inevitable, though it is an order of magnitude less than in either Meade and Celestron instruments.

The TSC-225 comes with Takahashi's EM-200 German equatorial mount on a wooden tripod. Built into the mounting are a neatly designed polar-alignment scope and an accurate 12-volt DC dual-axis drive corrector. Also included with the package are a 25-mm orthoscopic eyepiece, a 2-inch star diagonal with a 1.25-inch adapter, and a 7 × 50 straight-through finder. Available options include a fan that draws filtered ambient air through the tube assembly to aid in cooling.

Takahashi's TSC-225 reigns as the finest Schmidt-Cassegrain in production today, beating its American-made competitors on every count except two:

slow focal ratio and price. The former is important for astrophotographers but not others, whereas the latter hits everyone where it hurts: the wallet. The TSC-225 costs about three to four times more than either a Meade or Celestron 8-inch SCT!

The Choice Is Yours

With so many telescopes and so many companies from which to choose, how can the consumer possibly keep track of everything? Admittedly, it can be difficult, but hopefully Tables 4.1a through g will help a little. They list all the manufacturers mentioned above, telling you at a glance what companies make which kinds of telescopes.

Once you find brands and models that interest you, review the earlier comments about your choice(s). Then contact all of the appropriate manufacturers for catalogues, brochures, and anything else they are willing to send you. With literature in hand, give everything a thorough once-over but resist the temptation to buy impulsively based on color pictures of the telescopes being used by attractive models! Instead, continue to the back of the catalogue, where you should find specification sheets, for they will tell the true story. Study all the data (not just the pretty model) carefully.

There are many things to be on the lookout for when telescope shopping. If you are thinking about buying a refractor, make certain that all the optics are fully coated with a thin layer of magnesium fluoride to help reduce lens flare and increase contrast. As mentioned earlier, multiple coatings are the best. For reflectors and catadioptrics, check to see if their mirrors have enhanced-aluminum coatings to increase reflectivity and produce brighter images as well as protective overcoatings of silicon monoxide. Find out if the telescope comes with more than one eyepiece. Is a finderscope supplied? If so, how big is it? Though a 6×30 finder might be alright to start with, most observers prefer at least an 8×50 finder. If the telescope does not come with a finder, then one must be purchased separately before the instrument can be used to its fullest potential. Next, take a long, hard look at the mounting. Does it appear substantial enough to support the telescope securely, or does it look too small for the task? Remember all that we have gone over in this chapter up to now, and, above all, be discriminating.

Without a doubt, the best way of getting solid information about many kinds of telescopes is by joining a local astronomical society. Chances are good that at least one member already owns the telescope that you are considering and will happily share personal experiences, both good and bad. Plan on attending one of the club's observing sessions, or star parties as they may be called. At these functions, members bring along their telescopes and set them up side by side to share with each other the excitement of sky watching. To find the club nearest you, check either the annual "Guide to Astronomy" in *Astronomy* or the yearly "Resource Guide" in *Sky & Telescope*. You can also

Table 4.1a **Binoculars**

Manufacturer	7x35	7x42	7x50	8x40	8x56	9x63	10x42	10x50	10x70	11x80	12x50	14x70	14x100	15x80	16x70	16x80	20x80	20x100	25x100	25x150	40x150
Adlerblick		X					X														
Bausch & Lomb	X							X													
Celestron		X	X		X	X	X	X		X	X			X			X				
Edmund			X					X											X		
Fujinon			X						X						X						X
Minolta	X		X	X			X														
Miyauchi																	X	X		X	
Nikon	X		X	X				X													
Orion			X	X	X		X	X		X	X	X	X			X	X		X		
Parks	X		X							X							X		X		
Swift	X		X	X			X			X							X	X			
Unitron		X	X	X			X	X		X							X				
Zeiss, Carl		X	X		X																

97

Table 4.1b **Achromatic Refracting Telescopes**

Manufacturer	3.1″	3.5″	4″	5″	6″	8″	10″	15″
Celestron	X		X					
D&G				X	X	X	X	X
Meade	X							
Orion	X							
Pentax	X							
Unitron	X		X	X				
Vixen	X	X						

Table 4.1c **Apochromatic Refracting Telescopes**

Manufacturer	3″	3.3″	3.5″	3.7″	4″	5″	6″	7″	8″
Astrophysics					X	X	X	X	X
Celestron					X				
D & G						X			
Meade					X	X	X	X	
Pentax		X		X	X				
Takahashi	X				X	X	X		
Tele Vue					X				
Vixen			X						

Table 4.1d **Normal-Field Newtonian Reflecting Telescopes (NFTs)**
(focal ratio ≥ f/7)

Manufacturer	4.25″	4.5″	6″	8″	10″	12.5″	14.5″
Celestron		X					
Coulter				X			
Meade		X	X				
Orion		X	X				
Parallax	X		X	X	X	X	X
Parks			X				
Star-Liner			X	X			
Vixen		X					

contact a local museum or planetarium to find out if there is an astronomy club in your area.

Look through every instrument at the star party. Bypass none, even if you are not considering that particular kind of telescope. When you find one that you are considering, speak to its owner. If the telescope is good, he or she will brag just like a proud parent. If it is poor, he or she will be equally anxious to steer you away from making the same mistake. Listen to the wisdom of the owner and compare his or her comments with the advice given in this chapter.

Next, ask permission to take the telescope for a test drive so that you may judge for yourself its hits and misses. Begin by examining the mechanical in-

Table 4.1e Rich-Field Newtonian Reflecting Telescopes (RFTs)
(focal ratio < f/7)

Manufacturer	4.25"	6"	8"	10"	12.5"	13.1"	14.5"	15"	16"	17.5"	18"	20"	24"	25"	30"	32"	36"
Celestron		x															
Coulter			x	x	x					x							
Edmund	x																
JMI									x		x		x				
Jupiter					x			x			x						
Meade			x	x					x								
Obsession											x	x		x	x	x	x
OGS			x	x	x		x		x			x	x				
Orion			x	x	x												
Parallax	x	x	x	x	x		x		x								
Parks		x	x	x	x				x								
Safari											x						
Sky Designs							x				x	x					
Star-Liner			x	x			x										
Starsplitter					x		x		x		x	x		x	x		
Tectron						x					x			x	x		
Torus					x		x		x		x	x	x				

Table 4.1f Cassegrainian Telescopes

Manufacturer	6"	8"	10"	12.5"	14.5"	16"	20"	24"
OGS			x	x	x	x	x	x
Parks	x	x	x	x		x		
Star-Liner		x	x	x	x	x	x	x

Table 4.1g Catadioptric Telescopes

Manufacturer	Schmidt-Cas								Maksutov-Cas				Maksutov-Newt
	4"	5"	7"	8"	9"	10"	11"	14"	3.5"	6"	7"	12"	5.7"
Celestron		x		x			x	x					
Ceravolo													x
INTES										x			
Meade	x			x		x							
Questar									x		x	x	
Takahashi					x								

tegrity of the mounting. Tap (gingerly, please) the mounting. Does it vibrate? Do the vibrations dampen out quickly or do they continue to reverberate? Try the same test by rapping the mounting and tripod or pedestal. How rapidly does the telescope settle down?

Working your way up, check the mechanical components of the telescope itself. Does the eyepiece focusing knob(s) move smoothly across the entire

length of travel? If you are looking at a telescope with a rack-and-pinion focusing mount (by far the most common variety), does the eyepiece tube stop
when the knob is turned all the way, or does it separate and fall into the tube?
Do you find the side-mounted finderscope easily accessible?

When you are satisfied that the telescope performs well mechanically, examine its optical quality. By this time, no doubt the owner will have already
shown you a few dazzling objects through the telescope, but now it is time to
take a more critical look. One of the most telling ways to evaluate a telescope's
optical quality is to perform the star test outlined in Chapter 8. It will quickly
reveal if the optics are good, bad, or indifferent.

How should you buy a telescope? Some manufacturers only sell factory-
direct to the consumer, whereas others have networks of retailers and distributors. This latter case is certainly in the consumer's best interest, for it allows
price comparison as well as instant access to the instruments themselves,
rather than settling for a lot of words and some pretty pictures in catalogues.

When it comes time to purchase a telescope, shop around for the best deal
but do not base your choice on price alone; be sure to compare delivery times
as well! Some of the more popular telescopes, such as those from Celestron,
Tele Vue, and Meade, are available from dealer stock for immediate delivery.
If you want to get into amateur astronomy fast (and who doesn't like immediate
gratification), then these brands are certainly worth a look. At the opposite end
of the telescope spectrum are other companies whose delivery times can stretch
out to weeks, months, or even more than a year! Consult Appendix A for a list
of distributors or contact the manufacturer for your nearest dealer.

Once you decide on a telescope model, it is best, if possible, to purchase
the telescope in person. Not only will you save money in crating and shipping
charges, but you will also be able to inspect the telescope beforehand to make
sure all is in order and as described. Though most manufacturers and distributors strive for customer satisfaction, there is always the possibility of trouble
when merchandise is mail ordered.

"Don't Worry . . . the Check Is in the Mail"

It was a problem I had with a well-known source of astronomical equipment
that led to this section of the book. Briefly, an eyepiece I ordered from this
company proved defective; there was a small chip in one of the inner lens
elements. I called its offices and was to told to return it for replacement or
refund. I opted for replacement. Two weeks went by, but nothing happened. I
called again and was assured that the faulty eyepiece had been received and
that a new eyepiece would be sent to me by week's end. Fourteen more days
went by, and still nothing. I called the company again, at which time the owner
assured me that the eyepiece had been sent out a few days earlier and that it
should reach me any day. Two more weeks elapsed, and with no eyepiece in
hand, I wrote to the owner demanding an explanation. Within a week, I re-

ceived a refund check and a surprisingly nasty letter stating that the package containing my returned eyepiece arrived damaged because of my negligent packing (earlier, they had told me it arrived fine). Because of my "attitude," the company did *not* want me as a customer again! The company's name? Sorry, I cannot mention it, but I can say that it is NOT listed anywhere in this book. It is hoped that it will be made conspicuous by its absence!

Happily, this unfortunate experience is the exception, not the rule. The vast majority of astronomical supply companies are owned and operated by competent, friendly people. They want happy customers (remember, a happy customer is a repeat customer) and guard their good reputations jealously. Most are willing to bend over backwards to see that a problem is resolved to the customer's satisfaction. But what can the consumer do if he or she is dissatisfied with a manufacturer or distributor?

Begin on the right foot. Before returning a defective piece of merchandise, always speak to the manufacturer first about the problem. For this initial contact, a telephone call should suffice, but be certain to write down the name of whomever you spoke to. Request instructions for the most expeditious way to return the item for replacement or refund. Conform to the directions precisely, but to protect yourself, always follow up the conversation with a letter addressed to the representative to whom you spoke. In it, repeat the nature of the problem as well as the desired outcome. Send the letter by certified mail, return receipt requested, and keep a copy for your records.

Allow the company a reasonable length of time to respond to your complaint, typically two to four weeks. If, after that time, a satisfactory resolution has not been reached, write to the company again and inquire as to the delay. Clearly, you are anxious to resolve the dilemma and would appreciate immediate action on their part. State that you expect a response within a given period of time, say ten business days. Once again, send the letter by certified mail, return receipt requested, and keep a copy for your records.

Allow two more weeks to elapse. If there is still no response, call the company and find out who the owner is. Write to him or her directly, recounting all that has happened since the item was ordered. Mention the person's name with whom you have been dealing and state that you demand immediate satisfaction. Be polite, but be direct and precise. Again, send the letter—you guessed it—by certified mail, return receipt requested, keeping a copy for your records.

By now, the predicament should have been resolved, but if it has not, then it's time to take action. The major astronomical periodicals, such as *Sky & Telescope* and *Astronomy*, do not have in-house consumer advocates, yet they do take an active interest in consumer satisfaction with all who advertise in their magazines. Write to each of them with your complaint, being certain to send a copy to the president/owner of the offending company. In addition, send a copy of the letter to the Astronomical League (their address is 2112 Kingfisher Lane East, Rolling Meadows, IL 60008). The League is also interested in customer satisfaction and may offer assistance. If you suspect mail fraud, also contact the local postmaster or postal inspector or write to Consumer Advocate, Customer Services Department, Postal Service, 475 L'Enfant Plaza West S.W., Room 5910, Washington, D.C. 20260–6320; the telephone number is (202) 245–4550.

Most consumer advocates recommend charging all mail-orders to a major credit card; do not use a check or money order, if possible. Using a credit card gives you certain powers that are not available any other way. On the back of every credit card's monthly statement, there are instructions that clearly describe steps to be taken in the event of a consumer problem. Usually, the card requires that the consumer describe in specific detail the exact nature of the problem and provide copies of all receipts and documentation. The charge will then be put in contest until the problem is resolved. If a charge is contested, the consumer is not responsible for any interest that may accrue as a result. When a final determination is made, either a credit will be issued to the charge account or the balance plus interest will be due.

Contesting a charge should be viewed as a last-resort measure. Unfortunately, many consumers abuse this privilege, seeing it as a way to get an extended warranty for free. This is not its intention and just dilutes the system's effectiveness. Only put a charge in contest when a bona fide problem exists and the vendor refuses to cooperate. For instance, just because you decided that you don't like an item anymore is not reason enough to contest a charge, but poor quality or workmanship is. See the difference?

Honest Phil's Used Telescopes

What if you took the pop quiz in the last chapter and found out that the best telescope for your needs was, say, an 8-inch Schmidt-Cassegrain, but you cannot afford to spend $1,200 to get the instrument you want? What can you do? If you needed to buy a car but could not afford to buy a new one, the odds are that you would check the used car classified advertisements, right? If it works for cars, then why not for telescopes? All other things being equal, an old telescope such as my 23-year-old 8-inch Newtonian seen in Figure 4.8, will work just as well as a new telescope as long as it was treated kindly. Each month, both *Sky & Telescope* and *Astronomy* magazines have columns devoted to readers' classified advertisements, while *The Starry Messenger* is a monthly periodical devoted solely to used telescopes, binoculars, and accessories. Bargains are found inside every issue.

Look through the classifieds to see if anything strikes your fancy. Pay special attention to where the buyer is, because you will have to get the telescope from there to your home. If possible, restrict your search to an area that is within a day's drive so that you can check out the telescope firsthand instead of relying on a stranger's word. One person's treasure is another person's junk!

What should you look for in a used telescope or binoculars? Essentially the same things you would look for in a new one. You want to check the in-

Figure 4.8 *A blast from the past: the author's Criterion Dynascope RV8 (purchased new in 1971) piloted by his daughter Helen (also purchased new, but in 1984).*

strument both optically and mechanically. Inspect the instrument for any damage or mishandling. Are the optics clean? Did the owner store the telescope adequately? All of these are important considerations when deciding if the binoculars or telescope will be a worthy companion.

The following list might help you find a prince among the frogs by listing an inventory of my choices for the ten best telescopes of yesteryear. Bear in mind that prices may vary greatly for the same instrument depending on its condition. If a telescope is in absolutely pristine condition, then the asking price will probably be higher. On the other hand, if the instrument was dropped down one too many flights of stairs by the owner, then it might be listed at a bargain price. Shop carefully and shop wisely.

1. Astro-Physics 6-inch f/12 Super Planetary apochromatic refractor
2. Brandon 3.7-inch f/7 apochromatic refractor
3. Cave 6-inch f/8 Newtonian reflector or larger
4. Celestron C100, 4-inch f/10 achromatic refractor
5. Celestron Super C8 + , 8-inch f/10 SCT
6. Criterion Dynascope RV-6 (6-inch f/8) or RV-8 (8-inch f/7) Newtonians
7. Coulter CT-100, foldable 4.25-inch f/4 Newtonian
8. Fecker Celestar 6-inch Newtonian reflector
9. Optical Craftsmen Connoisseur 6-inch f/8 Newtonian reflector or larger
10. Quantum 4, 4-inch Maksutov-Cassegrain

Congratulations, It's a Telescope!

Be the telescope new or be it used, resist the urge to uncrate your baby immediately once you get it home. Instead, read its instruction manual from cover to cover. When you are done, read it again. Absorb all the information it has to offer. If you have any questions, call the dealership where you bought the instrument. Though they may not know the answer, they should at least have the manufacturer's phone number (if not, check Appendix A). Some companies even have technical assistance lines set up for just such an emergency.

Now, assemble the telescope per the instructions. Do everything slowly and deliberately. I know the anxiety is terrible, but take your time. Check the collimation to make certain that the optics are properly aligned with each other. If they are not, consult the owner's manual. If a part doesn't fit together properly the first time, do not try to force it. Instead, put everything down, sit back, take a deep breath, and smile. Remember, the universe has been around for billions of years; it will still be there when you get the telescope together!

With the telescope assembled and the skies finally clear (why is it always cloudy whenever you get a new telescope?), take your prize outside for first light. Pick out something special to look at first (I always choose Saturn, if it's up) and enjoy the view! By following all of the steps here as well as the other suggestions found throughout the chapters yet to come, you will be well on your way toward a fantastic voyage that will last a lifetime.

5

The Eyes Have It

Have you ever tried to look through a telescope without an eyepiece? It doesn't work very well, does it? Sure, you *can* stand back from the empty focusing mount and see an image at the telescope's focal plane. But without an eyepiece in place, the telescope's usefulness as an astronomical tool is greatly limited, to say the least.

Most telescopes come with at least one eyepiece, giving the novice a place to start. Though they may be adequate to begin with, these standard-issue eyepieces frequently offer only mediocre performance and image quality. Most new telescope owners quickly hunger for more.

Until recently, eyepieces (or oculars, if you prefer) were thought of as almost second-class citizens whose importance was considered minor compared to a telescope's prime optic. With few exceptions, many eyepieces of yore suffered from tunnel vision by producing narrow fields of view regardless of magnification as well as an assortment of aberrations. The 1980s, however, saw a revolution in eyepiece design. In the place of their lackluster cousins stood advanced optical designs that brought resolution and image quality to new heights. With the possible exception of selecting the telescope itself, picking the proper eyepiece(s) is probably the most difficult choice facing today's amateur astronomers.

Although eyepieces are available in all different shapes and sizes (as you can tell from Figures 5.1a and 5.1b), the discussion here will begin with a few generalizations. Figure 5.2 shows a generic eyepiece with its components labelled. Regardless of the ocular's internal optical design, the lens element(s) closest to the observer's eye is always referred to as the *eye lens*. The lens element(s) that is farthest from the observer's eye (that is, the one facing inward toward the telescope) is called the *field lens*. A *field stop* is usually mounted just beyond the field lens at the focus of the eyepiece, giving a sharp edge to the

Figure 5.1   *A selection of premium eyepieces from (top) Orion and (bottom) Tele Vue. Photos courtesy of Orion Telescope Center and Tele Vue, Inc.*

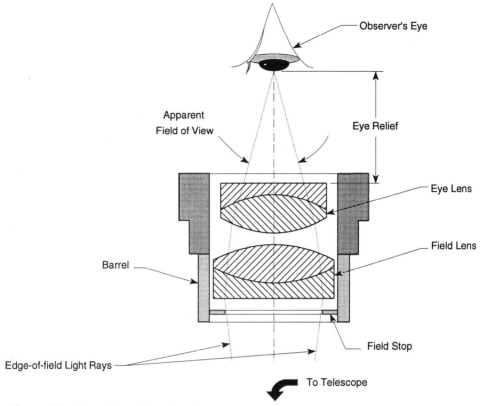

Figure 5.2 *A generic eyepiece showing components.*

field of view while preventing peripheral star images of poor quality from being seen.

Although the eyepiece must be sized according to the diameter of the eyepiece optics, the barrel (the part that slips into the telescope's focusing mount) is always one of three diameters. Most amateur telescopes use 1.25-inch diameter eyepieces, a standard that has been around for years. At the same time, most less expensive, department-store telescopes are outfitted with 0.965-inch oculars. Finally, the astronomical community has seen a recent boon in a whole new breed of giant eyepieces with 2-inch barrels.

Before looking at specific eyepiece designs, we must first become fluent in Eyepiece-ese, terms that describe an eyepiece's characteristics and performance. There are surprisingly few. Perhaps most important of all is *magnification* (or maybe I should say "lack of magnification.") As previously outlined in Chapter 1, magnification is equal to the focal length of the telescope divided by the focal length of the eyepiece. Therefore, the longer the telescope's focal length, the greater the magnification from a given eyepiece.

Wouldn't it be nice if there were one specific magnification value that would work well in every telescope for every object in the sky? Sadly, this is simply not the case. For certain targets, such as widely scattered star clusters or nebulae, lower powers are called for. To get a good look at the planets or smaller deep-sky objects (for example, planetary nebulae and smaller galaxies), higher powers are required. If the magnification is too high for a given telescope, then image integrity will be sacrificed.

Just how much magnification is too much and how much is just right? A good rule of thumb is to use only as much magnification as needed to see what you are interested in looking at. If you own a smaller telescope (that is, 7 inches or smaller in aperture), the oft-repeated rule of not exceeding $60\times$ per inch of aperture is suggested. This means that a 7-inch telescope can operate at a maximum of $420\times$, but remember this value is *not* cast in stone. It all depends on your local atmospheric conditions and the instrument. Given excellent optics, you may be able to go as high as $90\times$ or even $100\times$ per inch on some nights, while on others, $30\times$ per inch may cause the view to crumble.

On the other hand, larger telescopes (instruments with greater than 8-inch apertures), especially those with fast focal ratios, can seldom meet or exceed the $60\times$-per-inch rule. Instead, they can handle a maximum of only $30\times$ or $40\times$ per inch.

The choice of the right magnification is one that must be based largely on past experience. If you are lacking that experience, do not get discouraged, for it will come with time. For now, use Table 5.1 as a guide for selecting the maximum usable magnification for your telescope.

Notice how the table caps out at $300\times$ for all telescopes beyond 8 inches. Experience shows (and there are some reading this now who I am sure will disagree vehemently) that little is gained by using more than $300\times$ to view an object regardless of aperture. Only on those rare nights when the atmosphere is at its steadiest (that is, when the stars do not appear to twinkle) can this value be bettered.

Table 5.1   **Telescope Aperture versus Maximum Magnification**

Telescope Aperture		Magnification	
(In.)	**(cm)**	**Theoretical ($60\times$/inch)**	**Practical**
2.4	6	144	100
3.1	8	186	125
4.25	11	255	170
6	15	360	240
8	20	480	300
10	25	600	300
12.5	32	750	300
14	36	840	300
16	41	960	300
18	46	1080	300
20	51	1200	300
25	64	1500	300
30	76	1800	300

Another important consideration when selecting an eyepiece is the size of its *exit pupil*, which is the diameter of the beam of light leaving the eyepiece and traveling to the observer's eye, where it enters the observer's own eye pupil. You can see the exit pupil of a telescope or binocular by aiming the instrument at a bright surface, such as a wall or the daytime sky. Back away and look for the little disk of light that appears to float in the air just behind the eye lens.

How can you find out the size of the exit pupils produced by your eyepieces in your telescope? Easily, using either of the two formulas shown below:

$$\text{Exit Pupil} = D/M$$

where

D = the diameter of the telescope's objective lens or primary mirror in millimeters

M = magnification

or

$$\text{Exit Pupil} = F_e / f$$

where

F_e = the focal length of the eyepiece in millimeters
f = the telescope's focal ratio (its f-number)

Let's look at an example. An 8-inch f/10 telescope (focal length = 80 inches or 2,000 mm) equipped with a 40-mm eyepiece produces a magnification of $50\times$. Using either of the above methods, it can be seen that the eyepiece returns an exit pupil of 4 mm. The same eyepiece provides $40\times$ in a 14-inch f/4.5 instrument. Although we might expect this combination to produce a wonderful wide-field view, the exit pupil grows to roughly 9 mm across, too large to be useful because it will waste some of the telescope's light and produce a so-called dead spot in the center of the field. In this latter case, a shorter-focal-length eyepiece is called for.

Knowing the diameter of the exit pupil is a must, for if it is too large or too small, the resulting image may prove unsatisfactory. Why? The pupil of the human eye dilates to about 7 mm when acclimated to dark conditions (though this figure varies from one person to the next and shrinks as you age). If an eyepiece's exit pupil exceeds 7 mm, then the observer's eye will be incapable of taking in all the light that the ocular has to offer. Many optical authorities would be quick to point out that an excessive exit pupil wastes light and resolution. This is not necessarily the case for the owners of refractors. Say you own a 4-inch f/5 *refractor* and wish to use a 50-mm eyepiece with it. The resulting exit pupil from this combination is 10 mm. Though the exit pupil is technically too large, this telescope-eyepiece would no doubt provide a wonderful low-power, wide-field view of rich Milky Way starfields when used under dark skies.

The key phrase in that last sentence is *when used under dark skies*. If the same pairing was used under mediocre suburban or urban sky conditions, then the contrast between your target and the surrounding sky would suffer greatly. This is due to the fact that the eyepiece is not only transmitting starlight but also skyglow (light pollution)—*too much* skyglow.

What about using a 50-mm eyepiece with a 4-inch f/5 *reflector*? Sorry, but this is not a good idea. With conventional reflectors as well as catadioptric instruments, obstruction from the secondary mirror will create a noticeable black blob in the center of view when eyepieces yielding exit pupils greater than 8 mm or so are used. With these telescopes, it is best to stick with eyepieces of shorter focal lengths.

On the other hand, if the exit pupil is too small, then the image will be so highly magnified that the target may be nearly impossible to see and focus. Just as there is no single all-around best magnification for looking at everything, neither is there one exit pupil that is best for all objects under all sky conditions. It depends on what you are trying to look at. Table 5.2 summarizes (rather subjectively) my personal preferences.

Though magnification gives some feel for how large a swath of sky will fit within an eyepiece's view, it can only be precisely figured by adding another ingredient: the ocular's *apparent field of view*. Nowadays, most manufacturers will proudly tout their eyepieces' huge apparent fields of view because they know that big numbers attract attention. Unfortunately, few take the time to explain what these impressive figures actually mean to the observer.

The apparent field of view refers to the eyepiece field's edge-to-edge angular diameter as seen by the observer's eye. Perhaps that statement will make more sense after this example. Take a look at Figure 5.3a. Peering through a long, thin tube (such as one from a roll of paper towels), the observer sees a very narrow view of the world—an effect commonly known as tunnel vision. This perceived angle of coverage is known as the apparent field of view. To increase the apparent field of view in this example, simply cut off part of the cardboard tube. Slicing it in half (Figure 5.3b), for instance, will approximately double the apparent field, resulting in a more panoramic view of things.

In the world of eyepieces, the apparent field of view typically ranges from a cramped 25° to a cavernous 80° or so. Generally, it is best to select eyepieces with at least a 40° apparent field because of the exaggerated tunnel-vision effect

Table 5.2   **Suggested Exit Pupils for Selected Sky Targets**

Target	Exit Pupil (mm)
Wide star fields under the best dark-sky conditions (e.g., large star clusters, diffuse nebulae, and galaxies)	6 to 7
Smaller deep-sky objects; complete lunar disk	4 to 6
Small, faint deep-sky objects (especially planetary nebulae and smaller galaxies); double stars, lunar detail, and planets on nights of poor seeing	2 to 4
Double stars, lunar detail, and planets on exceptional nights	0.5 to 2

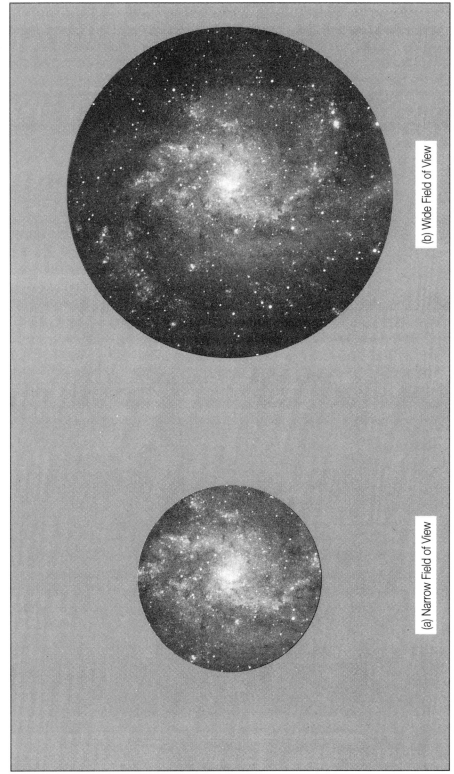

(a) Narrow Field of View

(b) Wide Field of View

Figure 5.3 *Simulated view through eyepieces with (a) a narrow apparent field of view and (b) a wide apparent field of view. Photograph of M33 by George Viscome (14.5-inch f/6 Newtonian, Tri-X film in a cold camera, 30-minute exposure).*

through anything less. An apparent field in excess of 60° gives the illusion of staring out the porthole of an imaginary spaceship. The effect can be really quite impressive!

Eyepieces with the largest apparent fields of view do not come cheaply. Some, especially the long-focal-length models, are quite massive both in terms of weight and cost. Typically, they must be made from large-diameter lens elements and may be available in 2-inch barrels only, restricting their usefulness to medium- and large-aperture telescopes only. Some are so heavy that you may have to rebalance the telescope whenever they are used! More about this when specific eyepiece designs are discussed later in this chapter.

By knowing both the eyepiece's apparent field (typically specified by the manufacturer) and magnification, we can calculate just how much sky can squeeze into the ocular at any one time. This is known as the *true* or *real field of view*, and can be approximated from the following formula:

$$\text{Real Field} = F/M$$

where

F = the apparent field of view

M = magnification

To illustrate this, we can continue with the earlier example of an 8-inch f/10 telescope and a 40-mm eyepiece. It has already been found that this produces $50\times$ and a 4-mm exit pupil. Suppose this particular eyepiece is advertised as having a 45° apparent field. Dividing 45/50 shows that this eyepiece produces a real field of 0.9°, almost large enough to fit two Full Moons edge to edge.

Another term that is frequently encountered in eyepiece literature but rarely defined is *parfocal*. This simply means that the telescope will not require refocusing when one eyepiece is switched for another in the same set. Without parfocal eyepieces, the observer may lose a faint object during the second focusing if the telescope is accidentally moved while refocusing. Please keep in mind that even when an eyepiece is claimed to be parfocal, that does not mean that it is *universally* parfocal. Just because two eyepieces of the same optical design, say a 26-mm from Brand X and a 12-mm from Brand Y, each claim to be parfocal, they are most likely not parfocal with each other. At the same time, two eyepieces of different optical designs may be manufactured by the same company and declared parfocal, but they are not likely to be parfocal with each other.

Finally, a well-designed eyepiece will have good *eye relief*. Eye relief is the distance from the eye lens to the observer's eye when the entire field of view is seen at once. Less expensive eyepieces may offer eye relief of only one-quarter times the ocular's focal length. This is much too close to view comfortably, especially if you must wear glasses. Modern designs permit eye relief nearly equal to the eyepiece's focal length, making observing more enjoyable.

Image Acrobatics

Since eyepieces can suffer from the same aberrations and optical faults as telescopes, it might be wise to list and define a few of their more common problems. Some have already been defined in earlier chapters, but now we will concentrate on their impact on eyepiece performance. (Of course, if any of the following conditions exist all of the time regardless of eyepiece used, then the problem likely lies with the telescope, not the eyepiece.)

If star images near the field's center focus at a different distance behind the eye lens than stars near the edge, then the eyepiece suffers from *curvature of field*. The overall effect is an annoying unevenness across the entire field of view.

Another flaw found in lesser eyepieces is *distortion*. Distortion is most readily detectable when viewing either terrestrial sights or large, bright celestial objects such as the Moon or Sun. This condition is usually characterized by a warping of the scene in a way that is similar to the effect seen through a fish-eye camera lens.

Chromatic aberration is nearly extinct in today's eyepieces thanks to the use of one or more achromatic lenses. Still, some less sophisticated eyepieces furnished as standard equipment with less expensive telescopes suffer from this ailment. It also may be present in eyepieces advertised as "war-surplus specials" (World War II, that is). If an ocular transmits chromatic aberration, the problem will be immediately detectable as a series of colorful halos surrounding all of the brighter objects found toward the edge of the field of view; the center of view is usually color free.

Spherical aberration has also been all but eliminated in most (but not necessarily all) eyepieces of modern design. If an eyepiece is free of spherical aberration, then a star should look the same on either side of its precise focus point. When spherical aberration is present, however, the star will change its appearance from when it is just inside of focus compared to when it is just outside of focus. This predicament is the result of uneven distribution of light rays at the eyepiece's focal point. Today, if spherical aberration is present, chances are it is being introduced by the telescope's prime optic (main mirror or objective lens) and not the eyepiece. At low powers (large exit pupils), it can also be introduced by the observer's own eye.

Just as with objective lenses and corrector plates, most eyepiece lenses are now coated with an extremely thin layer of magnesium fluoride. Coatings reduce flare and improve light transmission, two desirable characteristics for telescope oculars. A bare lens surface can reflect as much as 4% of the light striking it. By comparison, a single-coated lens reduces reflection to about 1.5% per surface.

A lens coated with the proper thickness of magnesium fluoride exhibits a purplish hue when held at a narrow angle toward a light. Top-of-the-line eyepieces receive multiple antireflection coatings, reducing reflection to less than 0.5%. Most multicoated lenses show a greenish reflection when turned toward a light.

Eyepiece Evaluation

Galileo, Kepler, Newton—they had it pretty easy when it came to selecting eyepieces. After all, they had few choices. Galileo used a single concave lens placed before the objective's focus. It produced an upright image, but the field of view was incredibly small and severely hampered by aberrations. Kepler improved on the idea by selecting a convex lens as his eyepiece. It gave a wider, albeit inverted view, but still suffered from aberrations galore. Progress in eyepiece design was slow in the early years.

Here, we will see just how far the art of eyepieces has progressed while evaluating which designs are best suited for the telescope you chose earlier.

Huygens

The first compound eyepiece was concocted by Christiaan Huygens in 1703 (just as with new telescope designs, an eyepiece usually bears the name of its inventor). As can be seen in Figure 5.4a, Huygens eyepieces contain a pair of plano-convex elements. Typically, the field lens has a focal length three times that of the eye lens.

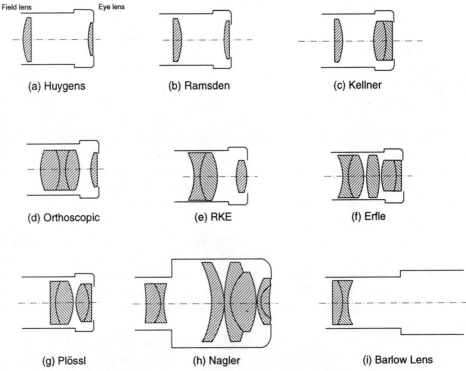

Figure 5.4   *The inner workings of several eyepiece designs from both yesterday and today. Eyepieces shown are (a) Huygens, (b) Ramsden, (c) Kellner, (d) Orthoscopic, (e) Edmund's RKE, (f) Erfle, (g) Plössl, and (h) Nagler. A typical Barlow Lens is shown in diagram (i).*

In the past, Huygens eyepieces were supplied as standard equipment with telescopes of f/15 and greater focal ratios. At longer focal lengths, these oculars can perform marginally well, though their apparent field of view is annoyingly narrow. In telescopes with lower focal ratios, however, image quality literally crumbles before our eyes. Spherical and chromatic aberrations abound, along with image curvature and an overall lack of sharpness, making Huygens eyepieces undesirable choices for most modern telescopes. No wonder so few companies still offer them.

Ramsden

Devised in 1783, this design was the brainchild of Jesse Ramsden, the son-in-law of John Dollond (you may recall him from Chapter 2 as the father of the achromatic refractor—small world, isn't it?). As with the Huygens, a Ramsden eyepiece (Figure 5.4b) consists of two plano-convex lenses. Unlike the earlier ocular, however, the Ramsden elements have identical focal lengths and are flipped so that both convex surfaces face each other.

In most cases, the lenses are separated by about two-thirds to three-quarters of their common focal length. This is, at best, a compromise. Setting the elements closer together improves eye relief but dramatically increases aberrations. Going the other way will decrease the design's inherent faults, though eye relief quickly drops toward zero. Therefore, like the Huygens, it is probably best to remember the Ramsden for its historical significance and pass it by in favor of other designs.

Kellner

It took over six decades of experimentation before an improvement to the Ramsden eyepiece was developed. Then, in 1849, Carl Kellner introduced the first achromatic eyepiece (Figure 5.4c). Based on the Ramsden, Kellner replaced the single-element eye lens with a cemented achromat. In his arrangement, the element closest to the observer's eye is made of flint, while the other is made of crown glass.

The Kellner's achromatic eye lens greatly reduces most of the aberrations common to Ramsden and Huygens eyepieces. Sometimes called Achromatic Ramsdens, they feature fairly good color correction and edge sharpness, little curvature, and apparent fields of view ranging between 40° and 50°. In low-power applications, Kellners offer good eye relief, but this tends to diminish as the eyepiece's focal length shrinks.

Perhaps the greatest drawback of the Kellner is its propensity for internal reflections. I have often heard of Kellners being referred to as "haunted eyepieces" because of their ghost images. This is especially noticeable when one is used to observing bright objects through a telescope of large aperture and short focal length. But thanks to the wonder of antireflection optical coatings, this effect is almost eliminated when Kellners are used with 8-inch and smaller telescopes.

Given all of their plusses and minuses, Kellners represent a good buy for budget-conscious owners of small-to-medium aperture telescopes who are in the market for a low-to-medium-power eyepiece.

Orthoscopic (Abbe)

Introduced in 1880 by Ernst Abbe, the orthoscopic eyepiece has become a perennial favorite of amateur astronomers. As shown in Figure 5.4d, it consists of a cemented triplet field lens matched to a single plano-convex eye lens. What results is close to a perfect eyepiece, harassed by neither chromatic nor spherical aberration. There is also little evidence of ghosting or curvature of field. Orthos offer wide, flat views with apparent fields between 40° and 50° and excellent eye relief. Color transmission and contrast are superb, especially when combined with today's optical coatings. With either a long focal length for low power, a short focal length for high power, or anywhere in between, orthoscopic eyepieces remain one of the best eyepieces for nearly all amateur telescopes.

RKE

From Edmund Scientific Company comes this fresh twist on the Kellner eyepiece. Instead of using an achromatic eye lens and a single-element field lens, the RKE (Figure 5.4e) does just the opposite. The computer-optimized achromatic field lens and single-element eye lens combine to outshine the Kellner in just about every respect. Actually, performance of the three-element RKE is most comparable to that of the four-element orthoscopic. Each has a moderate apparent field of view (45° for the RKE), good color correction, and image clarity. All work well at all focal lengths in all telescopes.

Erfle

The Erfle, the granddaddy of all wide-field eyepieces, was originally developed in 1917 for military applications. With apparent fields of view ranging between 60° and 75°, it was quickly embraced by the astronomical community as well. Internally, Erfles consist of either five or six elements; one variety uses two achromats with a double convex lens in between, while a second subspecies employs three achromats, as shown in Figure 5.4f.

Erfles give observers an outstanding panoramic view of the deep sky. The spacious view takes its toll on image sharpness, however, which suffers from astigmatism toward the field's edge. For this reason, Erfles are inappropriate for lunar and planetary observations or any occasion that calls for higher magnification. In low-power, wide-angle applications, though, they are very impressive.

Zoom

Zoom eyepieces combine a wide variety of focal lengths into one neat package. Typically, they range in focal length from about 7 mm to 21 mm or so. Sounds too good to be true? Unfortunately, it usually is. Most zoom eyepieces are compromises at best. For one thing, aberrations are frequently intensified in them, perhaps due to poor optical design or because the lenses are constantly sliding up and down in the barrel. Another problem is that their apparent fields of view are not constant over the entire range. The widest apparent fields occur at high power but shrink rapidly as magnification drops. Finally, many are not parfocal across their entire range, requiring you to refocus whenever the zoom eyepiece is zoomed. Premium-quality zooms that are parfocal and maintain a reasonably wide field at all magnifications can give impressive results, but you must be willing to pay for the performance. If you are not, avoid the urge to settle for cheaper imitations.

Plössl

One of the most highly regarded eyepieces around today, the Plössl (Figure 5.4g) features twin close-set pairs of doublets for the eye lens and the field lens. The final product is an excellent ocular that is comparable to the orthoscopic in terms of color correction and definition, but with even better eye relief and a larger apparent field of view. Ghost images and most aberrations are sufficiently suppressed to create remarkable image quality.

Though it was developed in 1860 by G. S. Plössl, an optician living in Vienna, Austria, the Plössl eyepiece took more than a century to catch on among amateur astronomers. Back in the 1970s, when I first became seriously engrossed in this hobby, I can recall hearing about a mysterious eyepiece called a Plössl. Try as I might, I could not find much published information on it, other than that I could not afford one! Perhaps the enigma of the Plössl was heightened by the fact that, back then, the only company offering one was Clavé in France, and its distribution was limited. Then, in 1980, the Plössl hit the big time. That year, Tele Vue Optics, Inc., introduced a line of Plössls that was to start a new eyepiece revolution among amateur astronomers. They were an instant success.

Taking aim at the trend set by Al Nagler (Tele Vue's owner and master optician), many companies now offer Plössls. All are quite good, too, but some are perceptibly better than others. I prefer not to generalize, but in this case, you pretty much get what you pay for. The more expensive Plössls typically maintain closer optical and mechanical tolerances while using multicoated optics, blackened lens edges, and antireflection threads. Little things like these can make the difference between seeing a marginally visible object and not.

Brandon

Offered exclusively by VERNONscope and Company, Brandon eyepieces have long been held in high esteem by discriminating amateur astronomers. Their patented four-element design yields sharp, crisp views with excellent image contrast across the entire field. Ghosting, prevalent in many lesser eyepieces, is effectively eliminated in the Brandons thanks to precise optical design and magnesium-fluoride coatings. Overall performance is comparable to Plössl and orthoscopic eyepieces for deep-sky objects but superior for the planets.

Brandon eyepieces come in six focal lengths: 48, 32, 24, 16, 12, and 8 mm. Of these, all have standard 1.25-inch barrels except the 48-mm, with has a 2-inch barrel. In addition, all feature foldable rubber eyeguards. But take heed: For those who plan on using screw-in filters, Brandon eyepieces do not have standard threads to accept other manufacturers' filters. Instead, they can only use VERNONscope filters.

Lanthanum LV

Like Edmund's RKE eyepieces, Lanthanum LV oculars are proprietary to their manufacturer, in this case Vixen Optics of Japan. Based on the Plössl, the Lanthanum LV adds a fifth lens element between the eye-lens and field-lens groups as well as an extra element(s) before the field lens, in effect creating a built-in Barlow Lens (these are discussed later in this chapter). Apparent fields of view are in the 45° to 50° range, similar to Plössls.

The added elements of the Lanthanum LV design help to overcome a problem common to many eyepieces. With most designs, as focal length shrinks, so does eye relief. As a result, short-focal-length eyepieces require the observer to get uncomfortably close to the eye lens, but that's not the case with these oculars. All Lanthanum LV eyepieces, ranging in focal length from 30 mm to 2.5 mm (the shortest focal length available) feature a long 20-mm eye relief. As a result, the observer can stand back from the eyepiece and still take the whole scene in, making life at the eyepiece much more comfortable. This is especially noteworthy for observers who must keep their glasses on while viewing.

Although eye relief is good, some respondents to my survey noted that image contrast and sharpness, especially near the edge of the field, are inferior to other oculars. So while these eyepieces do have something to offer certain observers, most will be better served by either orthoscopics, Plössls, or some of the super eyepieces that follow.

Super-Deluxe-Extra-Omni-Ultra-Maxi-Mega-Colossal. . .

Whatever happened to the words *standard* and *regular*? Are they still in the dictionary? Apparently not, judging by today's advertising. Every product, from refrigerators to pet food, is burdened with superlatives in some way. Top-of-the-line oculars are no different. Each is proclaimed by its manufacturer to be something extraordinary. And do you know what? They really are quite good.

These super-duper eyepieces first came on the scene to meet the demanding needs of amateurs using the increasingly popular Schmidt-Cassegrains and mammoth Newtonians. They also work equally well in refractors and longer-focal-length reflectors. Most use multicoated lenses made from expensive glasses to minimize aberrations as much as possible. With these eyepieces, the universe has never looked so good!

Not satisfied with the success of its Plössls, Tele Vue led the way in introducing oculars with extremely wide apparent fields of view. Tele Vue Wide-Field eyepieces use six elements to create a 65° apparent field that rivals the sharpness of narrower orthoscopics and Plössls. Spherical and chromatic aberrations as well as ghost images are also minimized.

A third line of oculars marketed by Tele Vue are its Nagler eyepieces. Naglers (Figure 5.4h) are characterized by a complex seven-element design that produces an 82° apparent field while correcting for astigmatism, chromatic aberration, spherical aberration, coma, and just about every other optical fault there is. A variation of these, the Nagler Type 2 eyepieces, utilizes an eight-element combination that takes a good idea and makes it even better.

Finally, Tele Vue recently introduced a true monster of an eyepiece: the Panoptic. The elements in the Panoptic combine to yield unparalleled low-power views with excellent eye relief through medium-to-large–aperture telescopes, making them ideal for observers who must wear eyeglasses. They come in four focal lengths (15, 22, 27 and 35 mm), each sporting a 68° apparent field of view and excellent correction. Performance of the Nagler and Panoptic eyepieces is just heavenly!

Many other companies offer wide-field eyepieces, though the optical layouts vary greatly. For instance, Meade Instruments has its Series 4000 eyepieces, made up of three different sets of oculars. Least expensive of the Series 4000 oculars are the Super Plössl eyepieces, which offer approximately 50° fields and excellent performance characteristics. This is actually a hybrid design that marries the Plössl to the Erfle.

The Meade Series 4000 family also includes two types of extreme-wide-field eyepieces, dubbed Super Wide Angle and Ultra Wide Angle. The Super Wide Angle oculars embody six elements that provide an impressive apparent field of 67°. As if that were not enough, the Ultra Wide Angle oculars yield an incredible apparent field approximately equal to the Naglers. They perform this magic by combining eight elements in five groups. Performance of all Series 4000 eyepieces is right up there with the best available.

Not to be left out, Celestron International markets Ultima eyepieces, available in a large variety of focal lengths, apparent fields, and optical designs. Its standard Ultimas (did someone say "standard?") use five elements to create fields between a slim 36° in the 42-mm model to a more acceptable 51° in shorter-focal-length versions. The Wide Field Ultimas feature a seven-element design for 70° apparent fields, whereas the Extra Low Power Ultimas have focal lengths from 45 mm to 80 mm (remember your exit pupil!) and apparent fields between 34° and 50°. All provide long eye relief and excellent correction against aberrations.

Another company to jump on the bandwagon is Orion Telescope Center. In addition to selling many of the previously mentioned brands of eyepieces, Orion also features UltraScopic and MegaVista oculars of their own design.

Ultrascopics use either five or seven elements to span a range of focal lengths from 35 mm to 3.8 mm, each with a 52° field. Orion's seven-element MegaVistas are competitive with Tele Vue's Wide Field and Meade's Super Wide Angle oculars. They come in focal lengths ranging from 40 mm to 10.5 mm, each offering an apparent field of view around 70°. Views are ghost free with pinpoint stars seen across the field.

As previously mentioned, all of these Super-Deluxe-Extra-Omni-Ultra-Maxi-Mega-Colossal eyepieces offer excellent image quality, color correction, and freedom of edge distortions. Their designers are to be congratulated for creating superb eyepieces for the backyard astronomer. On the negative side, however, I must point out that not only are their performances stellar but their prices are, too. Some of these may actually cost more than an entire telescope.

Are they worth the money? In all honesty, I have to say yes, especially if you own a fast (low focal ratio) telescope. But then again, if someone asked me what the finest car is, I probably would say either Mercedes Benz, Lexus, or some other upscale brand. Do I drive one of these? No, they cost more than my wallet can bear (although if I sell enough books . . .). I guess what I am trying to say is that while these top-of-the-line eyepieces are great if you can afford them, they are not absolutely necessary. Many hours of great enjoyment are yours with less costly eyepieces.

Barlow Lens

The Barlow Lens was invented in 1834 by Peter Barlow, a mathematics professor at Britain's Royal Military Academy. He reasoned that by placing a negative lens between a telescope's objective or mirror and the eyepiece, just before the prime focus, the instrument's focal length could be increased (Figure 5.4i). The Barlow Lens, therefore, is not an eyepiece at all, but rather an eyepiece amplifier.

Depending on the Barlow's location relative to the prime focus, the amplification factor can range up to about 3×. For example, remember our 8-inch f/10 telescope and 40-mm ocular? If we insert the eyepiece into a 2× Barlow Lens and then place both into the telescope, the combination's magnification will climb from 50× to 100×.

Why use a Barlow Lens? The first reason should be obvious. By purchasing just one more lens, the observer effectively doubles the number of eyepieces at his or her disposal. But the benefits of the Barlow go even deeper than this. Because a Barlow stretches a telescope's focal length, an eyepiece/Barlow team will yield consistently sharper images than an equivalent single eyepiece (provided that the Barlow is of high quality, of course). This becomes especially noticeable near the edge of view. Another important advantage is increased eye relief, something that short-focal-length eyepieces always have in short supply.

Telecompressors and Focal Reducers

Although the f/10 focal ratios of most Schmidt-Cassegrain telescopes have great appeal for observers who enjoy medium- to high-power sky views, their long focal lengths make it difficult to fit wide star fields into a single scene. For the amateur who craves a good view of larger objects such as the Andromeda Galaxy or the Orion Nebula, several manufacturers offer focal reducers. Basically working like a Barlow Lens in reverse, these add-on attachments cut a telescope's overall focal length, increasing its field of view.

Typical focal reducers (also known as *telecompressors*) fit between the back of a Schmidt-Cassegrain telescope and the star-diagonal eyepiece holder. Some models come with a star diagonal, while others use the star diagonal that came with the telescope. All have a universal thread, meaning that the Celestron telecompressor will fit a Meade telescope and vice versa and that aftermarket brands will fit both.

Although most reducers are great for shortening an instrument's focal length, they do nothing to correct for aberrations inherent in the telescope such as field curvature and may introduce a darkening effect called *vignetting* around the edge of the field. To help eliminate these scourges of photographers, Celestron offers a reducer-corrector. Not only does this attachment reduce the overall focal length of a telescope by 37% (from f/10 to f/6.3), but it also gives pinpoint star images across the entire field of view. Though greatly beneficial to all observers, this is especially attractive to astrophotographers, as the reduction in focal length also cuts down exposures by 2.5 times. Though designed with its own telescopes in mind, the Celestron reducer-corrector fits Meade SCTs as well. When teamed with an f/6.3 Meade SCT, the reducer-corrector yields a fast f/4 photographic system.

Coma Correctors

With the Dobsonian revolution of the 1980s, Newtonians with low focal ratios (RFTs) have become immensely popular—and with them, unfortunately, so has coma. You'll recall that coma is the aberration that causes stars near the edge of an eyepiece's field to look like tiny comet-like blobs instead of sharp points. Coma is an inborn trait of all Newtonians, though it only becomes objectionable when a telescope's focal ratio is f/5 or less.

To help combat the ills of coma, three companies, Tele Vue, Celestron, and Lumicon, presently manufacture coma correctors. Celestron and Lumicon employ two single lenses in their designs, while Tele Vue's model is based around a pair of achromats. To use one of the correctors, simply slip an eyepiece into its barrel, then place the pair into the telescope's eyepiece holder. This sounds suspiciously like a Barlow Lens, but rather than increasing power by two or three times, coma correctors limit their magnification effect to 15% or less. This means that while the view is greatly improved, any shrinking of the field size is negligible.

Though the Lumicon Coma Corrector is designed primarily for astrophotography, it works well for visual observations as well. It accepts both 1.25- and 2-inch eyepieces (an adapter is supplied for the former). The Celestron Multi-Purpose Coma Corrector also accepts 1.25- and 2-inch oculars and is equally effective for visual and photographic use. Finally, the Tele Vue Paracorr comes in two varieties: a 2-inch photo-visual model and a 1.25-inch visual-only version.

All coma correctors work well with nearly all eyepieces and are especially adept at providing stunning views of star fields and clusters. They are also effective at correcting off-axis astigmatism, another common problem of fast Newtonians. However, these correctors do not correct for bad optics. If your telescope has on-axis astigmatism, it is most likely caused by a poor-quality primary or secondary mirror. Most owners of f/4.5 or faster telescopes agree that the results are well worth their cost.

Reticle Eyepieces

For some applications, it can be useful (even necessary) to have an internal grid, or *reticle*, superimposed over an eyepiece's field of view. Reticle patterns, typically etched on thin, optically flat windows, come in a wide variety of designs depending on the intended purpose. The simplest reticles, two perpendicular lines that cross in the center of view, are most useful for aligning an equatorial mount to the celestial pole. Some reticles feature double crosshairs slightly offset from each other to create an open square in the center of view. These are especially beneficial when taking long-exposure astrophotographs. Peering through a guide telescope or an off-axis guider (discussed later), the photographer places a star in the central square. During the exposure, the photographer constantly checks that the star does not stray from the square, thereby ensuring that the telescope is tracking the sky properly. Other designs include bull's-eye targets and complex grid patterns. The former is also useful in astrophotography as well as during observing sessions and star parties when you are trying to pinpoint a hard-to-spot object for others to see. The latter reticle is handy when trying to judge the precise apparent size of an object or perhaps the angular separation of a double star.

Reticle eyepieces are frequently based on either the Kellner, orthoscopic, or Plössl eyepiece designs with focal lengths ranging between roughly 6 mm and 12 mm. In general, it is best to stick with the longer focal lengths as they tend to produce better image quality and definition. If more magnification is required than the eyepiece alone can deliver, use a Barlow lens.

Most reticle eyepieces come with strange-looking appendages sticking out one side of their barrels. These are illuminating devices used to light the reticle patterns for the observer. Ideally, an illuminator's light level should be adjustable so the reticle is just bright enough to be seen but not so bright as to overpower the object in view.

Illuminating devices use either small incandescent flashlight bulbs or light-emitting diodes (LEDs), both colored red. LEDs are preferred because they draw much less power than a conventional bulb. Power for the illuminator comes from either an external battery or telescope power panel (such as on some Meades and Celestrons) or from tiny internal batteries (usually type SR-

44 or equivalent). The former offer almost unlimited life to the illuminator, though dangling wires connecting the two may become entangled in the dark. The latter design is wireless, but there is always the chance that the batteries will wear out just when they are needed most.

Perhaps the most versatile illuminated reticle eyepiece on the market today is Celestron's Micro Guide, shown in Figure 5.5. Built around a 12.5-mm orthoscopic eyepiece, the Micro Guide features a laser-etched reticle that includes a bull's-eye style target for guiding during astrophotography and micrometer scales for measuring object size and distances such as the separations and position angles of double stars, and so on.

Another interesting development in the world of reticle eyepieces is Orion's PulsGuide illuminator. Rather than glowing steadily as other illuminators do, the PulsGuide alternately flashes the reticle on and off. The advantage here is that by not keeping the reticle on continuously, fainter stars may be spotted and followed with greater ease. Rate and brightness of pulse are both adjustable by the user. Note that the PulsGuide is not sold with a reticle eyepiece but only as a replacement for illuminators of other eyepieces. It is available in two models to fit into just about any illuminated reticle ocular sold today.

Although nice to have, reticle eyepieces are low on the list of priorities of most amateur astronomers unless they are contemplating telescope-guided astrophotography or advanced observing programs. The rest of us should probably direct our money elsewhere.

Figure 5.5   *Celestron's Micro Guide illuminated-reticle eyepiece. Photo courtesy of Celestron International.*

Pieces of the Puzzle

As an aid to putting all of the pieces to the eyepiece puzzle together, I've included Table 5.3 as a summary of this chapter's thoughts.

Now comes the moment of truth: Which eyepieces are best for you? It is always preferable to have a set of oculars that offers a variety of magnifications, because no one value is good for everything in the universe. Low power is best for large deep-sky objects such as the Pleiades star cluster or the Orion Nebula. Medium power works well for lunar sightseeing as well as for viewing smaller deep-sky targets such as most galaxies. Finally, high power is needed to spot subtle planetary detail or to split close-set double stars.

Table 5.3   **Eyepiece Recommendations**

	Refractor		Reflector			Catadioptric	
Type	≤f/10	>f/10	f/5 Newt.	f/8 Newt.	f/12 Cas.	f/10 SC	f/16 Mak.
Huygens	½★		½★	½★	★	½★	★
Ramsden	★	★	★	★	★	★	★
Kellner	★★^{L}	★★^{L}	★^{L}	★★^{L}	★★^{L}	★★^{L}	★★^{L}
	★^{M}	★^{M}	★^{M}	★^{M}	★^{M}	★^{M}	★^{M}
Orthoscopic	★★★	★★★	★★★	★★★	★★★	★★★	★★★
RKE	★★★	★★★	★★★	★★★	★★★	★★★	★★★
Erfle	★★★^{L}	★★★^{L}	★★★^{L}	★★★^{L}	★★★^{L}	★★★^{L}	★★★^{L}
Zoom	★★	★★	★★	★★	★★	★★	★★
Plössl	★★★	★★★	★★★	★★★	★★★	★★★	★★★
Brandon	★★★½	★★★½	★★★½	★★★½	★★★½	★★★½	★★★½
Lanthanum LV	★★½	★★½	★★½	★★½	★★½	★★½	★★½
Wide-Field	★★★★^{L}	★★★★^{L}	★★★★^{L}	★★★★^{L}	★★★★^{L}	★★★★^{L}	★★★★^{L}
Nagler	★★★★$^{M/H}$	★★★★$^{M/H}$	★★★★$^{M/H}$	★★★★$^{M/H}$	★★★★$^{M/H}$	★★★★$^{M/H}$	★★★★$^{M/H}$
Panoptic	★★★★$^{L/M}$	★★★★$^{L/M}$	★★★★$^{L/M}$	★★★★$^{L/M}$	★★★★$^{L/M}$	★★★★$^{L/M}$	★★★★$^{L/M}$
Super Plössl	★★★	★★★	★★★	★★★	★★★	★★★	★★★
Super Wide Angle	★★★★^{L}	★★★★^{L}	★★★★^{L}	★★★★^{L}	★★★★^{L}	★★★★^{L}	★★★★^{L}
Ultra Wide Angle	★★★★$^{M/H}$	★★★★$^{M/H}$	★★★★$^{M/H}$	★★★★$^{M/H}$	★★★★$^{M/H}$	★★★★$^{M/H}$	★★★★$^{M/H}$
Standard Ultima	★★★½	★★★½	★★★½	★★★½	★★★½	★★★½	★★★½
	★★★^{L}	★★★^{L}	★★★^{L}	★★★^{L}	★★★^{L}	★★★^{L}	★★★^{L}
Wide Field Ultima	★★★½	★★★½	★★★½	★★★½	★★★½	★★★½	★★★½
Extra Low Ultima	★★★^{L}	★★★^{L}	N/A	★★★^{L}	★★★^{L}	★★★^{L}	★★★^{L}
MegaVista	★★★★$^{L/M}$	★★★★$^{L/M}$	★★★★$^{L/M}$	★★★★$^{L/M}$	★★★★$^{L/M}$	★★★★$^{L/M}$	★★★★$^{L/M}$
Ultrascopic	★★★½	★★★½	★★★½	★★★½	★★★½	★★★½	★★★½

Notes:

1. *Eyepiece ratings are based on the following four-star scale: ★ = poor, ★★ = fair, ★★★ = good, ★★★★ = excellent, N/A = not applicable due to excessive exit pupil.*

2. *When performance changes according to magnification, the following superscripted abbreviations are used: ^{L} = low power only, ^{M} = medium power only, ^{H} = high power only. Be sure to check the exit pupil that will result from using a particular eyepiece in your telescope first.*

Table 5.4   **Four Telescope/Eyepiece Alternatives**

	Dream Outfit	Middle of the Road	Good and Cheap
4″ f/10 refractor	LP: 35-mm Panoptic	LP: 32-mm Plössl	LP: 25-mm RKE/Kellner
	MP: 19-mm Wide Field	MP: 12-mm Plössl	MP: 9-mm RKE/Kellner
	HP: 4.8-mm Nagler	HP: 6-mm Orthoscopic	HP: —
	Barlow Lens	Barlow Lens	Barlow Lens
8″ f/7 Newtonian	LP: 24-mm Wide Field	LP: 26-mm Plössl	LP: 26-mm Erfle
	MP: 14-mm Ultra Wide	MP: 12-mm Plössl	MP:—
	HP: 8.8-mm Ultra Wide	HP: 9-mm Orthoscopic	HP: 9-mm RKE/Kellner
	Barlow Lens	Barlow Lens	Barlow Lens
15″ f/5 Newtonian	LP: 22-mm Panoptic	LP: 24-mm Wide Field	LP: 26-mm Erfle
	MP: 12-mm Nagler	MP: 10-mm Plössl	MP: 9-mm RKE/Kellner
	HP: 7-mm Nagler	HP: 6-mm Orthoscopic	HP:—
	Barlow Lens	Barlow Lens	Barlow Lens
	Coma Corrector		
8″ f/10 Schmidt-Cassegrain	LP: 40-mm Wide Field	LP: 32-mm Plössl	LP: 25-mm RKE/Kellner
	MP: 16-mm Nagler	MP: 12-mm Plössl	MP:—
	HP: 9-mm Nagler	HP: 9-mm Orthoscopic	HP: 9-mm RKE/Kellner
	Barlow Lens	Barlow Lens	Barlow Lens
	Reducer/compressor	Telecompressor	—

In an ideal world, eyepiece quality would be our sole consideration, but in the real world, most of us also must factor in cost. As an aid to guide your selection, Table 5.4 offers three different possibilities for four of today's most popular telescope sizes. Each lists eyepieces according to the magnification they would produce (LP = low power, MP = medium power, HP = high power). The first is a dream outfit, where money is no object; the second offers a middle-of-the-road compromise between quality and cost; and the third represents the minimum expenditure required for a good range of eyepieces. Prices are not listed because they can change quickly and dramatically from dealer to dealer; be sure to shop around.

Keep in mind that these represent only three possibilities for each telescope. You are encouraged to flip back and forth between Tables 5.3 and 5.4 to substitute your own preferences, as it all boils down to this. While you may find this guide useful, the only way to learn which oculars are really best for you is to try them out first. Once again, I strongly recommend that you seek out and join a local astronomy club and go to an observing session. Bring your telescope along and borrow as many different types of eyepieces as possible. Take each of them for a test drive. Then, and only then, will you know exactly what is right for you.

6

The Right Stuff

Congratulations on making it through what some of you might have thought of as a telescope-and-eyepiece obstacle course. Though the range of choices was extensive, you should now have a fairly good idea about which telescopes and eyepieces best meet your personal needs.

A telescope alone, however, cannot simply be set up and used. First, it must be outfitted with other things, such as a finderscope, maybe some filters, a few reference books, a star atlas, a flashlight, some bug spray . . . well, you get the idea. So get out your wish list and credit card once again. It is time to go shopping in the wide world of telescope-related paraphernalia.

Let me just take a moment to calm your fears at the thought of spending more money on this hobby. Sure, it is easy to draw up a list of must-have items as you look through astronomical catalogues and magazines. Everything could easily tally up to more than the cost of your telescope in the first place, but is this truly necessary? Happily, the answer is no. Before you buy another item, you should first explore the accessories that are absolutely mandatory, those that can wait for another day, and the ones that can be done without altogether.

A quick disclaimer before going on: There are so many accessories available to entice the consumer that it would be impossible for a book such as the one you hold before you to list and evaluate every part made by every company. As a result, this chapter must limit its coverage to more readily available items. If, as you are reading this chapter, you feel that I have unjustifiably omitted something that you feel is the greatest invention since the telescope, then by all means share your enthusiasm with me. Write your own review and send it to me in care of the address listed in the Preface. I will try to include mention of that item in a future edition.

Finderscopes

After eyepieces (and arguably even before), the most important accessory in an amateur astronomer's bag of tricks is a *finderscope* (Figure 6.1), which is a small, low-power, wide-field spotting scope that is mounted piggyback on the telescope tube. Its sole purpose in life is to help the observer aim the main telescope toward the desired target. Although most telescopes come with finders, all finders are not created equal. What sets a good finder apart from a poor finder depends on its intended use. First, some words of advice for readers who will be using finders only to supplement the use of setting circles for zeroing in on sky targets. If this applies to you (and I truly hope it does not, especially after you read my editorial in Chapter 9), then the finder will probably be used only for locating brighter solar system objects, Polaris (to align the equatorial mount to the celestial pole), and perhaps a few terrestrial objects. In this case, just about any finder will do.

If, however, star hopping is your preferred method for locating sky objects (again, consult Chapter 9), then finderscope selection is critical to your success as an observer. The three most important criteria by which to judge a finder are magnification, aperture, and field of view. Finderscopes are specified in the same manner as binoculars. A 10×50 finder, for example, has a 50-mm aperture and yields $10 \times$. Most experienced observers (including yours truly) agree that the smallest *useful* size for a finderscope is 8×50, though finders, like binoculars, should be sized to match the sky conditions. Under typical suburban skies, an 8×50 finder (with a 6-mm exit pupil) will penetrate to about 8th

Figure 6.1   An 8×50 finderscope.

magnitude, which is roughly comparable with many popular star atlases. Rural astronomers, who generally enjoy darker skies, may prefer giant 10×70 and 11×80 finders that reveal stars a magnitude or two fainter but suffer from smaller fields of view. Meade Instruments; Orion Telescope Center; Roger W. Tuthill, Inc.; and Lumicon offer good selections of finders from which to choose.

The current rage in finderscopes is the *right-angle finder*. In these, a mirror- or prism-based star diagonal is built into the finder to turn the eyepiece at a 90° angle. Sure, a right-angle prism can make looking through the finder more comfortable and convenient, but is it really a good idea? To my way of thinking, no! There are two big drawbacks to right-angle finders. First, though the view through a right-angle finder is upright, the prism also flips everything left-to-right. This mirror-image effect matches the view through a telescope using a star diagonal (Schmidt-Cassegrain, refractor, etc.) but makes it very difficult to compare the field of view with a star atlas. By comparison, *straight-through finders* flip the view upside down but do not swap left and right. Personally, I find it easier to turn a star chart upside-down to match an inverted view than to turn it inside-out to match a mirrored image. (Special prisms, called *Amici* prisms, cancel out the mirroring effect but are supplied with relatively few finders.)

A straight-through finder also permits the use of both eyes when initially aiming the telescope. By overlapping the naked-eye view with that visible through the finder, an observer may point the telescope quickly and accurately toward the intended part of the sky. If you keep both eyes open when using a right-angle finder, one eye will see the sky and the other will see the ground!

Deep down inside, most observers agree that right-angle finders are poor substitutes for straight-through finders. At least I assume they must, given the incredible popularity of one-power finderscopes in recent years. By far, the most common of this new generation of sighting contraptions is the Telrad (Figure 6.2) by Steve Kufeld. The Telrad is described as a "reflex sight." Using a pair of AA batteries and a red light-emitting diode (LED), the Telrad projects a bull's-eye target of three rings onto a clear piece of glass set at a 45° angle. To sight through the Telrad, simply look through the glass window from behind. The glass acts as a beam splitter, reflecting the target rings to the observer's eye while at the same time letting the stars shine through from beyond. The brightness of the rings is controlled by a side-mounted rheostat that also acts as an on/off switch.

The Telrad is not meant to be used in lieu of a finder but only to supplement its use. The biggest advantage of the Telrad is that it allows easy aiming of a telescope without any need to flip or twist star charts. Its only disadvantage is that the window tends to dew over quickly in damp environs. But because it is not an optical device, the window may be wiped clear with a finger, sleeve, or paper towel. A more permanent solution to the dewing problem is to simply make a roof over the glass window by bending a large file card that has been painted black over the Telrad from side to side and secure it in place with masking tape.

Figure 6.2 *A one-power Telrad aiming device by Steve Kufeld, one of the most popular accessories among today's amateur astronomers.*

Surprisingly, only one other company—Tele Vue—markets a one-power sighting device: the Star Beam. Leaner and sexier than the gawky Telrad, which is made of black plastic, the slim, chromed-metal Star Beam works as well as its ugly-duckling cousin but costs about four times as much. Beauty is only skin deep, but cost cuts right to the bone, so the Star Beam has never achieved the popularity of the Telrad.

Finally, a word or two on finder mounts. Good finders are secured to telescopes by a pair of rings with six adjustment screws that allow the finder to be aligned precisely with the main instrument. Let the buyer beware, however, that many smaller finders (primarily 5×24 and 6×30 models) use single mounting rings with only three adjustment screws. These are notoriously difficult to adjust and even more difficult to keep in alignment. If there is a choice between a single ring or a pair of rings for mounting a finder, always select the pair.

Filters

For years, photographers have known the importance of using filters to change and enhance the quality and tone of photographs in ways that would be impossible under natural lighting. Today, more and more amateur astronomers are also discovering that views of the universe can also be greatly improved by using filters with their telescopes and binoculars. But there are so many from which to choose. Which is best? Some filters heighten subtle, normally invisible planetary detail, others suppress the ever-growing effect of light pollution, while still others permit the safe study of our star, the Sun. Which filter or filters, if any, are right for you depends largely on what you are looking at and from where you are doing the looking.

Light-Pollution Reduction (LPR) Filters

As all amateur astronomers are painfully aware, the problem of light pollution is rapidly swallowing up our skies. In regions across the country and around the world, the onslaught of civilization has reduced the flood of starlight to a mere trickle of what it once was. Are we powerless against this beast? Not entirely.

While Chapter 9 discusses the dilemma of light pollution in more specific detail, this section will look at some filters that may be used to help counteract it. First, let's briefly define light pollution: unwanted illumination of the night sky caused largely by poorly designed or poorly aimed artificial lighting fixtures. Rather than illuminating only their intended targets, many fixtures scatter their light in all directions, including up.

There are two kinds of light pollution: local and general. Local light pollution shines directly into the observer's eyes and may be caused by anything from a nearby streetlight to an inconsiderate neighbor. There isn't a filter made that will counteract this sort of interference. Light-pollution reduction filters are much more effective against general light pollution, or *sky glow*, which is the most destructive type of light pollution. It can turn a clear blue daytime sky into a yellowish, hazy night sky of limited usefulness.

Although modern technology (or its careless use) caused the problem in the first place, it also offers partial redemption. Many sources of light pollution shine in the yellow region of the visible spectrum, as indicated in Figure 6.3. Mercury-vapor lights, for instance, shine at 405 nm, 436 nm, and in the wide region between approximately 550 and 630 nm (*nm* is short for *nanometer*, a very small unit of measure; one nanometer is equal to 10^{-9} meters or 10 angstrom units). At the same time, many non-stellar sky objects (planetary nebulae, emission nebulae, and comets) emit most of their light in the blue-green region of the spectrum. Emission nebulae, for example, glow primarily in the hydrogen-beta (486 nm) and oxygen-III (496 nm and 501 nm) regions of the spectrum. In theory, if the yellow wavelengths could somehow be suppressed while the blue-green wavelengths were allowed to pass, then the effect of light pollution would be greatly reduced.

What exactly do light-pollution reduction filters do? A popular misconception is that LPR filters (Figure 6.4) make faint objects look brighter. Not true! LPR filters consist of thin pieces of optically flat glass that are coated in multiple, microscopically thin layers of special optical material; they are designed to block specific, undesirable wavelengths of light while letting others pass. The observer need only attach the filter to his or her telescope, usually either

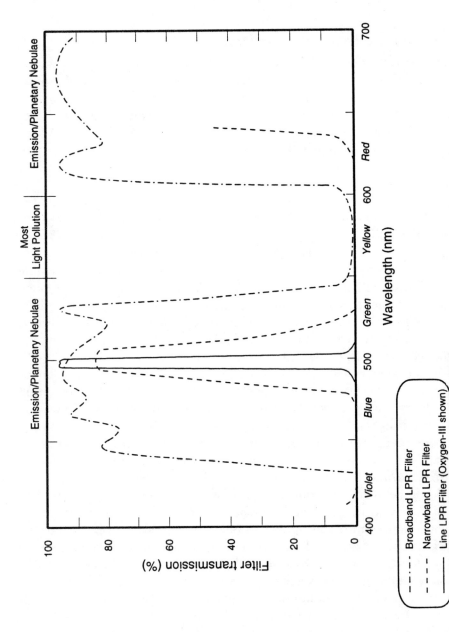

Figure 6.3 *Transmission characteristics of light-pollution reduction (LPR) filters. Chart based on data supplied by Lumicon, Inc.*

Figure 6.4 *An assortment of LPR filters. Photo courtesy of Lumicon, Inc.*

by screwing the filter into the field end of an eyepiece or, in the case of many Schmidt-Cassegrains, inserting it between the telescope and star diagonal. The net result is increased contrast between the object under observation and the filter-darkened background. Many deep-sky objects also shine in the deep red, hydrogen-alpha (656 nm) portion of the spectrum, a region that is all but invisible to the eye under dim light conditions. Still, most light-pollution reduction filters transmit these wavelengths as well for photographic purposes.

Light-pollution reduction filters (also called *nebula filters*) come in three varieties: *broadband*, *narrowband*, and *line* filters. The biggest difference is in their application. As shown in Figure 6.3, broadband filters pass a wide swath of the visible spectrum, from roughly 430 to 550 nm. Narrowband filters transmit wavelengths of light between about 480 and 520 nm. Line filters have an extremely narrow window, allowing only one or two specific wavelengths of light to pass.

No one type of filter is best for everything. Some enhance the object under observation, some have little or no effect, while others actually make the target fade or disappear completely! Which LPR filter is best under what circumstances remains one of the hottest topics of debate among amateur astronomers today. Table 6.1 may help sort things out. The filters are rated on a one- to four-star scale based on their performance on several different types of objects. One star indicates either negligible or negative results, two stars connote minimal positive effect, three stars indicate a noticeable improvement in appearance, and four stars denote a vast improvement in object visibility and/or detail.

Which filter or filters should you choose? If you are interested primarily in deep-sky objects and want to buy only one LPR filter, then go with either the DayStar or Orion Ultrablock. Though the results are close, these offer greater enhancement of emission and planetary nebulae than the other narrowband filters. Another good choice is Lumicon's O III filter, a remarkable

Table 6.1   **Light-Pollution Reduction (LPR) Filters: A Comparison**

	Open Clusters	Globular Clusters	Emission Nebulae	Reflection Nebulae	Planetary Nebulae	Galaxies	Comets
Broadband							
Celestron LPR	★	★	★★	★★	★★★	★	★★
Lumicon Deep Sky	★★	★★	★★	★★	★★★	★★	★★
Meade 908B	★★	★★	★★	★★	★★★	★★	★★
Orion Skyglow	★★	★★	★	★	★★★	★★	★★
Parks LPB	★★	★★	★★	★★	★★★	★★	★★
Narrowband							
DayStar	★★	★★	★★★★	★★	★★★★	★★	★
Lumicon UHC	★	★	★★★	★★	★★★★	★	★
Meade 908N	★★	★★	★★★	★★	★★★★	★	★
Orion Ultrablock	★	★	★★★★	★★	★★★	★	★
Line							
Lumicon H-Beta	★	★	★★★★	★	★	★	★
Lumicon O III	★	★	★★★★	★★	★★★★	★	★
Lumicon Swan Band	★	★★	★★	★★	★★★	★	★★★★

Notes:

1. *The 1.25-inch DayStar Nebula Filter reduces the apparent field of view of super wide-field eyepieces, thereby limiting the panoramic views those eyepieces are noted for. (The 2-inch version does not suffer from this drawback.)*
2. *Most observers find the Hydrogen-Beta filter to be of limited usefulness. It works well on extremely faint diffuse nebulae (in particular, the Horsehead Nebula, California Nebula, and Cocoon Nebula) but little else.*
3. *In general, broadband filters are much more useful for astrophotography than narrowband filters.*

accessory indeed. It will reveal nebulae that are impossible to see with any of the others. Though labelled "light-pollution reduction filters," the DayStar, Ultrablock, and O III produce excellent results even under the darkest sky conditions.

Reflection nebulae, clusters, and galaxies (all of which are illuminated primarily by starlight shining across the visible spectrum) seem to benefit little from any of the filters. A noticeable improvement does occur when using some of them in bright suburban or urban conditions, but they serve no useful purpose when used under darker suburban or rural skies. Finally, comets, because of their unique spectra, are best observed using Lumicon's Swan Band filter, designed with comets in mind.

The bottom line is this: Light-pollution reduction filters are wonderful assets for the amateur astronomical community. With them, observers can spot nonstellar objects that were considered impossible to find just a few years ago. But remember, no filter can deliver the beauty of the universe as well as a truly dark, light-free sky.

Lunar and Planetary Filters

Although light-pollution reduction filters have no positive effect on viewing subtle lunar and planetary features, there exists a wide variety of other (happily, cheaper) filters that do. Most seasoned observers agree that fine details on both the Moon as well as the five bright planets visible to the naked eye can be significantly enhanced by viewing through color filters. Though they cannot perform magic, color filters can make a definite difference. This improvement occurs for two reasons. First, they reduce *irradiation*, which is the distortion of a boundary between a lighter area and a darker region on the Moon's or a planet's surface. This effect usually is caused by either turbulence in the Earth's atmosphere or by the human eye being overwhelmed by the dazzling image. Of course, you don't want the object too dim to be seen well! Observers with small- and medium-aperture telescopes should stick with the lighter filters.

Filters also help to increase otherwise barely perceptible contrasts between two adjacent regions of a planet or the Moon by transmitting (brightening) one color while absorbing (darkening) some or all other colors. For instance, an observer's eye alone may not be able to distinguish between a white region and a bordering beige region on Jupiter. By using filters of different colors, the contrast between the zones may be increased until their individuality becomes apparent.

Which filters are best for which objects in the sky? Table 6.2 compares different heavenly sights with the results that may be expected when they are viewed through a variety of color filters. The filters are listed according to their *Wratten number* as well as their color. The Wratten series of color filters was created by Eastman Kodak and contains over 100 different shades and hues. Today, it is the industry's standard way to refer to a filter's precise color. (Note that the entire series was revamped not too many years ago. That is why some older books may refer to, say, a deep yellow filter as a K3 instead of the modern designation, No. 15.)

For those just starting out, choose basic colors, such as deep yellow (No. 15), orange (No. 21), red (No. 23), green (No. 58), and blue (No. 80A). Though you are free to use photographic color filters sold in camera stores, most amateurs prefer filters made especially for telescope eyepieces. Like light-pollution reduction filters, these color filters are designed to screw into the field end of eyepieces. Many telescope manufacturers and suppliers, including Orion, Optica b/c, Meade, and Celestron, offer a wide assortment of color filters.

Solar Filters

Monitoring the ever-changing surface of the Sun is an aspect of the hobby that is enjoyed by many. Before an amateur dares look at the Sun, however, he or she must be aware of the extreme danger of gazing at our star. **Viewing the Sun without proper precautions, even for the briefest moment, may result in permanent vision impairment or even blindness.** This damage is caused primarily by the Sun's blue and ultraviolet rays, the latter being the same rays

Table 6.2   Color Filters: A Comparison

Object	Filter	Result
Moon	Moon filter (Neutral density)	Reduces brightness of the Moon evenly across the spectrum, making observations easier without introducing false color.
	15 (Deep yellow)	Enhances contrast of lunar surface
	58 (Green)	Like deep yellow, green will enhance contrast and detail in some lunar features
	80A (Light blue)	Reduces glare
	Polarizer	Like the neutral density filter, reduces brightness without introducing false colors
Mercury	21 (Orange)	Helps to make the planet's phases clearly visible
	23A (Red)	Increases contrast of a planet against blue sky, aiding in daytime or bright twilight observation
	25 (Deep red)	Same as #23A, but deeper color
	80A (Light blue)	Improves view of Mercury against bright orange twilight sky
	Polarizer	Darkens sky background to increase contrast of planet; helpful for determining phase of Mercury
Venus	25 (Deep red)	Darkens background to reduce glare; some say they also help reveal subtle cloud markings
	80A (Light blue)	Improves view of Venus against bright orange twilight sky
	Polarizer	Reduces glare without adding artificial color (especially important for viewing the planet through larger telescopes)
Mars	21 (Orange)	Penetrates atmosphere to reveal reddish areas and highlight surface features such as plains and maria (best choice for small apertures)
	23A (Light red)	Same as #21, but deeper color (best choice for medium and large apertures)
	25 (Deep red)	Same effect as #23A, but deeper color (best choice of the three for very large apertures)
	38A (Deep blue)	Brings out dust storms on surface of Mars
	58 (Green)	Accentuates melt lines around polar caps
	80A (Light blue)	Accentuates polar caps and high clouds, especially near the planet's limb
Jupiter	11 (Yellow-green)	Reveals fine details in cloud bands
	21 (Orange)	Accentuates cloud bands
	56 (Light green)	Accentuates reddish features such as the Red Spot
	58 (Green)	Same as #56, but deeper color
	80A (Light blue)	Highlights details in orange and purple belts as well as white ovals
	82A (Very light blue)	Similar effect as #80A, though not as pronounced
Saturn	15 (Deep yellow)	Helps to reveal cloud bands
	21 (Orange)	Similar effect as #11, but deeper color
Comets	80A (Light blue)	Increases contrast of some comets' tails
Other Uses	15 (Deep yellow)	Helps block ultraviolet light when doing black-and-white astrophotography
	25 (Red)	Reduces impact of light pollution on long-exposure black-and-white photographs taken from light-polluted areas
	58 (Green)	Same as #25; works well for emission nebulae
	82A (Very light blue)	Suppresses chromatic aberration in refractors
	Minus Violet	Reduces impact of light pollution without dramatically distorting color; cheaper than broadband LPR filters

that cause sunburn. Although it may take many minutes before the effect of sunburn is felt on the skin, the Sun's intense radiation will burn the eye's retina in a fraction of a second.

There are two ways to view the Sun safely: either by projecting it through a telescope or binoculars onto a white screen or piece of paper or by using a special filter. A build-at-home telescope attachment for projecting the Sun is detailed in the next chapter, but for now, let's examine solar filters.

Sun filters come in a couple of different varieties. Some fit in front of the telescope, while others attach to the eyepiece. *NEVER* use the latter . . . that is, the eyepiece variety. They can easily crack under the intense heat of the Sun (focused by the telescope just like the Sun's image), leading tragically to blindness. My first telescope came with one of these so-called Sun filters. After only one use, I began to notice bright flashes of light whenever I looked toward the Sun. At first, I couldn't figure out what was wrong. But then, a closer inspection revealed that a hairline crack had developed in the filter, splitting it in half. I am one of the lucky ones; the filter did not shatter. If it had, the full intensity of sunlight would have rushed into my eye and quite possibly have caused permanent blindness.

The only safe kind of Sun filter is the type that fits securely in *front* of a telescope or binoculars; Figure 6.5 shows one example. In this way, the dangerously intense solar rays (and accompanying heat) are reduced to a safe level prior to entering the optical instrument. Be sure to use only specially designed Sun filters; DO NOT use photographic neutral-density filters, smoked glass, overexposed photographic film, or other makeshift materials that may pass invisible ultraviolet or infrared light.

The most popular manufacturers of safe solar filters are Thousand Oaks Optical Company and Roger W. Tuthill, Inc. Let's take a look at each. Thousand

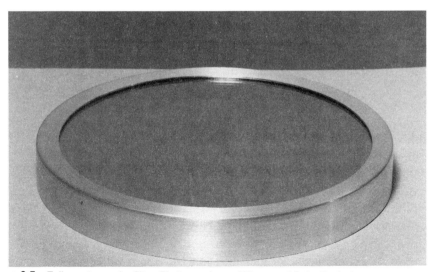

Figure 6.5 *Full-aperture solar filter. Photo courtesy of Thousand Oaks Optical.*

Oaks filters, made of parallel-plane glass, come in three varieties. Type 1 filters are triple-coated with a nickel-chromium alloy and come mounted in an aluminum cell. Type 2 filters are the same as Type 1, except that stainless steel is combined in the coating for added durability. Both reduce the Sun's brightness to 0.001% of its original intensity and are designed for visual and photographic use. Type 3 filters reduce the Sun's brilliance to 0.01% and are intended for photographic use *only*. All three Thousand Oaks filters produce yellow-orange solar images.

Tuthill has been selling the ever-popular Solar Skreen filters since 1973. (Celestron also offers Solar Skreen filters under its own brand name.) These filters are made from two tissue-paper-thin pieces of aluminized mylar that are sandwiched together. Like the Thousand Oaks Type 1 and 2 models, Solar Skreen filters reduce the Sun's brightness to 0.001% of its unfiltered intensity.

There are some pluses and minuses to Solar Skreen filters. One drawback is that they turn the Sun blue. This can be a little disconcerting at first, but most people get used to it after a while. But if you must see a yellow Sun (when doing color solar photography, for instance), add a Wratten No. 15 (deep yellow) or No. 21 (orange) filter to the eyepiece or camera. On the plus side, inch for inch, Solar Skreen filters are less expensive than their Thousand Oaks rivals.

Both Thousand Oaks and Solar Skreen filters may be purchased either already mounted in a cell for a specific telescope or unmounted. The mounted filters come in cells designed to slip over the front end of specific telescope models. Because these are full-aperture filters, the larger the telescope, the more expensive the filter will be.

Keep in mind, however, that a telescope's full aperture is *not* required for observations. (In fact, large-aperture instruments may produce an uncomfortably bright view regardless of the filter used.) To save a few dollars, purchase a small, unmounted filter and then mount it yourself using an off-axis mask similar to that detailed in Chapter 10 for viewing the Moon. Though the price will be lower, definition may be a little lower due to the smaller aperture.

Solar filters must be treated and stored with care lest they become damaged and unsafe to use. Regularly inspect the filters (especially the mylar type) for damage such as pinholes or irregularities in the coating by holding the filter up to a bright light. If a pinhole develops, it may not be cause to discard the filter just yet. Small holes may be effectively blocked off with a tiny drop of flat paint without image quality suffering too greatly. Gently dab a bit of paint over the hole using a tooth pick. If more serious damage is detected, then the filter must be replaced immediately.

All of the solar filters just discussed might be thought of as broadband filters, because they filter the entire visible spectrum (and you can't get much broader than that). There are also special narrowband solar filters that allow observers to see our star in a completely different light! These block all of the light from the Sun with the exception of one distinctive wavelength: 656 nanometers. This is one frequency of glowing hydrogen, called *hydrogen-alpha*. Viewing the Sun with an H-alpha filter allows the observer to monitor ruby-red solar prominences, bright white filaments, and intricate surface granulation.

Hydrogen-alpha filters typically consist of two separate pieces: an energy rejection filter (ERF) that fits over the objective to prevent the telescope from overheating and the hydrogen-alpha filter itself, which fits between telescope and eyepiece. The ERF also acts to stop down the telescope to f/30, necessary for the proper operation of the H-alpha filter. The filters themselves are available in several different, extremely narrow bandwidths (usually expressed in angstroms, where 1 angstrom equals 0.1 nanometer). The narrower the bandwidth, the higher the contrast but the fainter the image. Bandwidths of either 0.8 or 0.9 angstrom are the most popular because these offer the best compromise between brightness and contrast.

Unfortunately, H-alpha solar filters are expensive. DayStar Filter Corporation (the same company that manufactures the DayStar Nebula Filter) is the leading source for these special accessories. The most expensive is the University model, which, as the name implies, is geared toward the rigid requirements of professional institutions. The ATM filter, however, is designed with the serious amateur solar astronomer in mind. Both the University and ATM filters need to be heated in order to function properly and therefore require an external source of AC power. The least costly H-alpha filter of all DayStar models is the T-Scanner. Unlike the other two, the T-Scanner requires no external power supply and so is completely portable, making it even more attractive. Its only drawback is that the bandwidth passed is a little wider than the others, lowering contrast some. Still, the T-Scanner is the most popular hydrogen-alpha filter available today.

Lumicon also manufactures H-Alpha solar filters for the amateur marketplace. The Lumicon Solar Prominence Filter works on the same no-heater principle as the T-Scanner. Its 1.5 angstrom bandpass, wider than any of the DayStar filters, renders good views of prominences and sunspots but is not designed for viewing flares, filaments, and granulation. Complete with ERF prefilter and all required adapters and hardware, the Lumicon filter costs hundreds less than the T-Scanner, and it is an excellent value for anyone living within a budget who wishes to get into this phase of solar observing.

As Chapter 10 will point out later on, never use a finderscope to align a telescope to the Sun. Viewing through the unfiltered finder can cause permanent eye damage as surely as peering at the Sun with an unfiltered telescope. To align the telescope with the Sun, use the shadow technique also described in that chapter.

Other Eyepiece Accessories

Collimation Tools

A telescope will deliver only poor-quality, lackluster images unless it is in proper optical alignment, or collimation. It wasn't too many years ago that the only tool available to check collimation was the simple *sight tube*, which consists of a hollow pipe the same diameter as an eyepiece with a small central hole in an otherwise solid metal cap. When placed in a telescope's focusing mount, a sight tube is fine for checking the alignment of refractors, catadioptrics, and normal-focal-length Newtonians but is inadequate by itself for the critical collimation required for rich-field Newtonian reflectors.

To answer the outcry for better collimation tools, Tectron Instruments and AstroSystems, Inc., offer a modified sight tube as well as two other accessories to make the job of telescope collimation a little easier: the Cheshire eyepiece and the Autocollimator.

The sight tube is helpful for approximating the collimation of just about any telescope. At one end lies a small peephole used by the observer to check collimation; at the other, a pair of thin wire cross hairs serves as a reference for centering optical components. (Collimation procedures are outlined in detail in Chapter 8, as are instructions for making a sight tube at home.)

The Cheshire eyepiece (Figure 6.6), invented 70 years ago by Professor F. J. Cheshire at a British university, is intended primarily for in-the-field adjustments of fast Newtonians. Not an optical assembly, a Cheshire eyepiece is a variation on the sight tube with a part of one side cut out and a 45° mirrored surface inside. Tectron's version, again made from black-anodized aluminum, is durable enough to deliver years of service even with a fair amount of abuse. AstroSystem's interpretation is similar but can be combined with a sight tube for a handy all-in-one tool.

Figure 6.6   *A Cheshire eyepiece, a must for collimating fast Newtonian reflectors.*

The Autocollimator is the most sensitive of the three tools for collimating a fast Newtonian reflector. At first glance, it looks like a short sight tube—a hollow tube capped with a piece of metal that has a tiny hole drilled in the center. The difference becomes evident only when you look inside the Autocollimator. There, encircling the peephole and perpendicular to the eyepiece mount's optical axis, a flat mirror serves to reflect all light from the primary mirror back down the telescope.

Are all three necessary? The sight tube, whether purchased or built at home, is a MUST for anyone who owns a telescope and is concerned with optimizing its performance. I also consider the Cheshire eyepiece a requirement for owners of rich-field (RFT) Newtonian reflectors. Don't leave home without it! The Autocollimator, though a handy tool for fine-tuning the alignment of fast Newtonians, is not absolutely necessary.

Rubber Eyecups

Nothing is more annoying to an observer than having to shade a bright light from entering the corner of the eye while looking through a telescope. The distraction caused by glare seen out of the corners of the eyes can be enough to cause a faint celestial object to be missed. To help shield an observer's eye from extraneous light, many eyepieces sold today come with collapsible rubber eyecups, and they certainly can make a big difference!

Though all eyepieces do not come with built-in eyecups, most can be retrofitted with after-market versions that work just as well. Orion, Edmund, Tele Vue, and several other companies sell rubber eyecups designed either to fit their own particular lines of oculars or to offer a universal fit.

Binocular Viewers

Research has shown that when it comes to viewing the night sky, binocular (two-eyed) vision is definitely better than monocular (one-eyed) vision. Our power of resolution and ability to detect faint objects are dramatically improved by using both eyes. Some people enjoy up to a 40% increase in perception of faint, diffuse objects by viewing them with both eyes instead of just one. In addition, color perception and contrast enhancement also benefit.

Although few experts will argue against the benefits of using binoculars over a same-size monocular telescope, the advantages of binocular viewing attachments for conventional telescopes are not as clear. Binocular viewers (Figure 6.7) customarily either screw onto a telescope (as with Schmidt-Cassegrain telescopes) or slide into a telescope's focusing mount. Inside, a beam splitter cuts the telescope's light into two equal-intensity paths, sending the light toward two eyepiece holders.

Are these rather costly accessories worth the price? Well, yes and no. First, they must be used with two *absolutely identical* eyepieces, raising the total investment even higher. Even then, however, they do not work as well as true

Figure 6.7 *Binocular viewing attachment for telescopes. Photo courtesy of Celestron International.*

binoculars. This should probably come as no surprise when you stop and think about it. In essence, binoculars are two independent telescopes strapped together. Binocular viewers, on the other hand, must rely on the light-gathering power of a single telescope. As such, the images perceived through a telescope equipped with a binocular viewer will actually be *dimmer* than through the same telescope outfitted with a single, equal-magnification eyepiece.

Binocular viewers come into their own when viewing the brighter planets and, especially, the Moon. Our satellite seems to take on a three-dimensional effect (actually, it's an optical illusion) that cannot be duplicated with a traditional monocular telescope. In addition, an increase in planetary detail can also be expected with a binocular viewer thanks to the aforementioned improvement in subtle contrast and color perception.

Several companies, including Orion, Celestron, and Tuthill, offer binocular viewers. The best feature fully coated optics. If you buy one, make sure it has an adjustment for setting the eyepiece spacing!

Books, Star Atlases, and Computer Software

Not long ago, I read that no pastime has more new books published about it each year than amateur astronomy. While that's bad news for us authors (too much competition), it is good for the hobby. In fact, there are so many excellent books and periodicals available that it is difficult to draw up a short list, but below is my attempt.

Where are the best places to buy books on astronomy? Certainly not most bookstores! Most (especially the larger chains found in malls) carry few, if any, and what few they do have are usually tucked away in a dark corner, while books on astrology, UFOs, and Hollywood's latest scandals line the front rows. (Please be aware, however, that even if a bookstore does not carry a particular book, they can always order it.)

The best outlets for astronomy books are Sky Publishing Corporation (publishers of *Sky & Telescope* magazine) and Kalmbach Publishing Company (publishers of *Astronomy* magazine). Both carry all of the latest books available from leading publishers. Another popular source is the Astronomy Book Club, a division of Newbridge Communications (3000 Cindel Drive, Delran, New Jersey 08075). Members get to choose from the newest and best books at lower prices and earn bonus points for added savings (buy a specified number of books, get another at about 50% off).

Star Atlases

AAVSO Star Atlas; Scovil, C.; Sky Publishing Corporation, 1980. A highly detailed atlas covering the entire sky to 9th magnitude and showing all known variable stars (and there are thousands) with a brightness range of at least 0.5 magnitude and a maximum visual magnitude brighter than 9.5. It gives many accurate, labelled comparison stars to about magnitude 9.5, but it is weak on deep-sky objects.

Cambridge Star Atlas 2000.0; Tirion, W.; Cambridge University Press, 1991. An excellent first star atlas! The *Cambridge Star Atlas 2000.0* shows all stars down to magnitude 6.5 as well as 866 deep-sky objects. A list accompanies each colorful chart tabulating the finest objects found in that particular slice of sky.

Edmund Mag 6 Star Atlas; Dickinson, T., V. Costanzo, and G. Chaple; Edmund Scientific Company, Barrington, New Jersey, 1982. Another collection of star charts identifying stars down to magnitude 6.2 and hundreds of telescopic targets. Facing each chart is a list of interesting sights located within that area of sky. Also included is an excellent introduction to astronomy and observing.

Norton's 2000.0; Ridpath, I., et al.; John Wiley & Sons, 1989. Now in its 18th edition, *Norton's 2000.0* (formerly known as *Norton's Star Atlas*) is the grand-daddy of all modern star atlases. Data tables adjacent to each page list details for many of the objects found in that particular part of the sky. Extensive discussions on telescopes and observing techniques make this more a reference book of amateur astronomy than just a star atlas.

1000+; Lorenzin, T., with T. Sechler; Lorenzin, 1987. Like *Norton's*, *1000+* is much more than a mere star atlas. The book is centered around descriptions of more than 1,000 deep-sky objects, each described as seen through the author's 8-inch Schmidt-Cassegrain and plotted on the accompanying atlas, which shows stars to 6th magnitude.

Sky Atlas 2000.0; Tirion, W.; Sky Publishing Corporation, 1981. In a word: the *best*. The *Sky Atlas 2000.0* accurately plots all stars in the sky down to magnitude 8.0 and 2,500 deep-sky objects on 26 charts. The *Sky Atlas 2000.0* is available in three editions: an unbound field edition (showing white stars on a black background) and desk edition (black stars on white) and a spiral-bound deluxe color edition.

Uranometria 2000.0; Tirion, W., B. Rappaport, and G. Lovi; Willmann-Bell, 1987. The *Uranometria 2000.0* plots more than 330,000 stars to magnitude 9.5 and 10,000 deep-sky objects on hundreds of 9-by-12-inch charts, showing the sky in detail never before available to the amateur at a reasonable price. The *Uranometria 2000.0* is sold in two volumes; one covering the northern sky from declination $+90°$ to $-6°$ and the second including $+6°$ to $-90°$ (the overlap is intentional). This atlas is recommended for advanced amateurs only.

Periodicals

Astrograph; Box 2283, Arlington, Virginia 22202. A unique bimonthly periodical devoted solely to amateur astrophotography. Each issue contains useful articles on technique, equipment, results, and just about anything else of interest to celestial photographers.

Astronomy; P.O. Box 1612, Waukesha, Wisconsin 53187. North America's best-selling astronomical periodical, each issue of *Astronomy* includes several feature articles covering a wide range of topics, as well as a monthly star map, regular columns on events and objects of note in the sky, and book and product reviews.

Astronomy Now; 193 Uxbridge Road, London W12 9RA, United Kingdom. From England comes this popular-level monthly periodical. *Astronomy Now* closely resembles its transatlantic cousins *Astronomy* and *Sky & Telescope*, with each issue containing several interesting feature articles and highlighting sky events for the month to come.

ATM Journal; 17606 28th Avenue SE, Bothell, Washington 90012. Like the sorely missed magazine *Telescope Making*, *ATM Journal* offers amateur telescope makers a forum to display and describe homemade telescopes, observatories, and other accessories. If you are considering building your own telescope, this magazine will be of great interest.

Griffith Observer; Griffith Observatory, 2800 East Observatory Road, Los Angeles, California 90027. Published bimonthly, the *Griffith Observer* is sort of a miniature *Sky & Telescope* or *Astronomy*. Feature articles on anything from astronomical history to the latest discoveries may be found in each issue, as are a simple star map and sky data.

Sky & Telescope; P.O. Box 9111, Belmont, Massachusetts 02178. *Sky & Telescope* (like *Astronomy*) may be thought of as two magazines stapled together and separated by a monthly star map in the center. Before the centerfold are feature articles on up-to-the-month discoveries and the latest professional research. After the star map, regular departments highlight sky events occurring during the month and review homemade telescopes and accessories. *Sky & Telescope* is an excellent magazine for the amateur astronomer.

Sky Calendar; Abrams Planetarium, Michigan State University, East Lansing, Michigan 48824. The cheapest way to learn what's up in the sky is to subscribe to *Sky Calendar*. Published monthly, this single-sheet publication features a calendar-style format that highlights one or more interesting naked-eye sights for most nights of the month. On the reverse side is a monthly star map for constellation identification. I highly recommend this one for beginners.

Southern Sky; P.O. Box 3686, Weston Creek, ACT 2611, Australia From Australia comes this excellent new bimonthly magazine. Like *Astronomy* and *Sky & Telescope*, *Southern Sky* features articles focusing on both the latest astronomical discoveries as well as sky events worth viewing. Regular features include columns on deep-sky observing, astronomy clubs, astrophotography, antique telescopes, and a whole-sky star chart.

Stardate; University of Texas, McDonald Observatory, Austin, Texas 78712. This is a scaled-down version of *Astronomy* or *Sky & Telescope*. Each bimonthly issue features a lead article or articles as well as a star map and descriptions of celestial events for the month to come.

Starry Messenger; P.O. Box 6552, Ithaca, New York 14851. The original used equipment periodical, listing hundreds of classified advertisements for all sorts of telescopes, binoculars, and related accessories.

Annual Publications

Astronomical Calendar (annual); Ottewell, G.; Astronomical Workshop, Furman University, Greenville, South Carolina 29613. The abundantly illustrated *Astronomical Calendar*, published annually, describes celestial events visible around the world for the particular year of issue, making it one of the most useful publications in an amateur's library.

Observer's Handbook (annual); Percy, R., et al.; Royal Astronomical Society of Canada. Like the *Astronomical Calendar*, the *Observer's Handbook* details all of the predicted celestial events that will occur during its year of publication, with events for each month listed chronologically. Special departments describe in detail upcoming eclipses, meteor showers, and much more.

Introductory Books

Guide to Amateur Astronomy; Newton, J. and P. Teece; Cambridge University Press, 1988. *The Guide to Amateur Astronomy*, rich in information for both the budding stargazer and the seasoned amateur alike, is built around five major headings, including basic fundamentals, a guide to the night sky, tips and suggestions for advanced observing projects, astrophotography, and amateur telescope making.

Peterson's Field Guide to the Stars and Planets; Menzel, D., and J. Pasachoff; Houghton Mifflin, 1993. The *Field Guide to the Stars and Planets* is one of the finest introductory books on the science and hobby of astronomy. Though small in physical size, it contains more useful information than most books four times as large. The current edition includes a 7th-magnitude all-sky star atlas drafted by Wil Tirion.

Sky Observer's Guide; Mayall, N., et al.; Western Publishing, 1985. First published in 1959, the *Sky Observer's Guide* includes sections on observing the Sun, Moon, planets, comets, meteors, asteroids, satellites, deep-sky objects, and just about anything visible in small telescopes.

365 Starry Nights; Raymo, Chet; Prentice Hall, 1982. This book is an excellent introductory guide for the whole family. It features a different sky activity for each night of the year, building the reader's knowledge with the passage of time.

Observing Guides

(Note that while a few of these guides include finder charts, most require the use of a separate star atlas.)

Astronomical Objects for Southern Telescopes; Hartung, E.; Cambridge University Press, 1985. One of the best guides to the southern sky ever compiled. This book lists and describes hundreds of objects visible both above and below the celestial equator (focusing mainly on the latter) through amateur-size telescopes.

Burnham's Celestial Handbook, volumes 1, 2, and 3; Burnham, R., Jr.; Dover, 1978. Throughout its 2,138 pages, *Burnham's Celestial Handbook* breaks down the sky by constellation, covering just about everything in the known universe.

Complete Manual of Amateur Astronomy; **Sherrod, C.; Prentice Hall, 1990.** An excellent guide for all intermediate and advanced amateurs who want to pursue in-depth observing programs. Topics covered range from observing the planets to monitoring variable stars and astrophotography.

Exploring the Moon through Binoculars and Small Telescopes; **Cherrington, E.; Dover, 1983.** An essential guide for the devout lunatic (Moon watcher). Divided into 28 day-by-day chapters, the author describes the appearance and location of just about every major feature visible on our nearest neighbor.

The Messier Album; **Mallas, J. and E. Kreimer; Sky Publishing Corp., 1978.** Originally published in *Sky & Telescope* as a monthly series of articles from May 1967 to September 1970, the *Messier Album* is one of the best guides to all 109 members of the legendary Messier catalogue. Interesting facts, visual appearance, plus a chart and photograph are included for each object listed.

Observing Handbook and Catalogue of Deep-Sky Objects; **Luginbuhl, C. and B. Skiff; Cambridge University Press, 1990.** An excellent reference for intermediate and advanced amateurs. After brief introductory comments, the book describes over 2,000 objects visible in medium-size telescopes. Sketches and photographs of some of the objects are included, as are computer-generated charts for some of the more crowded fields.

Observing Variable Stars: A Guide for the Beginner; **Levy, D.; Cambridge University Press, 1989.** This book is written for the beginner who wants to do more than just look at the night sky. The many different kinds of variable stars and how to best observe them are described in detail, as are dozens of individual stars.

Seeing the Sky; **Schaaf, F., et al.; John Wiley & Sons, 1990.** This is a trilogy of books written to introduce beginners to the night sky. The first book in the series (*Seeing the Sky*) addresses the naked-eye sky, the second (*Seeing the Solar System*) probes our family of planets, while the third (*Seeing the Deep Sky*) contains much information for the deep-sky enthusiast.

Star-Hopping for Backyard Astronomers; **MacRobert, A.; Sky Publishing Corporation, 1993.** An excellent guide for novice astronomers yearning to learn and understand how to find objects with binoculars or telescopes. Fourteen expeditions across the sky let observers discover and conquer 160 deep-sky objects. Advice is also offered on equipment selection and observing technique.

Touring the Universe through Binoculars; **Harrington, P.; John Wiley & Sons, 1990.** Now, this is a book! (You expected me to say anything less?) *Touring the Universe through Binoculars* is unique among binocular books, as it was written for the intermediate and advanced amateur who has made a conscious decision to use binoculars for observing the night sky. The book contains information on observing the Moon, Sun, planets, minor members of our solar system, and over 1,100 deep-sky objects.

***Universe from Your Backyard*; Eicher, D.; Cambridge University Press, 1988.** From *Astronomy* magazine comes this collection of reprinted articles describing 690 of the finest deep-sky objects for backyard astronomers. Simple finder charts, photographs, and drawings combine to make this an excellent introduction to deep-sky observing.

***Webb Society Deep Sky Observer's Handbook,* volumes 1 through 8; Jones, K., et al.; Enslow, 1979–90.** A useful but specialized series of paperback observing guides for the serious deep-sky observer. Each volume focuses on a specific type of deep-sky object, giving up-to-date theories of formation, advice on observing technique, and a catalogue of at-the-telescope descriptions and drawings. Unfortunately, while written content is good, the sketches are poorly reproduced.

Astrophotography

***Astrophotography Basics* (Publication P150); Eastman Kodak; Kodak, 1988.** As the name implies, *Astrophotography Basics* provides just enough good, basic instruction for the beginner to capture heavenly bodies successfully on film. And best of all, it can be had for free (in single quantities only). Call the Kodak Customer Service hotline at 1-800-242-2424 and ask for a copy of Publication P150.

***Astrophotography for the Amateur;* Covington, M.; Cambridge University Press, 1991.** One of the finest books ever written for the amateur looking to break into the challenging field of astrophotography. It features good discussions on capturing just about everything in the sky on film, from the simplest star trails to advanced through-the-telescope techniques.

***A Manual of Advanced Celestial Photography;* Wallis, B., and R. Provin; Cambridge University Press, 1988.** The perfect book for anyone who has the basics down pat and is striving to join the ranks of the quasiprofessionals. Wallis and Provin discuss in great detail many topics that most other books skim over, such as gas-hypersenstization methods, advanced modus operandi at the telescope, and how to perform magic in the darkroom.

***Photography With Your Telescope;* Brown, S.; Edmund Scientific Company, 1966.** This is the book that launched many amateurs into astrophotography back in the 1960s and 1970s. Though times have certainly changed since then, it still contains a lot of good information on basic astrophotography for the beginner.

Telescope Making and Optics

Build Your Own Telescope; Berry, R.; Charles Scribner's Sons, 1985. An excellent first book on constructing telescopes. Includes complete plans for five refractors and Newtonian reflectors, ranging in aperture from 4 to 10 inches.

How to Build Your Own Observatory; Berry, R. (editor); Kalmbach Publishing, 1990. As a next step after building or purchasing a telescope, many amateurs decide to build their own observatory. This book, reprinted from the files of the now defunct but sorely missed magazine *Telescope Making*, contains many examples of amateur-made observatories.

How to Make a Telescope; Texereau, J.; Willmann-Bell, 1984.
Making Your Own Telescope; Thompson, A.; Sky Publishing, 1980.
Standard Handbook for Telescope Making; Howard, N.; Harper and Row, 1984 (out of print). These are three of the best books ever written about making a telescope. Each contains sufficient instructional details for the reader to build his or her own telescope from scratch. Chapters describe everything from mirror grinding, polishing, and testing to making your own eyepieces, finders, and mountings.

Telescope Optics: Evaluation and Design; Rutten, H., and M. van Venrooij; Willmann-Bell, 1988. An excellent reference book for the advanced amateur who wants to learn more about how telescopes, eyepieces, and related optics work.

References and Catalogues

Deep Sky Field Guide; Cragin, M., J. Lucyk, and B. Rappaport; Willmann-Bell, Inc., 1993. A companion for the advanced deep-sky observer, this itemized catalogue details all of the thousands of deep-sky objects plotted in the *Uranometria 2000.0* star atlas, arranged chart by chart.

Sky Catalogue 2000.0, volumes 1 and 2; Sinnott, R., and A. Hirschfield; Sky Publishing, 1985.
NGC 2000.0; Sinnott, R.; Sky Publishing, 1988. Think of these as telephone books to the universe. The former is a two-volume set, with Volume 1 listing information (positions, magnitude, spectral type, etc.) for all stars in the sky down to magnitude 8.0 (the same limit as the *Sky Atlas 2000.0*). Volume 2 provides much of the same information for thousands of deep-sky objects. The *NGC 2000.0* is an updated and expanded version of the original New General

Catalogue originally compiled by John Dreyer a century ago. Contains up-to-date data on all 13,226 NGC and IC (Index Catalogue) objects.

Miscellaneous Reading

***The Astronomical Scrapbook*; Ashbrook, J.; Sky Publishing, 1984.** This is a compilation of articles that appeared in *Sky & Telescope* magazine across nearly a quarter century. The late Joseph Ashbrook continues to delight readers with accounts of both famous and little-known astronomical visionaries, personalities, cranks, and hoaxes, such as the great Paris fiasco and the city on the Moon.

***History of the Telescope*; King, H.; Dover, 1979.**
***The Telescope*; Bell, L.; Dover, 1981.** Two reprints of wonderful theses on the origin of the telescope.

***Russell W. Porter*; Willard, B.; Bond Wheelwright Company, 1976.** This book should be required reading for anyone who has ever contemplated building his or her own telescope. Willard's well-researched text describes the fascinating life and times of Russell Porter, the founding force behind the hobby of telescope making back in the 1920s.

***Starlight Nights*; Peltier, L.; Sky Publishing, 1980.** An inspiring autobiography of one of America's most famous amateur astronomers of all time. This book details the love and devotion to the night sky that drove Peltier to discover six novae and 12 comets and make 132,123 variable star observations in a career that spanned six decades.

Computer Software

***Dance of the Planets*; Arc Scientific Simulations, P.O. Box 1955, Loveland, Colorado 80539.** Known for its unparalleled graphics, *Dance* will let you simulate eclipses; view planets, moons, and comets close up; and plot motions against the sky in either real or accelerated time. However, it has a rather steep learning curve and requires IBM-compatible; 640K RAM; hard drive; and EGA or VGA graphics; a math coprocessor is also required.

***EZ Cosmos*; Future Trends Software, Inc., 1508 Osprey Drive, Suite 103, DeSoto, Texas 75115.** This is an excellent sky-simulation program made even better with the introduction of a Windows version in 1992. Stars are shown to magnitude 9.5, along with thousands of deep-sky objects and all major solar system members. The latest version incorporates Level I of *TheSky* from Software Bisque (reviewed later in this chapter), adding a telescope interface, coordinate lines for right ascension and declination, and constellation boundaries. Requires IBM-compatible; 512K RAM; hard drive; and EGA, VGA, or Hercules graphics.

***Lodestar Plus*; Zephyr Services, 1900 Murray Avenue, Pittsburgh, Pennsylvania 15217.** Another nice planetarium program, Lodestar Plus accurately shows the sky for any date between 9,999 B.C. and 9,999 A.D. for any location on the Earth. Features over 9,000 stars to magnitude 7.8, all 7,840 NGC objects, and all major members of the solar system. Requires IBM-compatible; 512K RAM; hard drive; and CGA, EGA, VGA, or Hercules graphics. A math coprocessor is recommended.

***TheSky*; Software Bisque, 912 12th Street, Suite A, Golden, Colorado 80401.** Working within either DOS or Windows, *TheSky* is one of the finest sky-simulation programs around for the deep-sky observer. The Sun, Moon, and planets are plotted and animated against a myriad of colorful background stars. *TheSky* comes in either Level II or Level III, depending on the number of objects stored in its database. Requires IBM-compatible; 512K RAM; hard drive; and EGA, VGA, or Hercules graphics.

***Skyglobe*; KlassM Software, 284 142nd Avenue, Caledonia, Michigan 49316.** A great shareware planetarium program. The latest version shows naked-eye stars as well as the Sun, Moon, planets, and many deep-sky objects. It provides nice color graphics when used with an EGA or better monitor. Requires IBM compatible; 640K RAM; hard drive; and CGA, EGA, VGA, or Hercules graphics.

***Voyager II*; Carina Software, 830 Williams Street, San Leandro, California 94577.** Another outstanding sky-simulation program, but this one is for Apple Macintosh users. *Voyager II* displays 14,000 sky objects and may be used as either a general star map or detailed star atlas. Animation compresses time to show eclipses or planets orbiting the Sun. Requires 1 megabyte (1MB) RAM, hard disk, and System 6.05 (or higher) or 7.0.

Electronics for Telescopes

As with just about every other aspect of our lives, amateur astronomy is becoming more sophisticated thanks to tremendous advances in electronics and computerization. It is now possible for backyard telescopes to perform a variety of flips and whirls that were once only available at professional observatories. Here is a short review of some of the more popular electronic equipment on the market today.

Clock-Drive Correctors

Although most clock drives supplied with equatorially mounted telescopes track the sky accurately enough for short-duration visual observations, many astronomers (especially astrophotographers) find it necessary to correct for slight differences between tracking speeds and the stars' apparent motion across the sky. Thus, the need for clock-drive correctors was born.

Many of today's top-of-the-line telescopes (such as the Celestron Ultima or Meade LX models) use DC stepping motors to power their clock drives. These sophisticated instruments have correctors built right into their drive mechanisms that not only allow fine adjustment of the tracking rate but also permit relatively fast slewing across the sky. With these, the observer need only align the mounting to the celestial pole and turn on the drive. The telescope may then be aimed using a hand-held control box connected to the drive's power panel.

Telescopes such as Celestron's and Meade's base models, as well as most older instruments, use AC-powered synchronous motors in their clock drives and therefore require external drive correctors (Figure 6.8). Because the speed of synchronous motors will not vary with changes in voltage (above their threshold voltage, they run; below it, they stall), drive correctors alter the frequency of the supply voltage. Normally, 110 voltage is supplied at 60 hertz (60 Hz, or 60 cycles per second). Lowering this to, say, 50 hertz will slow a synchronous motor, whereas raising the value to 70 hertz will speed it up. By plugging a telescope's clock drive into a drive corrector and connecting the corrector to a power source, the telescope's rate of tracking may be adjusted.

Less expensive drive correctors use rheostats to adjust the output to the clock drive, while better drive correctors use three- or four-position switches. In the latter case, the switch will have separate settings for solar, sidereal, lunar

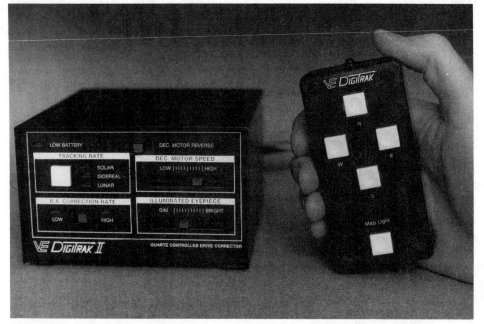

Figure 6.8 *Digitrak dual-axis clock-drive corrector. Photo courtesy of Vogel Enterprises.*

(only an approximate setting, because the Moon's daily motion will vary greatly depending on where it is in its orbit), and possibly King. (Named for Edward King, an early twentieth-century astronomer at Harvard Observatory, the *King rate* averages the changes in tracking speed according to an object's elevation above the horizon. Its purpose is to take into account atmospheric refraction— not a consideration for most applications.) Correctors also have momentary switches (preferably mounted on separate hand-held control boxes) for speeding up or slowing down the drive motor just in case a quick tracking tweak is needed.

Drive correctors come in two basic styles: single-axis or dual-axis. The simpler single-axis correctors only adjust the speed of the right ascension clock-drive motor, whereas dual-axis correctors also control the declination slow-motion motor. Add-on motor assemblies may be purchased either from a telescope's manufacturer or from suppliers of after-market accessories. For instance, Jim's Mobile Industries (JMI) sells the Motodec, an easy-to-mount 12-volt DC unit that attaches directly to the mounting's declination slow-motion knob. Though the motor must be used in conjunction with a dual-axis drive corrector (it does not have its own power source), a slip clutch allows free manual movement of the telescope with the power either on or off. Motodec may be used with JMI's own drive correctors or with any other brand featuring dual-axis control and standard plug-in connections. Another important feature to look for when investigating drive correctors is how they are powered. Some only plug into a 110-volt AC wall socket, and others only use 12-volt DC, such as from a car's battery. The best correctors have the flexibility to use either AC or DC, making them as useful in the backyard as they are in the wilderness.

Several companies manufacture drive correctors for the amateur market. Toward the less expensive end of the range are the Model 43 by Meade Instruments and the Mighty Mite by Vista Instruments. Both of these single-axis correctors operate from 12-volt DC only. The Meade corrector represents a good value but unfortunately does not come with a separate hand controller. Instead, all of the controls are inconveniently mounted on the front panel of the unit's main power box. The Mighty Mite lets users vary the speed of synchronous drive motors up to 33% above and below their nominal rate simply by pushing "slow" and "fast" buttons on the supplied handheld control box. Somewhat more expensive is Lumicon's quartz-reference dual-axis drive corrector, which comes with a handheld controller for easy adjustments.

More expensive models feature digital control for greater tracking precision. For instance, Lumicon sells a digital quartz-reference dual-axis drive corrector. It also features a declination-braking feature and plug-in jacks for map and setting circle illuminators.

Vista offers the AC- or DC-powered Programmable Drive Corrector, available in three models for either single- or dual-axis control. This unique design lets the user preset the tracking rate precisely, a nice improvement over other correctors that force users to play hit-or-miss with rheostats or those that use three- or four-position switches.

Mototrak quartz-controlled drive correctors from JMI enjoy some interesting and useful features, including dual-axis operation, optional declination motor, and electronic Motofocus focusing control. Motofocus mounts onto a telescope's focusing mount to permit fine adjustment of the focus, handy for high-powered observations or photography but also annoying at times because it means another loose wire hanging off the telescope. All three functions are controlled by a single hand-unit for greatest ease.

Vogel Enterprises manufactures the Digitrak line of drive correctors, some of the most highly regarded correctors around. All Digitrak correctors feature quartz control for greater precision and a choice of power inputs, either DC only, AC/DC, or an internal rechargeable battery.

A drive corrector is a must for long-exposure astrophotography, but if all you want to do is run the clock drive from a car battery, then you really don't need one. Instead, save some money by purchasing an AC/DC Inverter. Inverters plug into a standard 12-volt automobile cigarette lighter to produce 110-volt AC. Though the current is too low to power an appliance, it is adequate to run a clock drive. Inverters are available from several astronomical suppliers (such as Orion, Vogel, Tuthill, and JMI) as well as from many auto parts stores (including the catalogue giant J.C. Whitney, Inc., 1917–19 Archer Avenue, P.O. Box 8410, Chicago, Illinois 60680) and camping outlets (the latter two usually having the lowest prices).

Rechargeable Batteries and Portable Power Packs

There will undoubtedly come times when neither a wall outlet nor a car is available for powering a clock drive. In cases such as these, if the telescope cannot be brought to the power, then the power must be brought to the telescope. Not long ago, this would have meant lugging around a small wet-cell battery (such as from a motorcycle), an inconvenient and potentially dangerous chore.

Many amateurs are still under the impression that heavy wet cells are needed to run clock drives. This is simply not true. Today, technology in battery design has advanced to the point where lightweight rechargeable cells can power an AC clock drive (using an intermediate drive corrector or voltage inverter) for an entire evening on a single charge. (As with drive correctors, many of today's leading telescopes that use DC stepping motors in their clock drives feature built-in power supplies. These supplies run off common flashlight batteries yet still deliver enough power to turn the drive. In these cases, separate power sources are not needed.)

While battery output is rated in voltage, charge life is expressed in *ampere-hours;* the higher the amp-hour value, the longer the battery's charge life. To calculate how long a battery will last on a single charge, first find out the electrical demand of the device to be powered, stated as a voltage, current, and wattage. The required voltage is almost always printed right on the apparatus, but wattage and current may not be. If the wattage is also noted, simply divide it by the voltage to obtain the device's current demand. For instance, a 12-volt motor requiring 6 watts of power will have a current demand of 0.5 amperes. If neither wattage nor current is specified, they must be measured using a meter.

Let's say you want to purchase a rechargeable battery to run your telescope's clock drive. Connect everything as you normally would but insert an ammeter between the corrector/inverter and battery. Turn the drive on and check how many amps it is drawing from the battery. For the sake of discussion, we will say that it draws exactly 1 amp at 12 volts (chances are it will be less). Divide the rechargeable battery's amp-hour rating by the measured load in amps to find the battery's expected life. In the example above, a 5–amp-hour battery will power this setup for 5 hours while a 12–amp-hour battery will last 12 hours, and so on. If more than one accessory is to be powered by the battery, be sure to measure and add the load of each. For instance, both the Tuthill No-Du Cap and the Orion Dew Zapper contact heater (each described later) draw about 1 amp at 12 volts, raising the total load to 2 amps. In this second example, the batteries noted above will last 2.5 and 6 hours, respectively, before requiring recharging.

Perhaps the most popular rechargeable battery used by amateur astronomers is the 12-volt Carefree Porta-Pac. The Porta-Pac, available with either a 5– or 12–amp-hour capacity, features a built-in cigarette lighter receptacle for plugging in drive correctors, antidew accessories, or any other item that can run off 12-volt DC. The permanently sealed lead-acid battery is designed to be both spill-proof and maintenance-free and will provide years of unfailing service as long as a charge is maintained. This last point is an important one to remember, for if allowed to drain completely, the battery will not recharge properly.

Before buying a battery from a mail-order outlet, check the Yellow Pages under "Batteries." Many communities have retail stores that offer a broad selection of batteries for all occasions and may have just the one to fit your needs at a lower cost than you would pay elsewhere.

Motorized Focusing Devices

Apart from poor sky conditions, the greatest hurdle to overcome when making high-powered observations is achieving a sharp focus. As magnification rises, precise focusing becomes extremely critical. At the highest powers, even the slightest turn of the focusing knob will vibrate most telescopes, blurring the image. Though not required for observations made at low or medium power, electric focus motors are handy for the crucial focusing required for high-power visual and photographic observations.

Although some premier telescopes come with motorized focusing as standard equipment, others sell motorized focusers as options. Most of these are dedicated units designed to fit only one or two specific telescope models. As an after-market solution, Jim's Mobile Industries (JMI) offers the Motofocus electric focusing motor, with models to fit many of the most popular telescopes. The Motofocus attaches to the telescope without drilling and allows either coarse or fine adjustment of the focusing. The more adventuresome might even wish to make their own motorized focusers. The next chapter illustrates what one amateur telescope maker accomplished for less money than a comparable store-bought unit would have cost.

Computerized Telescope-Aiming Systems

One of the biggest complaints that amateur astronomers have had for years is how useless setting circles are that come with many equatorial mounts. Because of their small size and gross calibration, the circles serve as little more than decoration. All this changed with the invention of electronic digital setting circles, which make it possible to aim a telescope accurately to within a small fraction of a degree.

The first digital setting circles for amateur telescopes were introduced around 1980 by Roger W. Tuthill, Inc. Called the Celestial Navigator, it attached to the right ascension and declination axes of an equatorial mount to provide a digital readout of exactly where the telescope was aimed. It was a boon to observers who preferred using setting circles for locating objects in the sky. (It also revealed that many equatorial mounts did not have perpendicular axes; in fact, some were off by several degrees!)

Just as with almost anything electronic, digital setting circles have become smaller, easier to use, cheaper, and more plentiful in the past couple of decades. Using them could not be easier. Forget polar-aligning (though it is still usually required to use the clock drive); just aim the telescope at two of the many stars stored in memory, tell the unit which ones they are, and the built-in algorithm does the rest. This ease of operation means that the setting circles can be attached to just about any kind of telescope mount—either equatorial or altazimuth. Don't know what to look at? That's okay, because many of these units come with a whole catalogue of thousands of sky objects from which to choose. Simply move the telescope until the LED prompt announces you have hit the preselected target, look in the eyepiece, and there it is.

Here is a quick rundown of the more popular digital setting circles on the market today. The Lumicon Sky Vector is a neat little unit that mounts unobtrusively to just about any telescope. Two encoders attach to the instrument's axes and are connected to the brain of the outfit, as it were, by means of a pair of thin wires. Once properly secured and calibrated, the digital setting circles will automatically keep track of the passage of time as well as where the telescope is aimed. The LED reads to within an accuracy of 10 arc-minutes in declination and 1 minute of right ascension. More sophisticated models of the Sky Vector come with libraries of either 1,500 or over 8,000 objects stored in memory, depending on the model.

JMI's DSC-MAX, NGC-miniMAX, NGC-MAX, and SGT-MAX and Celestron's Advanced Astro Master work in pretty much the same way as the Sky

Vector. The Celestron unit is designed specifically for Celestron telescopes supported on either fork or Super/Great Polaris equatorial mounts, though it may be adapted to other brands and mounting styles with a drill (be careful!) and a little creativity. It also features an 8,000+ object library from which the user may browse.

Simplest of the four JMI models is the DSC-MAX, which provides accurate telescope positions but does not have a built-in library of objects. Next up, the NGC-MAX (Figure 6.9) holds data on more than 8,000 deep-sky objects (the cheaper mini-MAX includes 1,950 objects in its memory). The SGT-MAX provides a direct link to a computer that, when used with the software provided, enables the user to watch a monitor as the telescope is moved across the real sky. Like Lumicon's Sky Vector, the JMI MAX family is designed to fit almost all popular telescopes around and is also adaptable to homemade instruments.

Tuthill offers a new generation of Celestial Navigator, now known as the Celestial Navigator Mark VI, which features a memory that contains information on 4,500 deep-sky objects and may be used with many German or fork equatorial mounts as well as Dobsonian alt-azimuth systems. The Mark VI is

Figure 6.9 *NGC-MAX computerized telescope-aiming system. Photo courtesy of JMI, Inc.*

also available with an RS-232 port for attaching your telescope with a personal computer. Appropriate software, such as *TheSky* or *EZ Cosmos* (version 4 or later) is required to complete the telescope-to-computer system.

Similar to many of the above units is Orion's new Sky Wizard, which comes in three varieties. Model 1 features all of the planets and Messier objects, Model 2 adds an additional 2,000 deep-sky objects, while Model 3 has a 10,000-object database. As sold, one must also purchase a separate installation kit to secure the Sky Wizard onto a telescope mount; as with the others, it works equally well with both equatorial and alt-azimuth mounts.

One of the most sophisticated computer-aided devices is the Starport, sold by Opto-Data. The Starport features a touch-sensitive screen that acts as both an electronic star atlas and telescope guidance system. The heart of the contraption is a memory that stores the positions of all stars to 8th magnitude, 10,000 deep-sky objects, and all planets. The user touches the screen to advance through the menu, leading toward whatever object has been selected. The screen then displays the field of view with the target highlighted. When hooked up to a telescope (any mount, equatorial or alt-azimuth), the Starport will follow the instrument through the sky until the quarry is found.

All of these computerized telescope-aiming systems require that a pair of encoders be mounted to the telescope's axes. Most of the models offer customized packages for attaching to fork-mounted Celestron or Meade Schmidt-Cassegrains simply by unscrewing and replacing one screw on both axes. However, other telescopes may not have it so easy, requiring some drilling and tapping in order to fit the encoders to their mountings. I strongly urge you to contact the manufacturer to find out what is required to attach the encoders to your telescope before you purchase any of the units mentioned above.

Another important question for readers considering the purchase of one of these units for their 8-inch Schmidt-Cassegrain telescope is this: Can the telescope swing into its storage position without hitting the right ascension encoder? The encoders engage the right ascension axis at its pivot point, smack dab in the middle of the fork. If the encoder sticks up too high, then the front of the telescope tube will not clear, thus forcing the user to detach the encoder whenever the instrument is disassembled for storage in its footlocker. Only the JMI encoders permit telescope storage without the need for encoder removal.

Back in the 1980s, one manufacturer advertised that an observer could see "100 galaxies per hour" with its computer-aided digital circles. Sure, they are a great way to increase your productivity as an observer, if productivity is only measured in terms of sheer numbers. But since when are we out to break the land-speed record? The idea is not to whiz across the universe at warp speed but rather to get to know the universe intimately. Without becoming too philosophical, I cannot recommend digital setting circles, especially if you are just starting out. It would be like giving a calculator to someone who does not know how to perform simple multiplication. The calculator will give the right answer, but without it the person is lost. Unless you are involved in an advanced program, such as variable-star observing or hunting for supernovae in other galaxies, resist the impulse until you finish the section of Chapter 9 entitled "Finding Your Way."

Smile . . . Say Pleiades

Chapter 10 concludes with an introduction to the art of astrophotography, one of the most popular pastimes of amateur astronomers. As most soon discover, however, there is a lot more to taking a good picture of the night sky than they might have suspected at first. Patience is the most important requirement of the astrophotographer, followed closely by equipment. It's tough to bottle the former for sale, but there are lots of companies looking to sell you the latter.

Cameras

While successful astrophotographs can be taken with many different types of cameras, most amateurs prefer the 35-mm single-lens reflex (SLR) camera. Single-lens reflexes allow the photographer to look directly through the lens of the camera itself, a critical feature for aligning the image, especially when photographing through a telescope (with most other cameras, the photographer is viewing through a separate viewfinder). SLRs also offer the maximum flexibility in terms of film and lens availability, both spoken about later.

Not all 35-mm SLRs are suitable for astrophotography. To be usable, a camera must have a removable lens with manually adjustable focus; provisions for attaching a cable release to the camera and the camera to a tripod; a manually set, mechanical shutter with a "B" (bulb) setting; mirror lockup; and interchangeable focusing screens. Unfortunately, few of today's 35-mm SLRs fit this bill. In an attempt to attract more weekend photographers, most camera manufacturers offer cameras with automatic everything, from focus to exposure to flash control. All of these are nice for taking pictures of the family picnic, but they are of no use to astrophotographers. Quite to the contrary, the long exposures required for astrophotos (usually measured in minutes, even hours) will quickly drain the power from expensive camera batteries. When that happens, the camera shuts down and becomes useless until a fresh set of batteries is inserted. So which cameras are best for astrophotography? Table 6.3 lists several excellent alternatives, both past and present.

Expensive does not necessarily mean *better for astrophotography*. All of these cameras will work well for wide-field constellation shots as well as through-the-telescope photos of the Moon and Sun (the latter requiring safety precautions outlined in Chapter 10). The differences in cameras will become more apparent when taking long-exposure telescopic shots. Here, the benefit of interchangeable focusing screens and mirror lockup will become apparent. Most subjects photographed through telescopes are very faint, making it difficult to line up and focus the shot when viewing through most standard focusing screens. A simple ground glass screen will provide the brightest possible images, a great aid in focusing and composing. Mirror lockup is essential for

Table 6.3 **Suggested Cameras for Astrophotography**

Camera Model	Nonbattery Bulb setting	Manual Lens	Interchangeable Viewscreen	Mirror Lockup
Today's Best				
Canon F-1	Y	Y	Y	N (older ones did)
Contax S2	Y	Y	N	N
Leica R6.2	Y	Y	Y	Y
Nikon F3HP	Y	Y	Y	Y
Nikon FM2	Y	Y	Y	N
Olympus OM-4T	Y	Y	Y	N
Pentax K1000	Y	Y	N	N
Pentax LX	Y	Y	Y	Y
Ricoh KR-5 Super II	Y	Y	N	N
A Few of Yesterday's Best				
Canon FTb	Y	Y	N	Y
Minolta SRT-101	Y	Y	N	Y
Miranda G	Y	Y	Y	Y
Nikon F and F2	Y	Y	Y	Y
Olympus OM-1 and OM-2	Y	Y	Y	Y

reducing *mirror slap*, which occurs every time the shutter is tripped and the camera mirror pivots out of the way. By swinging the mirror out of the way before the shutter is opened, most vibration will be eliminated, thereby reducing the chances for blurred images.

Lenses

Just as all cameras are not suitable for photographing the sky, neither are all lenses. But before an educated choice can be made, the photographer must first determine what he or she wants to photograph. For wide-field photography, either with the camera attached to a fixed tripod or guided with the stars, the standard lens may be all that is needed. Most 35-mm single-lens reflex cameras are supplied with lenses of 50- to 55-mm focal length. These cover an area of sky 28°×40°. If a wider field is desired, then either a 28-mm or 35-mm lens would be a good choice. They cover 50°×74°, and 40°×59°, respectively. On the other hand, if a magnified view is what you want, try an 85-mm, 135-mm, or 200-mm telephoto lens, with 16°×24°, 10°×15°, and 6.9°×10.3° fields of view, respectively.

In general, an astrophotographer wants to use as fast a lens as possible, because the faster the lens, the shorter the required exposure. Photographers refer to the speed of a lens just as astronomers talk about the focal ratio (the f-number) of a telescope. They mean the same thing. Recall that the focal ratio of a telescope is determined by dividing its focal length by its aperture.

Quite simply, the faster the lens, the lower the f-number. Holding the focal length constant, the only way to lower the focal ratio is to increase the lens's aperture. For instance, most 50-mm and 55-mm camera lenses have f-ratios between f/2 and f/1.2, resulting in apertures that range between about 1 and 2 inches. The larger the aperture is, the greater the light-gathering power of the lens, and therefore the shorter the required exposure will be.

Years ago, lens quality varied dramatically from one company to the next. But today, design and manufacturing procedures have been so perfected that most lenses will produce fine results (of course, every photographer thinks his or her lenses are best). Flare and distortion, two of the biggest problems in older lenses, have been all but eliminated thanks to optical multicoatings. In general, most lenses made by reputable companies (all those listed above, as well as those by Vivitar and Tamron, to name a few) are fine for astrophotography.

A BIG exception to all this is the zoom lens. Though extremely popular for everyday photography, zoom lenses almost always produce inferior results to their fixed-focal-length counterparts. For one thing, they are almost always slower (of higher focal ratio) than fixed lenses, making longer exposures necessary. Secondly, and most important, the optical quality of inexpensive zoom lenses is almost always inferior to that of fixed lenses. Instead of a zoom lens, it is best to select two or three lenses of different focal lengths, such as a 28-mm, 50-mm, and 135-mm.

Film

The film industry has seen advances in the past two decades the likes of which have never been witnessed before. Without a doubt, the wide variety of films that are readily available today is a great boon to astrophotography. In fact, the selection is so vast that it can leave the photographer confused. "Which film shall I use?" is a question often uttered even by veterans.

To help alleviate some of this bewilderment, Table 6.4 lists most currently popular films. The table is broken into three broad categories by film type (black-and-white negative, color negative, and color slide) and then sorted by film speed, or ISO value. ISO is the modern equivalent to the older ASA designation. Basically, the greater the numerical ISO value, the faster the film records light. For instance, an ISO 400 film will record the same amount of light in one-quarter the time required by ISO 100 film. High–ISO-value films allow shorter exposures and therefore less chance of photographer error (accidentally kicking or hitting the telescope, tracking error due to polar misalignment, etc.).

Why, then, would anyone consider using slower films? If a frame of film is studied close up, it will be found to be made up of a pebbly surface called *grain*. It is one of those irrefutable laws of nature that fast film has larger grain structure than slower films. Larger grain means lower resolution and therefore poorer image quality.

Another reason to use slower film is something called *reciprocity failure*. This is one of those terms bounced around by most photographers but probably

Table 6.4   ***Film Comparisons***

Black-and-White Negative		
Film	**ISO**	**Grain Structure**
Kodak Tech Pan 2415	25[1]	Extra fine
Kodak T-Max 100	100	Very Fine
Kodak Plus-X	125	Fine
Kodak Tech Pan 2415 (hypered)	200	Very fine
Kodak T-Max 400	400	Moderate
Kodak Tri-X	400	Coarse
Kodak T-Max 3200	3200	Very coarse

Color Negative			
Film	**ISO**	**Grain Structure**	**Sky color (background)**
Kodak Ektar 25	25	Extra fine	Can vary depending
Kodak Ektar 125	125	Fine	on printing. Ask lab
Fujicolor Super G 400[1]	400	Moderate	to print photos so
Kodak Ektar 1000[1]	1000	Coarse	that the sky comes
Fujicolor Super G 1600	1600	Coarse	out black.
Konica SR-G 3200[1]	3200	Very Coarse	

Color Slide			
Film	**ISO**	**Grain Structure**	**Sky color (background)**
Kodachrome 25	25	Extra fine	Neutral
Fujichrome 50	50	Fine	Beige
Fujichrome Velvia	50	Very fine	Beige
Ektachrome 50HC	50	Fine	Beige
Kodachrome 64	64	Very fine	Cyan
Ektachrome Elite 100	100	Very fine	Neutral
Fujichrome 100[1]	100	Fine	Green
ScotchChrome 100	100	Fine	Greenish
Kodachrome 200	200	Fine to moderate	Magenta
Ektachrome 200	200	Moderate	Cyan
Agfachrome 200	200	Moderate	Bluish
Fujichrome 400D	400	Moderate	Greenish
Ektachrome Elite 400	400	Moderate	Neutral
ScotchChrome 400	400[2]	Moderate	Reddish
Ektachrome P800/1600	400[2]	Moderate to coarse	Neutral
Fujichrome P1600D	400[2]	Moderate to coarse	Cyan
Agfachrome 1000RS	1000	Coarse	Reddish
ScotchChrome 1000[1]	1000	Coarse	Green

Notes:
1. Also available hypersensitized.
2. May be push-processed to ISO 800, 1600, or 3200.

understood by few. As an illustration, think back to the example of ISO 100 versus ISO 400 film. The films differ in ISO value by a ratio of 1 to 4, while their speed differs by a ratio of 4 to 1. (Remember, ISO 100 film requires four times the exposure as ISO 400 film to record the same scene.) Because these two ratios are reciprocals of each other, this concept is called *reciprocity*.

Reciprocity does not remain constant as exposure times increase; instead, the film's ability to record light falters and finally ceases. After that, no further light buildup will occur regardless of length of exposure—hence the term reciprocity failure.

There are ways to diminish a film's reciprocity failure while also increasing its ISO speed. The most popular technique is called *gas hypersensitizing*. Here, film is baked for days in a special oven containing a combination of nitrogen and hydrogen called *forming gas*. Though some amateurs prefer to cook their own, gas-hypersensitized film may be purchased from many mail-order sources, such as Lumicon. Users should note that not all films respond well to hypering and that the hypering process has an effective shelf life of about a month. All hypered film should be sealed in an airtight package (such as the plastic film can it came in) and stored in a freezer before and after use.

Other ways to improve film performance is to chill it to about $-15°F$ using a special cold camera or to preflash it. These methods are quite advanced, however, and will therefore be left to the experts. Consult the selection of books on astrophotography listed earlier in this chapter for details.

In addition, Table 6.4 comments on the trueness of the films' colors. You may be surprised to learn that color films do not necessarily record the same subject in the same colors. With color print film, the final color can be adjusted in the printing process, but with color slide film, what you see is what you get.

Tripods

If you will be affixing your camera to a tripod (as opposed to shooting through a telescope), then pay close attention to the tripod you will be using. Many less expensive tripods sold in department stores and other mass-market outlets are just not sturdy enough to support a camera steadily for any length of time. If the tripod is shaky, then the photographs will be hopelessly blurred. It makes no sense mounting a camera outfit costing hundreds or even thousands of dollars on a cheap tripod!

Here are a few things to look for when purchasing a camera tripod. First, the legs should be extendable so that the camera may be raised to a comfortable height. Make certain, however, that the tripod remains steady when fully extended. Sturdier models feature braces that bridge the gap between the tripod's legs and the center elevator post. Next, take a look at the foot pads. Better tripods have convertible pads that feature both a rubber pad for use on a solid surface as well as a spike for softer surfaces like grass or dirt. (A tip: When using a tripod on sand, place a plastic coffee can lid under each foot for added rigidity.)

Of all the tripods made, most photographers agree that Bogen makes the sturdiest. For instance, the Bogen model 3036 is sturdy enough to hold a 4-inch refractor even with its legs fully extended. Lesser tripods would collapse under such a load. Other brands worth considering are Tiltall, Star-D, Velbon, Vivitar, and Slik.

Camera-to-Telescope Adapters

For astrophotography through the telescope, adapters bring the two together. For prime-focus photography (if the term is foreign to you, skipping ahead to Chapter 10's introduction to astrophotography would be in order now), the most common way to affix a camera to a telescope is a two-piece T-ring/adapter combination. The T-ring attaches to the camera in place of its lens, while an adapter attaches to the telescope. The ends of the adapter and T-ring are then screwed together to form a single unit.

Different cameras require different T-rings. For instance, one for a Minolta will not fit a Canon. Likewise, different adapters are required for different telescopes. In the case of most catadioptric telescopes, an adapter called a *T-adapter* screws onto the back of the instrument in place of the visual back that holds the star diagonal and eyepiece. Most refractors and reflectors, on the other hand, use a different item called a *universal camera adapter*, which is inserted into the eyepiece holder.

Positive-projection astrophotography, commonly used when shooting the planets or lunar close-ups, requires that an eyepiece be inserted between the lensless camera and the telescope. Most camera adapters, such as the one

Figure 6.10 *Camera-to-telescope adapter system. Photo courtesy of Orion Telescope Center.*

shown in Figure 6.10, come with an extension tube for this purpose. The eyepiece is inserted into the tube, and the tube then screwed in between the adapter and T-ring.

Celestron, Meade, Questar, and some other telescope manufacturers offer camera adapters that custom-fit onto their telescopes. After-market brands, often less expensive, are also available. For instance, Orion Telescope Center and Optica b/c sell several different adapters to fit most popular telescopes. None are supplied with camera T-rings, which must be purchased separately. Better photographic supply stores carry T-rings for most common single-lens reflex cameras, as do many astronomical mail-order companies.

Off-Axis Guiders

For long-exposure, through-the-telescope astrophotography, the photographers have no choice but to visually monitor their telescopes' tracking. To do this, they must peer through either a side-mounted guide scope or an off-axis guider. Mounting a guide scope onto the side of the main instrument can be both clumsy and expensive. Their weight can actually deform the main telescope's tube, causing image degradation and guiding errors. These factors have led many astrophotographers to choose the latter method.

An off-axis guider looks like the letter T, with two hollow tubes attached to each other at a 90° angle. The main body of the guider fits between the telescope and camera, while the perpendicular leg contains a tiny prism that is used to divert a small amount of starlight toward an illuminated-reticle eyepiece that fits into the guider. To use the guider, plug or screw it into the telescope's eyepiece holder between the telescope and the camera and align the crosshairs of the eyepiece with a star. During the exposure, the photographer monitors the guide star through the reticle eyepiece to make sure it never leaves the crosshairs.

Off-axis guiders are available from a number of different sources. The simplest (and therefore, cheapest) feature rigid prisms, whereas more expensive models come with prisms that may be rotated around the field. In practice, the latter is easier to use because the freedom of prism movement permits a much wider choice of guide stars. The Orion UltraGuider, Celestron Radial Guider, Lumicon Easy-Guider, and the Multi-Star Guider by Northern Lites are among the best off-axis guiders for the amateur astrophotographer.

Focusing Aids

Focusing a telescope for the eye alone is easy, but achieving a sharp focus when viewing through a camera viewfinder can be quite another matter. Although standard focusing screens work well under bright conditions with fast lenses, they produce dim, ill-defined images when attached to a telescope for nighttime use. Many astrophotographers swap their camera's standard focusing screen for a clear ground-glass matte screen. Special viewfinder screens called Intenscreens, available from Beattie Systems, Inc., deliver images up to four times as bright as standard screens. Unfortunately, they can also cost many times more than standard screens, but they do work. (Some camera light meters that take readings off the focusing screen might need their ASA/ISO speed dials adjusted to compensate for the Intenscreen's extra brightness.)

Even with a clear matte screen in place, getting a crisp image is still tough to do. Three focusing devices that have won favor among astrophotographers are the SureSharp and PointSource, both by Spectra Astro Systems, and Celestron's Multi Function Focal Tester. The SureSharp, introduced in 1987, consists of two parts: a modified T-ring that attaches to either a telescope's camera adapter or off-axis guider and a conical housing containing a Ronchi grating (a clear glass window with evenly spaced opaque, etched lines) at the *exact* distance as the film plane of a 35-mm SLR camera. With the telescope pointed at a 4th-magnitude or brighter star, look at the grating through an opening in the housing. If the star is out of focus, it will appear as a disk crossed by dark gray bands. Turning the focus, the number of bands decreases until, at the point of focus, the lines disappear. When the target star is focused precisely, the disk of the star is small enough either to pass between the opaque lines (the disk appearing completely illuminated) or to be completely blocked by a line (causing the disk to disappear).

The PointSource works on the same principle as the SureSharp. Screw the PointSource into any camera adapter that uses a T-ring for mounting onto a camera. Attach the camera adapter to the telescope, look at the Ronchi grating, and focus the telescope as described above. The biggest advantage of the PointSource is that it can be used with just about any T-ring and may be shared with different camera brands.

Celestron's Multi Function Focal Tester (MFFT-55) also screws onto a standard T-adapter in place of a camera. Just aim the telescope at a bright star and focus its image on the ground-glass screen of the MFFT-55. That's all there is to it. Remove the MFFT-55 (carefully, so as not to disturb the focusing), attach the camera, and the telescope is set for in-focus photography. Instructions also explain how the MFFT-55 can be used for a variety of other tests, such as telescope collimation and checking the squareness of the camera adapter to the optical axis. The Celestron MFFT-55 lists for more than either of the others, but deep discounts can cut that price sharply.

If you are interested in photographing bright, extended objects like the Sun, Moon, planets, or maybe the brightest deep-sky objects, then one of these focusing devices may not be needed. To get a sharp focus the first time, every time, make a mask for your telescope by cutting a circular piece of cardboard the same diameter as its aperture. Now cut two smaller circles in the mask directly opposite each other. (Two-inch diameters work well for 8-inch and larger telescopes.) Make certain that they are not cut off by the telescope's secondary mirror (if it's a reflector or catadioptric). With the mask attached in front of the telescope, attach the camera, aim at the target, and look through the viewfinder. If the telescope is out of focus, you will see two images. Turn the focusing in or out until the two slowly blend into one. When only a single image is seen, the telescope is properly focused. It works with all telescopes, but only for *extended* objects like the Moon and planets, not stars. A commercial version of this device, called Kwik Focus, is marketed by P&S Skyproducts.

CCD Cameras

The pursuit of astrophotography has changed dramatically in the past decade. The availability of super-sensitive charge-coupled devices, or CCDs for short, now makes it possible to take photographs of the Moon, planets, or deep-sky objects using exposures many times shorter than is possible with conventional film. Exposures as short as 30 seconds with a CCD camera, such as the model shown in Figure 6.11, will record the same detail as a half-hour exposure using conventional film. (Estimates indicate that the most common amateur-marketed CCD cameras have an equivalent ISO speed value of 20,000!) In addition, the effects of light pollution can be eliminated by computer processing, making it possible to capture great deep-sky images from within cities.

Two of the most popular CCD cameras offered to amateurs are the Lynxx PC by SpectraSource Instruments and Santa Barbara Instrument Groups' (SBIG) ST-4. Both attach to the prime focus of the telescope, just like a regular camera. Images can then be displayed on a computer monitor as they are recorded as well as stored and processed on a personal computer for outstanding results.

The CCD's field of view depends on the size of the CCD chip itself. Both the Lynxx PC and ST-4 use the same 0.1-inch square TC-211 chip supplied by

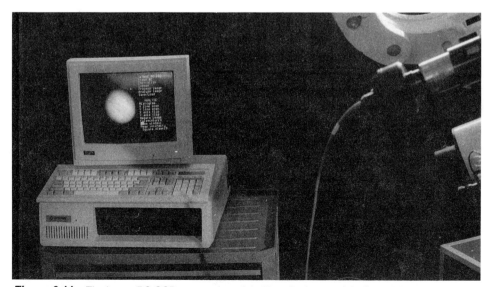

Figure 6.11   *The Lynxx-PC CCD camera in action. Note the image of Jupiter on the computer screen.*

Texas Instruments, resulting in very small fields. For instance, with an 8-inch f/10 SCT, the field of view is only 4 arc-minutes square.

Although both models use the same CCD chip, most agree that the Lynxx produces superior images, and this fact can be traced to several of its features. Most important of all is that the Lynxx PC uses a 12-bit A/D converter for storing and processing the images. A 12-bit converter gives a greater dynamic range in image brightness than the 8-bit converter used in the ST-4. These pros are partially offset by a couple of inherent cons. The 12-bit converter also means that an image taken with the Lynxx PC takes up about 50% more memory space to store and is slower to process than the same picture recorded with the ST-4.

The Lynxx 2000 is a new CCD camera recently introduced by Spectra-Source Instruments. With an expanded format of up to 336×488 pixels, the Lynxx 2000 offers five times the spatial resolution of previous Lynxx CCD systems. The associated software is available for both IBM-compatible and Macintosh computers.

The new SBIG ST-6 has received rave reviews from all who have tried it. In his extensive review of the ST-6 in the October 1992 issue of *Sky & Telescope*, Dennis di Cicco wrote, " . . . The ST-6 surpasses even my wildest expectations . . . it's pretty impressive to have images of 13th-magnitude Pluto jump off the computer screen after a 10-second exposure. . . ."

Image quality with the ST-6 is outstanding, approaching that of photographic film quality, thanks to 16-bit data. Using Texas Instruments' TC-241 CCD, the ST-6 covers a 16-by-12-arc-minute area with an 8-inch f/10 telescope, an increase of almost 800% over the Lynxx PC and ST-4 CCDs. Even more amazing is the CCD's ability to cut through light pollution and moonlight to produce exceptional results on nights when most deep-sky astrophotographers would stay inside and read a good book.

The only apparent drawback to the ST-6 is its overall size. Though it weighs only 2.25 pounds, the camera's head is large enough to interfere with the mounting of fork-mounted Schmidt-Cassegrain telescopes. With the ST-6 coupled to an 8-inch Meade or Celestron, the telescope cannot be pointed farther north than about 65° declination when viewing from a latitude of approximately 42° north. Other styles of equatorial mountings are not faced with this problem.

All SpectraSource and SBIG CCD cameras work in black and white only, with no color CCD cameras suitable for long exposures anticipated in the foreseeable future. Yet color CCD photography is possible, thanks to a variation on the tricolor photographic technique introduced by James Maxwell in 1861. Three separate exposures are made with red, green, and blue filters over the camera and are then combined electronically with computer software to produce a true-color image. SBIG offers a color filter assembly and the necessary software to turn the ST-6 into a color camera. The results, seen in just about any recent issue of *Sky & Telescope* or *Astronomy*, are outstanding!

While still more expensive than 35-mm SLR cameras, CCD cameras will undoubtedly drop in price just as CCD-based home video camcorder prices have fallen as the technology becomes cheaper to produce. Of course, all require a personal computer and high-resolution monitor (VGA or better), adding even more to the start-up cost. Yet many amateurs think this is a small price to pay for the outstanding quality that can be achieved with CCDs. (One postscripted piece of advice for anyone thinking of buying a CCD camera: Make sure that the camera's tracking system is compatible with your telescope drive's electronics. Not all are!) To learn more about this rapidly changing field of amateur astronomy, consider subscribing to *CCD Astronomy*, a new quarterly periodical from Sky Publishing Corporation.

The Great Outdoors

An area that few manuals of amateur astronomy address is the environment around the observer. Sure, most books complain about excessive light pollution and the need for good sky conditions, but there is so much more to enjoying the night sky than just the sky.

Baby, It's Cold Outside

The old saying that "clothing makes the man" (excuse me, *person*) is certainly true in astronomy. Nothing can take the enjoyment out of observing faster than physical discomfort. Though usually not a problem during summer, it certainly can be at other times of the year. Even the sturdiest telescope mount will wobble if the observer using it is shivering!

It goes without saying that the clearest nights occur after a high-pressure weather front sweeps the atmosphere of clouds, haze, and smog. Unfortunately, the clear atmosphere also causes the Earth to lose a great deal of the heat that it has built up during the day. Many amateurs decide to sit these cold nights out, but by doing so, they are missing some of the clearest skies of the season. Others try to brave the cold wearing their usual overcoat and a thin pair of gloves but soon return inside, teeth chattering and fingers numb. Is this any way to enjoy the wonder of the universe?

Most hardy souls agree that layering clothes works best. For temperatures above 20°F, wear (from the inside out) a tee shirt, flannel shirt, sweater, and parka on top, while underwear, long underwear, and heavy pants round out the bottom. In colder temperatures or when the wind is howling, replace the sweater with a one-piece worksuit. These provide a solid, windproof barrier between you and the cold, cruel world.

Though these items should keep you warm enough in moderately cold conditions, they also can make you stiff as a board. The multilayered look can make it difficult to pick up that pencil that was dropped on the ground—or even to bend for a peek through the eyepiece! Happily, today there is a better way.

Thanks to modern synthetic fabrics, it is now possible to stay outside even in subzero temperatures in relative comfort and with full freedom of movement. Increasing numbers of observers are joining other outdoor enthusiasts who wear clothing made of advanced materials such as Dupont's Thermax and Thinsulate. Both have amazing heat-retention properties yet are thin and light enough to permit the wearer to bend with ease.

The best selections of cold-weather apparel are found at either local sporting goods retailers or in national mail-order catalogues such as L.L. Bean, Campmor, and Damart. Unlikely as it may sound, I have gotten much of my cold-weather clothing either from local bike shops or from two mail-order outlets: Bike Nashbar and Performance Bicycle Shop. (I'm an avid bicyclist by day.)

Keeping the extremities warm is the most critical part of your cold-weather assault. In less extreme temperatures, a pair of thick socks and work boots for the feet, a hat for the head, and a pair of gloves for the hands should do the trick. Under colder conditions, the head is best protected with a silk or wool balaclava. Looking like a full-face ski mask, a balaclava is thin enough to hear through yet warm enough that it may make a heavy hood or hat unnecessary.

For the hands, try a pair of ski mittens stuffed with Thinsulate or a similar material. Unfortunately, though they are warmer than gloves, mittens can make it difficult even to focus the telescope. Some mittens come with a thin insulating glove that may be worn separately, which is handy when changing eyepieces but still a problem when taking notes or making drawings. One other possibility is to wear a separate thin glove inside each mitten. Once again, many bicycle outlets sell thin winter gloves that work quite well.

Nothing is more painful than frozen feet. I have seen some people walk out in 10° weather wearing a heavy parka, a down vest, long underwear, thick hood, heavy gloves, and a pair of sneakers! They didn't last long.

Work boots and so-called moon boots have excellent heat-retention qualities, but only when used with a good pair of socks. The two-pair strategy usually works best. Wear a thin pair on the inside and a thicker, thermal pair on the outside. For truly frigid weather, even the best insulated gloves and boots may not do the trick. While some outdoorsfolk use heated hand warmers that run on cigarette lighter fluid, I get nervous keeping an incendiary device near clothing. (Do you smell something burning?)

No doubt about it, battery-powered socks will keep your feet warm, yet they fail on several different accounts. First, all those I have tried warmed my feet like a microwave oven warms a bowl of leftover spaghetti—unevenly. Invariably, part of my foot was too hot, while another part was too cold. Then, too, you must decide what to do with the battery pack. In most cases, it must be strapped onto your leg, adding to the discomfort. Some people love electric socks, which is fine, but I prefer another approach.

A safer means are nontoxic chemical hand- and feet-warming pads, such as those manufactured by Mycoal Warmers Company in Japan and sold in North America by Grabber of Concord, California. Once removed from their packaging and exposed to the air, these nontoxic heaters maintain a temperature of about 140°F for several hours. (I have found the advertised claim that they last "seven hours or more" to be a little optimistic. From actual use, four or five hours is probably a better estimate.) Afterwards, simply discard the pad. It is important to note, however, that the warmer must be wrapped in a cloth or other protective material before use. Burns can result if these heaters are left in direct contact with the skin for long. Other companies make similar heat pads, but I have found those by Mycoal to be the best.

Don't Bug Me!

All that talk about cold weather makes me long for spring and warmer nights. But of course, with warmer weather come things that go buzz in the night. Mosquitos, gnats, and blackflies can prove more annoying than the cold. Can anything be done to ward off these nighttime pesties? Different solutions, ranging from voodoolike rituals to toxic chemical brews, have been advanced over the years with varying degrees of success. For instance, there's an old wives' tale that claims spreading rancid butter over the skin will fend off flying insects. While not true, it is a great way to keep other astronomers away! Others believe that if a little spritz of insect repellent works pretty well, then a lot should work wonders. Not exactly sound logic, as many hikers will attest—once the ill effects wear off, that is.

The best way to avoid insect bites is a combination of long-sleeved clothing, an observing site that is high and dry, and a good insect repellent. Studies show that the most effective brands use N, N-diethyl-3-methylbenzamide, better known as DEET. Most commercial repellents specify a DEET concentration of no more than 25 to 30% because higher amounts are potentially harmful. Common repellents that use DEET include Cutter™, Off™, Deep-Woods Off™, and 6-12 Plus™.

Exercise caution whenever applying an insect repellent. DEET may be applied directly to the skin and clothing, but be sure to use it far from your telescope or other equipment. Though it works well at warding off insects, DEET also acts as a wonderful solvent when sprayed onto vinyl, plastic, painted surfaces, and optical coatings! DEET should not be used on infants or young children.

A repellent growing rapidly in popularity is actually not a repellent at all. For years, Avon has been selling Skin-So-Soft™ lotion and bath oil as a way to promote youthful skin. Maybe that's why a lot of astronomers look so young, because many people swear it works as a mosquito repellent, too. Tests in *Consumer Reports* and others have found that it works marginally, if at all. Maybe that's why folklore says that to be effective, Skin-So-Soft must be applied repeatedly. In fact, in his book *Medicine for the Outdoors* (Little, 1986), author Paul Auerbach states that Skin-So-Soft must be applied 11 to 28 times as often to work as well as DEET.

Still More Paraphernalia

Flashlights

Every astronomer has an opinion of what makes a flashlight astronomically worthy. Some prefer pocketable penlights, others like focusable halogen models, while a few favor dual-bulb models (providing a built-in replacement just in case one burns out). Most agree that the best are small enough to fit into a pocket but large enough not to get lost at night. White or brightly colored housings are also preferred to black or dark models, because they are easier to find if dropped.

Regardless of the style or design, a flashlight must be covered with a red filter to lessen its blinding impact on an observer's night vision. There are many different ways of turning a white light red. Some of the more common methods include painting the bulb red with fingernail polish or using red tissue paper or transparent red cellophane from stationery and party supply stores. These tend to chip, tear, or fade with time, forcing repeated filter renewal or replacement. A more permanent solution is to use red gelatin filter material sold in art supply or camera stores. One or two layers of Wratten gelatin filter No. 25 (yes, the same classification system as eyepiece filters) work especially well. But perhaps the most versatile red filter material is sold by auto parts stores as repair tape for car taillights. Sold in rolls of several feet, its adhesive backing makes it ideal for sticking onto a flashlight.

Far better are flashlights that use an LED instead of a conventional bulb. These are growing in popularity thanks to their deep, pure red color and low power drain. Many astronomical suppliers sell LED "astronomy flashlights," with one of the favorites being the Starlite by Rigel Systems. It features an adjustable-brightness LED and runs on a single 9-volt battery (which can be a little tricky to replace). In general, LED flashlights are a little pricey, but for the frugal astronomer, the next chapter includes plans for a complete LED flashlight that can be made for about $10, as well as a plan to retrofit any flashlight with an LED for about $5.

Dew Caps and Dew Guns

One of the most frustrating things that we, as amateur astronomers, are forced to deal with is dew. The formation of dew on a telescope objective, finder, or eyepiece can end an observing session as abruptly as the arrival of a cloud bank.

Dew forms on any surface whenever that surface becomes colder than the dewpoint temperature, which varies greatly depending on both air temperature and humidity. To illustrate this, picture a can of cold soda. In the refrigerator, the can's exterior is dry because its surface temperature is above the dewpoint of the surrounding refrigerated air. Now, take the can out of the refrigerator and leave it at room temperature. Almost immediately its surface becomes laden with moisture because the can is now colder than the warmer air's dew-point. Under a clear sky, objects radiate heat away into space and soon become colder than the surrounding air.

Nothing, neither telescopes, binoculars, eyepieces, star atlases, nor cameras, is impervious to the assault of dew, but there are ways to slow the whole process down (no, one is NOT to give up and go inside—although there have been times that I was tempted!). One option is to install a *dew cap* on the telescope, which is a tube extension that protrudes in front of an objective lens or corrector plate. It shields the glass from wide exposure to the cold air, thus slowing radiational cooling. Binoculars, refractors, and catadioptric telescopes stand to gain the most from dew caps because their objectives and corrector plates lie so near the front of the telescope tube.

Reflectors usually do not need a dew cap because their primary mirrors lie at the bottom of the tube, which itself acts as a dew cap. The only exceptions to this are if the secondary mirror dews over (only in exceptionally damp conditions) or if the reflector has an open-truss tube. The former situation can be slowed by installing battery-powered resistors akin to those discussed in the next chapter. For the latter, many amateurs find it beneficial to wrap the truss with cloth, effectively shielding the mirror from radiational cooling.

Dew caps merely slow the cooling of an objective lens or corrector plate, thereby only delaying the formation of dew. Depending on the humidity, this is enough. But to be effective, a dew cap must extend in front of the objective or corrector at least one and a half times (preferably two to three times) the telescope's diameter. Although most refractors built this century are supplied with dew caps, most binoculars and catadioptrics are not (the lone exception being the Questar Maksutovs). Owners must therefore either purchase or construct dew caps. Chapter 7 details how to make dew caps easily at home using simple materials, but for now let us examine what is available commercially.

Surprisingly few companies offer dew caps. Of those available, most are made of molded plastic and designed to slip on and off the telescope as needed. For instance, Orion Telescope Center supplies Flexshield dew caps. Made of a thin, durable material called Kydex, Flexshields are sold as flat rectangles. They are formed into tubes by simply wrapping the ends around and sealing the full-width, permanently attached Velcro™ closure. After the observing session, simply pull the Velcro™ seal apart, and the Flexshield lies flat for easy storage. Flexshields come in sizes to accommodate telescopes from 5 to 14 inches in diameter.

Roger W. Tuthill, Inc., also offers a complete line of No-Du dew caps (Figure 6.12). The No-Du caps are rigid cylinders made out of a plastic derivative, making them more difficult to store than the Orion caps. Each is lined with black felt, which is claimed to increase the cap's effectiveness. Tuthill's larger dew caps also feature built-in cylindrical heating elements. By warming *ever so slightly* the corrector plate and the air inside the cap by convection, dewing can be effectively prevented even in high humidity. The cap's heating element

Figure 6.12   *Tuthill No-Du Cap.*

emits between 10 and 20 watts of heat, depending on telescope size, and requires a 12-volt DC power source, such as a car battery, DC power supply (useful only if a 110-volt AC outlet is nearby), or a rechargeable battery with at least 5 (preferably 10 or more) ampere-hour capacity. No-Du caps are available for telescopes ranging from 4 to 14 inches in diameter, but heated versions are sold only for 8-inch and larger scopes.

Orion makes a different style of contact heating element called the Dew Zapper for 8- and 10-inch Schmidt-Cassegrains. The Dew Zapper wraps around the front end of the telescope tube to heat the corrector plate by conduction to just above the air's dewpoint. Although it may be used independently of a dew cap, the Dew Zapper is more effective with a cap in place. Both 110-volt AC and 12-volt DC versions are available; the former delivers up to 20 watts of heat, while the latter supplies about 10 watts at 1 amp.

Some find the Dew Zapper inconvenient to use, as it is held onto the telescope's tube by means of a large elastic band. A better solution would have been to add mating Velcro™-type strips on either end of the heating coil, keeping it a one-piece unit. Another problem common to both the Dew Zapper and the No-Du cap is that they require external power sources, which means each comes with a long electric cord. Cords are easy to snag and trip over, especially at night, but because of the required power draw, there seems to be no way around this.

Right about now, if you listen carefully, you can hear the traditionalists in the crowd jumping up and down, screaming that all this is blasphemy. Many amateurs believe that imparting *any* heat to a telescope will cause optical distortion. True enough, heat will upset the delicate figures of optics. This is why some telescopes must be left outside for an hour or more before observing to let the optics adjust to the outdoor temperature. In practice, however, the disturbance from contact heaters will be minimal as long as the heating is done in moderation. By definition, the dewpoint can only be less than or equal to the air temperature, never greater (under most clear-sky conditions, the dewpoint is going to be much less than the air temperature). The purpose of a contact heater is not to raise the telescope's temperature above that of the air, only above the dewpoint. Therefore, the telescope should never feel warm to the touch, and overheating should never become a problem.

A third way to wage war against dew is to use a low-power hair dryer or heat gun. By blowing a steady stream of warm air across an optical surface, dew may be done away with, albeit temporarily. If an AC outlet is within reach, a small portable hair dryer at its lowest setting makes a good dew remover. If you are out in the bush away from such amenities, then use a heat gun designed to run off a car battery (DC heat guns draw too much power to use with rechargeable batteries). A wide variety of sources sell basically the same 12-volt heat gun at a wide variety of prices for a wide variety of uses. While Orion Telescope Center sells it as a "dew remover gun," auto parts stores call it a "windshield defroster," and camping equipment outlets offer it as a "mobile hair dryer." Call them what you will, all are pretty much the same. The gun puts out about 150 watts of heat, not exactly enough to dry hair or defrost a windshield but adequate for dew removal from an eyepiece.

Although I never go observing without one, heat guns do have some drawbacks. First, though fine for undewing finderscopes and eyepieces, their small size limits their effectiveness for objectives and corrector plates. (If the lens or mirror is much larger than about three inches, it is likely that the entire surface will not be cleared before a portion becomes fogged again.) A second shortcoming is that, sadly, the dew will return as soon as the surface cools below the dewpoint, making it necessary to halt whatever you are doing and undew the optic all over again.

If dew is a big problem where you observe, then consider getting all three gadgets, that is, a dew cap, contact heater (or heated dew cap), and heat gun. Never wait for dew to form; always turn the heater on before observing begins. Put the dew cap on (remember the finder, too) and have the heat gun holstered at the ready to clear any fog that may form on eyepieces. By following this three-step program, optical fogging should be minimized, if not eliminated.

Observing Chairs

It is a well-proven fact that faint objects and subtle detail will be missed whenever an observer is fatigued or uncomfortable. Yet many amateur astronomers spend hours outside at their telescopes without ever sitting down. Observing is supposed to be fun, not a marathon of agony!

Observing chairs help relieve the stress and strain associated with hours of concentrated effort at the telescope. The best chairs have padded seats and telescoping posts so that they may raised and lowered with the eyepiece (not always possible with long-focus refractors or large-aperture Newtonians).

Musician's stools or drafting-table chairs make excellent observing chairs. Check the offerings at local music shops and drafting/art supplies stores. For the astronomical market, Orion Telescope Center makes two observing stools that differ little from piano stools. The standard model adjusts between 19 and 25 inches, while the deluxe chair raises between 21 inches and 27 inches above the ground.

Another alternative is an innovative chair made from polished metal tubes and a padded seat sold by AstroSystems, Inc. The seat can be set at any height between just 9 inches to 36 inches, making the chair usable with most amateur telescopes. When stored, the AstroSystems Observing Chair folds to $4'' \times 16'' \times 47''$. While more flexible than the Orion stools, the AstroSystems chair is twice Orion's price.

Telescope Mounts

When the Beach Boys made the song "Good Vibrations" famous, they certainly were not singing about telescope mounts! One of the biggest complaints that my informal survey of telescope owners revealed is dissatisfaction with the mountings that are supplied. In an effort to remedy this situation, many amateurs seek to retrofit their instruments with substantially larger, sturdier support systems. Some of the finest are manufactured by Astro-Physics. Although designed for its fine refractors, Astro-Physics equatorial mounts can be fitted to many other instruments. Chapter 4 carries a complete discussion. In addition, here are a few other companies that manufacture some excellent telescope mounts.

Some of the world's finest telescope mounts are manufactured by Hollywood General Machining Company and carry the Losmandy name on their sides. Losmandy products include a number of add-on features to make an existing mount a little easier to use as well as complete mounts for a wide variety of telescopes. Especially impressive is the G-11 mount, previously noted in Chapter 4's discussion of the Celestron CG-11 SCT. The G-11 may be purchased separately and is ideal for telescopes up to about 60 pounds. It comes with a well-designed dual-axis drive system with a built-in quartz corrector and periodic error correction as well as an integrated polar-alignment scope.

Another giant in the telescope-mounting industry is Edward R. Byers Company. Like Losmandy, Byers mountings are famous for their rigidity and precise tracking ability. Chiefly a build-to-suit custom house, Byers features mountings designed primarily for medium-to-large aperture amateur telescopes. Byers also manufactures some of the finest clock drives available on the amateur market. Especially popular among serious astrophotographers is its Celestron 14 Star-Master retrofit drive system. The Byers system completely replaces the stock drive that comes with the C14 with a precisely machined worm-and-worm-gear assembly. Due to the extent of machining required to retrofit the drive, the customer must disassemble the C14 fork mount and send the base to Byers, where all the modifications are completed in their factory.

Two companies that offer equatorial platforms for Dobsonians are Jupiter Telescope Company and Equatorial Platforms. The Jupiter system is supplied as an option with its Juno line of telescopes and is described in Chapter 4. The Equatorial Platforms mounting, similar in design to that of Jupiter, is solidly made of finely crafted wood and well-machined components. The stationary bottom board is replaced by a board that pivots about a polar axis. Once polar-aligned with the motor (either an AC-powered synchronous or DC-powered stepper) switched on, the telescope will track the stars just like an equatorially mounted instrument. This ability does not come cheaply; in fact, the mounting may cost more than the telescope itself.

Binocular Mounts

After purchasing a pair of giant binoculars, most people suddenly come to the realization that the binoculars are too heavy to hold by hand and must be attached to some sort of external support. The favorite choice is the trusty camera tripod, but as was discussed earlier, not all tripods are sturdy enough to do the job. While all of the brands mentioned above are strong enough to support binoculars, most are too short to permit the user to view anywhere near the zenith comfortably (the bigger Bogens being an exception to this).

What are the alternatives? There's basically only one: Use a special mount designed specifically for the purpose. Some unusual designs have come and gone in the past, but a few are available today that work quite well. There are also a few that are poorly designed, affording little advantage to holding the glasses by hand.

One of the better binocular mounts around is the Steadi/Scan (Figure 6.13a) made by Safari Telescopes. Using the same principle as a Dobsonian telescope mount, the Steadi/Scan pivots binoculars in both altitude and azimuth. To get the silky smooth movement, the Steadi/Scan uses black-anodized aluminum riding on pads of Teflon™. Mounted directly onto a camera tripod, the Steadi/Scan offsets the binoculars from the center of the tripod, making it much easier to view objects that are high above the horizon. Note that to make full use of the Steadi/Scan, a tall tripod and/or a chair is required. The Steadi/Scan is sturdy enough to support up to 80-mm glasses.

The Sky Hook (manufactured by Star Bound and shown in Figure 6.13b) looks like what we engineers call a *kluge*—that is, a design that just cannot and will not work. Constructed of chromed steel, the Sky Hook looks more like one of those suitcase holders found in many hotel rooms than it does a piece of astronomical equipment! But don't laugh; it works! The binoculars are held by a cantilevered hooklike arm that moves up and down and back and forth on a pair of round tracks. Once everything is set up (about a five-minute operation),

Figure 6.13a *The Steadi/Scan Binocular mount from Safari Telescopes.*

Figure 6.13b *The Sky Hook binocular mount, demonstrated by the daughter of its creator, Bob Miles. Photo courtesy of Star Bound.*

the binoculars move smoothly in all directions. To use the Sky Hook, the observer need only seat him- or herself in a lawn chair for comfortable access to the entire sky.

Most of the following binocular mounts are based on a design that works on the same principle as swing-arm desk lamps. Everyone has seen the kind of lamp that may be raised and lowered to any position while the light itself remains aimed at a constant angle. In this case, the light has been replaced by binoculars, permitting observers both short and tall to view comfortably without having to re-aim the glasses each time. (For a complete description and plans for a homemade version of this style of binocular mount, see Chapter 7.)

One company to adopt this design is Astronomical Innovations, which sells the BM-100 binocular mounting. Though massive in appearance, it proves wobbly when supporting giant glasses. The mounting may be supported either on a heavy-duty (repeat, *heavy-duty*) photographic tripod or on that same outfit's Porta-Pier.

Astronomical Innovations also sells the MM-50 Mini Mount, which does little more than raise and slightly offset the binoculars from the tripod, again making a tall tripod a must. The design is adequate for lighter $7\times$ and $8\times$ glasses, but not for larger models.

Vista Instruments offers the Binocular Guider, also based on the parallelogram design. The Vista Binocular Guider, made primarily from aluminum channels, is strong enough to support only lightweight binoculars. Any pair larger than about 10×50 causes flexure and is just too much for the mount to hold steadily.

Most expensive of the current crop of binocular mounts is the GrandView, a beautifully crafted mount and tripod made from furniture-grade oak. Sturdy enough to support binoculars up to eight pounds, the GrandView can accommodate observers from 38 to 80 inches tall. Of course, this level of craftsmanship does not come cheaply, but the quality is unparalleled.

Vibration Dampers

These make great stocking stuffers for any amateur who owns a tripod-mounted telescope. One of the biggest problems facing amateurs using such telescopes is how to combat the vibrations caused by anything and everything from nearby automobile traffic or another person moving about to the wind.

While nothing can guarantee shake-free viewing, Celestron has introduced a simple little gadget that can make a big difference. Vibration dampers (Figure 6.14) are made of a hard outer resin and soft inner rubber polymer, isolated from each other by an aluminum ring. Simply place one suppressor under each tripod leg. Any vibration will be absorbed by the inner polymer before it can be transmitted to the telescope. Such a simple yet effective idea—why didn't someone think of this before?

Velcro™: The Unsung Hero of Amateur Astronomy

What a wonderful invention. Velcro™ (actually a brand name for a hook-and-loop fabric-fastening system) takes two completely incompatible substances

Figure 6.14   *Vibration damper pads. Photo courtesy of Celestron International.*

and makes them stick together. It works great for holding flashlights to clip-boards, small LED illuminators to setting circles, pencils to telescope tubes, and just about anything else you can think of. Best of all, Velcro™ can be pulled apart and reused again and again.

Tool Kit and Spare Parts

Have you ever had your telescope set up in the middle of a field far from home when, suddenly, you discover a piece missing? It may be only a ten-cent nut or a screw, but without it the telescope simply will not function. Sure, there are plenty of spares in the garage at home, but none are with you.

If you have been an amateur astronomer for some time, then the scenario above has undoubtedly happened to you. (If not, don't worry; your day is com-ing.) That is why it is so important to plan ahead for just such an emergency. Compile a list of parts and tools needed to put your telescope together and bring along plenty of spares. The seasoned amateur knows how important this can be for those middle-of-nowhere repairs.

Although most spare parts can be gotten from your local hardware store, a few special items may be difficult to track down. One place to look for those hard-to-find items is Small Parts, Inc., of Florida. They have just about every-thing for the do-it-yourselfer, from nuts, bolts, and threaded rods to couplers, knobs, gears, and pulleys. Best of all, they do not require huge minimum orders as so many other mail-order hardware suppliers do. A copy of their free cata-logue is yours for the asking (their address is listed in Appendix A).

The list of available accessories for today's amateur astronomer could run on for pages and pages, but I must cut it off somewhere. To help put all of this in perspective, Table 6.5 lists what the well-groomed amateur astronomer is wearing these days. Some of the items may be readily purchased from any of

Table 6.5   **Must-Have Accessories for the Well-Groomed Astronomer**

Commercial	Homemade
• Narrow-band LPR filter	• Eyepiece and accessory tray attached to telescope
• Amici prism for right-angle finder	
• *Sky Atlas 2000.0*	• Adjustable observer's chair (see plans in Chapter 7)
• Velcro™	
• A good red flashlight	• Two old briefcases; one for charts, maps, and flashlights, the other lined with high-density foam for eyepieces, filters, etc.
• Warm clothes	
• Hand and foot warmers	
• Bug repellent (Cutters™, Deep-Woods Off™)	• A backlit clear-plastic clipboard for sketching (see plans in Chapter 7).
• Guidescope or off-axis guider (long-exposure photography only)	
	• Observation record forms
• Drive corrector (if telescope comes equipped with a clock drive)	• Cloaking Device (see Chapter 9)
• Camera-to-telescope adapter	

a number of different suppliers; others can be made at home. A few may even be lying in your basement, attic, or garage right now.

The most important accessory that an amateur astronomer can have is someone else with whom to share the universe. Though many observers prefer to go it alone, there is something special about observing with friends. Even though you may be looking at completely different things, it is always nice to share the experience with someone else. If you have not already done so, seek out and join a local astronomy club. For the names and addresses of clubs near you, consult either *Astronomy*'s or *Sky & Telescope*'s annual guides to clubs, planetaria, observatories, and conventions. They may be found in the June and September issues, respectively. If you have an astronomy buddy, then you have the most important accessory of all, one that money cannot buy!

7

The Homemade Astronomer

Amateur astronomers are an innovative lot. While manufacturers offer a tremendous variety of telescopes and accessories for sale, many hobbyists prefer to build much of their equipment themselves. Indeed, some of the finest and most useful equipment is not even available commercially, making it necessary for the amateur to go it alone.

Many books and magazines have published plans for building complete telescopes. Rather than reinvent the wheel here, I thought it might be fun to include plans for a variety of useful telescope- and binocular-related accessories. Here are ten such projects, ranging from the very simple and inexpensive to the advanced and costly. Their common thread is that they were created by amateurs to enhance their enjoyment of the universe. These projects are just a few samples of the genius of the amateur telescope maker.

Light-Pollution Shields

Many of today's eyepieces and binoculars come with built-in collapsible rubber eyecups to keep stray light from entering the observer's eyes. Under most conditions, they work quite well. However, if light from a nearby source, such as a streetlight, passing cars, or a neighbor's porch light, is severe, eyecups alone are just not enough. Many amateurs have resorted to planting high hedges around their observing sites just to ward off such offenders.

Another way to tackle the problem is to make a simple light-pollution shield from black cloth. Figures 7.1 and 7.2 show two variations of the same theme. The first, made by yours truly, simply consists of a wide piece of black double-knit cloth nailed between two 8-foot sections of 1-by-2-inch pine. Each piece of pine has two holes drilled in it to match up with ¼-20 hanger bolts

Figure 7.1 *Fence-mounted light-pollution shield.*

permanently screwed into stockade-fence posts. Oversized shoulder washers and wing nuts allow the shield to be erected and dismantled quickly. So far, the shield has survived wind gusts up to 20 miles per hour without pulling the fence down! (If wind proves to be a problem, cut a few U-shaped slits into the cloth. This will let wind pass through the shield.) Table 7.1 lists all the items needed to construct the shield.

For hobbyists who enjoy astronomy on the go, Chris Bayus of Flint, Michigan, has come up with a clever self-supporting light shield. In his design, a black terry cloth shield is supported by a framework of inexpensive PVC piping. The frame takes only a couple of minutes to slip together before each use.

Table 7.2 lists all of the parts needed to make the Bayus light shield. Bring a copy of the list to your local plumbing or home supply store. Note that the specific pipe lengths probably will not be available but instead will have to be cut by hand at home using either a PVC pipe cutter or a hack saw. (The dimensions shown are only suggestions and can be changed to meet your specific situation. If your telescope is especially long or if you are tall, then you will want to make the frame larger.)

With everything cut to size, position the pieces on the ground in the pattern shown in Figure 7.3. Though parts of the frame ultimately will be glued

Figure 7.2 *Chris Bayus's light-pollution shield. Photo courtesy of Chris Bayus.*

Table 7.1 **Fence-Mounted Light Shield—Parts List**

Description	Quantity
Black cloth	Sized to suit your needs
Pole, 8′ long (1″×2″ lumber)	2
Hanger bolt, ¼-20 thread × 2.5″ long	4
Wing nut, ¼-20	4
Shoulder washer, 0.25″	4
Fence	As required

together to make setup easier, it is best to assemble the pieces before gluing just to make sure everything fits correctly. Add the diagonal braces and raise the frame. Is it straight, or a little crooked? Now is the time to fine-tune all lengths. Once satisfied, take the frame apart and begin bonding some of the pieces together as indicated in Figure 7.3. To the 11 pieces of 3.5-foot pipe, glue the two 90° elbows, two of the T-couplings, the straight couplings, and the end caps. Lay two of the remaining T-couplings on the ground. To each, glue a 1.5-foot pipe to the left side, a 1-foot-5-inch pipe to the right side, and a 2-inch pipe to the bottom. After the adhesive sets, cement a 45° elbow to the end of the 2-inch pipe, with the elbow angled down parallel to the T-coupling. Lastly, to the remaining 3-foot pipe join the last T-coupling to one end and the cross coupler to the other. Make certain the pieces all line up correctly, as PVC glue is tough to get apart once it sets.

Table 7.2 **Portable Light Shield—Parts List**

Description	Quantity
PVC piping (all 1 inch inside diameter)	
42″ long	11
36″ long	1
18″ long	2
17″ long	2
2″ long	2
T-coupling	5
90° elbow	2
45° elbow	2
Straight Coupling	2
Cross	1
End Cap	5
PVC glue	As required
Velcro™ pads (1″ square)	12
Black cloth (Bayus used bathrobe material)	5.5 yards
U-bolts, ¼-20 thread x 1.5″, 4″ long	2

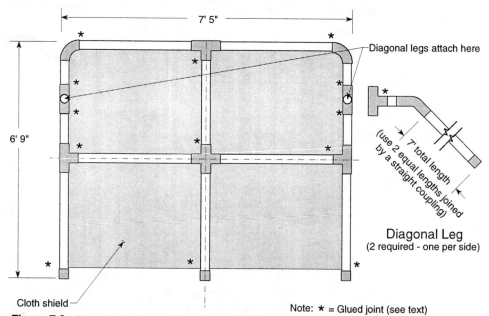

Figure 7.3 *Plans for the Bayus light-pollution shield.*

The time of reckoning is at hand. Reassemble the frame on the ground. If it was made correctly, all of the pieces should slip together easily. Use sandpaper if some of the connections are too tight. Add the two rear-facing diagonal braces and raise the frame. How does it look?

Cut the cloth in half widthwise so that you have two pieces approximately 45 inches by 99 inches each. Sew them back together to create a single piece approximately 90 inches by 99 inches. Fold over the edges and stitch in a 2-inch hem all the way along. Returning to the PVC frame, add adhesive-backed squares of Velcro™-style hook-and-loop material to the front of each corner, the middle of the top beam, the center of the cross, and at the bottom of each leg. Lay the cloth over the frame and mark where it covers the Velcro™. Sew mating Velcro™ material onto the cloth so that the shield will stick to the frame when assembled.

To use the shield, first assemble the framework on the ground. Attach the diagonal braces to the downward-facing 45° elbows to keep everything from tipping over and raise the frame. In calm weather, the shield is self-supporting, while in windy conditions it will need to be staked to the ground. That's where the U-bolts come in. Simply push one over each of the two back support legs and into the ground. Properly staked, the light shield can survive winds up to roughly 20 miles per hour. Finally, attach the cloth curtain to the mating hook-and-loop pads on the frame members.

When not in use, disassemble the shield for easy storage. The cloth and frame can be kept in a tent carrying case or any other suitably sized bag. To store his, Bayus prefers an old golf bag.

If you are plagued by either nearby light pollution or gusts of wind, then by all means consider making one of these shields or come up with a design of your own. It can turn an otherwise unusable yard into an enjoyable observing site.

An Observing Throne

Anyone who has ever spent more than, say, half an hour at the eyepiece of a telescope knows that it can be a painful experience. The constant bending over to look through the eyepiece can put a tremendous strain on an observer's neck, back, and legs. To help alleviate some of the discomfort, many amateur astronomers use commercially made observing chairs or stools. As mentioned in their review in the last chapter, these seats are fine for use with Cassegrain-style telescopes. Unfortunately, most do not have the range in height required to follow the eyepieces of long-tubed refractors and Newtonian reflectors, as these instruments swing between horizontal and vertical positions.

To complement his homemade 10-inch f/5.5 Newtonian, John Stanbury, an architect from Wellesley, Massachusetts, designed and handcrafted the observing chair shown in Figure 7.4. Not just an observing chair, mind you, Stanbury's elegant design is more of an observing *throne*. His motivation came after reading J.B. Sidgwick's classic *Amateur Astronomer's Handbook* (Dover Publications, Inc., 1971) and after examining the chair used with Harvard College Observatory's 15-inch refractor.

To make it easy to store, transport, and set up, the Stanbury chair is designed along the same lines as a folding stepladder. Simply open the chair as you would a small ladder, set it down on the ground, and it's ready for use. To change the height of the seat, lift the front of the seat board, slide it up or down as desired, and gently release it. What could be simpler? Table 7.3 is an itemized bill of materials for the chair.

The four vertical A-frame legs are made from 1-by-3-inch lumber, each measuring 48 inches. As shown in Figure 7.5, the inside of each of the chair's two front legs have 0.75-inch-wide longitudinal grooves running their entire length. These grooves act as guides for the seat itself, described later. The

Table 7.3   Observing Throne—Parts List

Description	Quantity
Leg, 48″ long (1″x3″ lumber)	4
Seat, 10″ x 17″ (0.75″)	1
Seat side support, 6″x10″ (0.75″)	2
Upper cross brace, 6.5″x10″ (0.75″)	2
Front cross brace, 4″x13″ (0.75″)	2
Rear cross brace, 4″x16″ (0.75″)	2
Hardwood dowel, 16.5″ (1″ diameter)	1
Spacer, 3″ diameter (0.5″)	2
Hardwood dowel, 13″ (0.75″ diameter)	1
Steel rod, 13″ (0.375″ diameter)	1
Brass folding brace	2

Figure 7.4   John Stanbury's observing chair. Photo courtesy of John Stanbury.

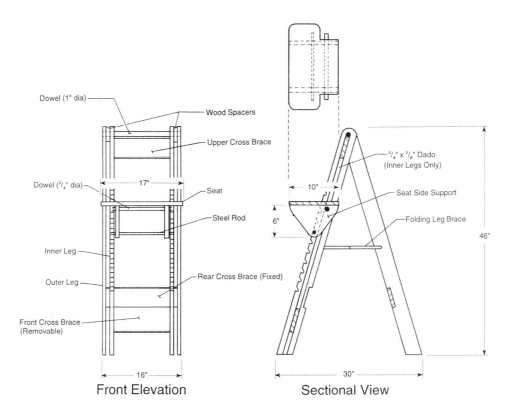

Figure 7.5 *Plans for the Stanbury observing chair.*

grooves were cut with a dado attachment on a table saw, though Stanbury notes a handheld router can also be used.

Cut the matched pairs of notches into the front legs as shown on the drawing. These notches engage the seat and support the weight of the observer. Each measures 0.5-inch deep by 0.75-inch high and is tilted at a slight angle to let gravity assist in holding the seat in place.

Three cross braces, also made from 0.75-inch-thick stock, add lateral rigidity to the chair and are essential to the design. They are glued and screwed into place as shown. Note the cutouts behind the bottom pair of notches in the front legs. They match the location of the lower brace plate, permitting the frame to fold together without the plate hitting the backs of the front legs.

Hinging the A-frame at the top is a 16.5-inch-long, 1-inch-diameter hardwood dowel. To prevent the dowel from slipping, it is fastened to the outer legs with wire brads. The inner legs must rotate freely on the dowel so that the chair can be folded for storage when not in use. Two 0.5-inch wooden spacers separate the pairs of legs to keep them from rubbing together and possibly marring the wood finish. Finally, two brass folding braces similar to those found on stepladders span the front and back legs. When opened, these braces stop the legs from folding out from under the chair.

The seat is made from a 10-by-16-inch piece of 0.75-inch-thick hardwood veneer plywood. Two triangular braces, both cut from the same plywood stock as the seat, serve as reinforcing gussets. Glue and screw the seat board and the gussets together.

A hardwood dowel and a steel rod, each approximately 13 inches long, extend out from the wooden gussets and secure the seat to the A-frame. The 0.375-inch steel rod engages in the pairs of notches to support the weight of the observer. The hardwood dowel, 0.75 inch in diameter, extends from the seat into the matching longitudinal grooves along the front legs, holding the seat to the A-frame. To change the height of the seat, simply lift its front edge to disengage the lower dowel from the notches. With the rod acting as a pivot, move the seat to the desired height and lock the lower dowel into the corresponding notches.

Whether you choose to buy one or make your own, an observing chair is one of the most important accessories for the amateur astronomer. Being seated in a relaxed position is sure to make your time at the eyepiece more productive and much more enjoyable.

An Illuminated Clipboard

One of the best habits an amateur astronomer can develop is maintaining a logbook of descriptions and drawings of all observations made with a telescope, binoculars, or the unaided eye. But if you have ever tried to write or draw at the eyepiece, then you know that this is not an easy task. In addition to a pencil, paper, clipboard, and red-filtered flashlight, you need about three hands to hold everything. This sounds more like a juggling act in a circus. And to make matters worse, all this has to be done in the dark!

Greg Bohemier of Berkshire, New York, has come up with a clever idea that is sure to make the job of recording observations a little easier. Rather than trying to balance a flashlight under his chin or grip its barrel with his teeth, Bohemier has designed and constructed a backlit clipboard (Figure 7.6) that anyone can replicate from readily-available parts.

Table 7.4 lists all you need to build one for yourself. To help locate the right components, each item is cross-referenced with its Radio Shack part number if applicable.

Figure 7.7 shows both the plans and schematic for the board. The clipboard itself is a snap to make out of 0.1875- to 0.25-inch clear plastic. Some better-stocked office supply stores sell clear plastic clipboards, or you may choose to make your own. Many hardware stores stock sheets of clear plastic (usually under the trade name Plexiglas), as do window and auto glass shops. They may even cut it to size for free. If neither of these sources are available, then look under "Plastics" in the telephone book's Yellow Pages. The shop you visit may have a piece of scrap for the taking. Cut the plastic sheet to whatever size you wish, though the plan here calls for 9 by 12 inches. A coping saw or a

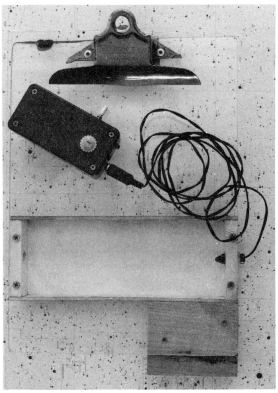

Figure 7.6   *Greg Bohemier's illuminated clipboard.*

Table 7.4   **Illuminated Clipboard—Parts List**

Description	Part Number	Quantity
Light-emitting diode, red	Radio Shack #276-041	2
Light-emitting diode, yellow	Radio Shack #276-021	1
Case, 4"x2"x1.1875"	Radio Shack #270-220	1
Battery connector	Radio Shack #270-325	1
Resistor, 180-ohm	Radio Shack #271-014	1
10-K Potentiometer	Radio Shack #271-1715	1
Switch	Radio Shack #275-1571	1
Wire, 20-gauge (exact gauge not critical)		As required
Clipboard		1
Plastic sheet, clear		1

fine-toothed sabre saw produces the best results, but do not strip off the protective film from the plastic (if so supplied) until afterwards to prevent scratching. Next, remove the spring-loaded chrome clip from a regular (opaque) clipboard by drilling out the two rivets that hold it in place. Position the clip toward the top center of the plastic board, marking and drilling the positions of the clip's two mounting holes. Attach the clip to the plastic board with stainless-steel screws.

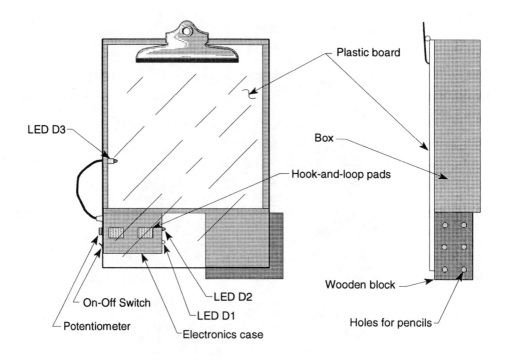

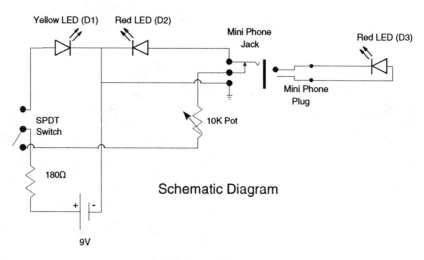

Figure 7.7 *Plans and schematic for the Bohemier illuminated clipboard.*

The most complicated part of this project is constructing the light source. For this, Bohemier recommends an observing light created by Robert Gravel of Westfield, Massachusetts. As with many amateur astronomers today, Gravel prefers LEDs over conventional flashlight bulbs, and with good reason. LEDs draw much less power from batteries than bulbs do, so the batteries will last longer.

The circuitry and battery are mounted in a plastic chassis independent from the clipboard itself. This allows the added flexibility of being able to use the light separately. Notice that the circuit uses three LEDs: one yellow (D1) and two red (D2 and D3). The chassis-mounted yellow LED may be used instead of a regular (unfiltered) flashlight whenever a brighter light is required, such as when setting your telescope up, tearing it down, or when searching for that eyepiece you just dropped (ouch). One of the two red LEDs is used at all other times. The chassis-mounted red LED (D2) is useful for looking at star charts and other references. The second red LED (D3), attached to a length of wire that plugs into the jack on the case, affixes to the clipboard's light box, described later. The LEDs are selected by flipping the single-pole-double-throw switch, while the brightness is controlled by the potentiometer.

Follow the diagram carefully when soldering the components together. Take extra care when soldering the LEDs, as they can be damaged if accidentally overheated. Prior to soldering the wire from D3 to the jack, cut a 1-inch-long section of tube from a ballpoint pen and thread the wire through. Finally, insulate the LED leads from each other with electrical tape or shrink sleeving to prevent electrical shorting. Once the circuit is assembled, hook up a 9-volt transistor-radio battery and turn the switch on. Do the LEDs light? If none of them work, first make sure the battery is good. If it is, then recheck all the solder connections. If only one or two light, the faulty LED may have been damaged during soldering and must be replaced.

With all LEDs working properly, place LED D3 inside the ballpoint pen tube so that its tip just sticks out one end. Fill the ballpoint pen tube with epoxy and set it aside to dry. Once the glue is set, wrap the outside of the plastic tube with a piece of adhesive-backed Velcro™ (the fuzzy side) around the outside.

Returning to the clipboard, drill and countersink about half a dozen equally spaced 0.125-inch holes around its edge. Using 0.125-inch Masonite or the equivalent, some wood glue, and some rust-proof wood screws, make a box with the same outside dimensions as the plastic clipboard. Pine slats, 0.5 by 0.875 inch, can be used for the sides thereby making the box 0.875-inch deep. Paint the inside of the box white and the outside any color you wish (if at all). Place the clipboard over the open top of the box so that the corners line up and attach the clipboard to the box using either brass or chrome-plated wood screws and some glue. Once the glue has set, drill one or more holes into the side of the light box to accommodate the LED. These holes should be just big enough to hold the Velcro™-wrapped LED snugly. Finally, stick adhesive-backed Velcro™ strips to the back of the plastic chassis and the clipboard so the two may be easily attached to each other if you wish.

Finally, add a wooden block that is drilled to hold pencils. No doubt about it, the Bohemier clipboard makes recording notes at the eyepiece a lot easier.

An Astronomer's Flashlight

Speaking of LEDs, many amateur astronomers have switched from using flashlights with standard bulbs to ones equipped with LEDs instead. The trouble is that LED-equipped "astronomer's flashlights" cost quite a bit more than their bulb-equipped counterparts. But why spend more money than you have to? Here's a way to convert any two-battery flashlight into an LED light for less than $5 (not including the cost of the flashlight itself). The plan shown in Figure 7.8 comes from David Kratz of Poquoson, Virginia, and is a modification of an idea from Bob Deen and Randy Hammock, employees of NASA's Jet Propulsion Laboratory. The required parts are listed in Table 7.5.

The choice of LED is up to you. The first listed in Table 7.5 (#276-066A) is best if you are going to use the light to read star charts or write notes. The second (#276-086) is brighter and best if you are using the flashlight to find your way while walking at night. The increased brightness of the latter also comes in handy when setting up or tearing down equipment. (Brightness can also be increased by using a lower-value resistor, but do not use less than 50 ohms.)

Regardless of your choice of LED, begin by removing the regular bulb from the flashlight. Wrap it or another burned-out bulb in several layers of paper towels and then break it by squeezing the glass with a pair of pliers. *Carefully* unwrap the bulb and discard the paper towel along with the shards of glass. Holding the lip of the bulb with a pair of pliers, use a small screwdriver to clean out the remaining glass from the metal socket. (Some form of eye

Table 7.5 **Astronomer's Flashlight—Parts List**

Description	Part Number	Quantity
Flashlight, two-cell (3 volts)	Just about any will do	1
Light-emitting diode, red	Radio Shack #276-066A	1
OR		
Light-emitting diode, red	Radio Shack #276-086	1
Resistor, 100-ohm, 0.25-watt	Radio Shack #271-1311	1

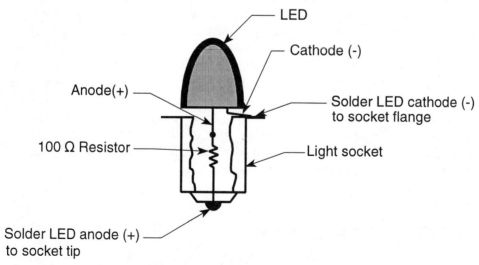

Figure 7.8 Bob Deen and Randy Hammock's LED flashlight bulb replacement.

protection, such as safety glasses, is recommended for this step.) Using a soldering iron, heat the conductive tip of the socket to remove the central wire of the bulb's filament. Finally, heat the slit on the side where the filament's other half is soldered to the socket and remove the wire.

Solder one side of the resistor (it doesn't matter which side) to the anode (+) lead of the LED. The package will show which lead is the anode—probably the longer of the two. The resistor should be very close to the LED itself, or the unit will not fit back into the flashlight. To hold the resistor and LED together for soldering, make a hook on each lead. Hook and crimp them together, then solder. Afterwards, bend the soldered leads so that they are straight and remove any excess with wire cutters.

Bend the LED's cathode lead at a 90° angle and stick the LED-resistor assembly into the light socket. Make sure the resistor lead goes through the central hole of the socket, while the LED's cathode lead should line up with the side slit. Carefully push the resistor into the socket as far as it will go. Solder the leads and again clip off any excess. Now simply insert the new LED assembly into the flashlight and flip on the switch. That's all there is to it: a perfect astronomer's flashlight.

Dealing with Dew

One of the most frustrating things that can happen to an amateur astronomer is having the optics of his or her telescope, binoculars, or camera suddenly become covered with dew on an otherwise perfectly clear night. As mentioned in the previous chapter, dealing with dew is not an easy task. But with a combination of a dew cap and a little warmth from a heating element, lens and mirror fogging can be effectively eliminated.

Soup cans work fine for finderscopes, camera lenses, and smaller binoculars. Line one end with self-adhesive foam weatherstripping, available from any hardware store. The foam will provide a secure fit while preventing the can from scratching the lens barrel. The interior of the can should be painted flat black or lined with black paper or felt to minimize any stray reflections.

Use aluminum flashing to make dew caps for telescopes and larger binoculars. First, measure the outer circumference of the lens barrels. Add about 10% for a safety factor plus an extra inch for overlap. Cut a piece of flashing to this length. Wrap the ends around so that one overlaps the other by that extra inch. Using adhesive-backed Velcro™, affix the loop half to one side of the flashing and the mating hook material to the opposite end so that when the dewcap is wrapped around to form a cylinder, the Velcro™ material holds the cap together. Now either paint the inside of the cap with flat black paint or, better yet, line it with black adhesive-backed felt or other cloth. Finally, add enough weatherstripping around the inside end to ensure a snug fit over the front end of the telescope tube.

If dewing is a serious problem, then a dew cap alone will not do the trick. In cases such as these, the observer has no choice but to use a heated dewcap.

Costly commercial models were reviewed in Chapter 6, but many amateur telescope makers have fabricated their own at home for about $5. As noted in that
earlier review, the purpose of a heated dew cap is to warm the air around the
optic (lens, corrector plate, or mirror) *just enough* to keep dew or frost from
forming. Too much heat will cause image distortion; too little will not prevent
fogging. Just how much heat is enough?

The amount of electrically produced heat delivered per second is expressed
in terms of watts and is known as power. Trying to find just the right number
of watts to keep dew off a telescope is a process of trial and error. Many things
will influence the final number, such as size of telescope and dampness. In
general, an 8-inch telescope (such as a Celestron or Meade Schmidt-Cassegrain) will require between 2 and 3 watts to prevent the corrector from fogging
over. One to 1.5 watts should work well for finderscopes, binoculars, camera
lenses, and refractors up to about 3 inches aperture.

Anyone can make a simple heating device that runs off a 12-volt DC power
source (such as a car battery or a rechargeable cell) and some inexpensive
resistors. Electrical resistance is measured in terms of ohms. After guessti-
mating how much heat a telescope will need to remain dew-free, use the following formula to calculate the corresponding resistance:

$$\text{Resistance (in ohms)} = \frac{\text{volts}^2}{\text{watts}}$$

Let's imagine you want to build a heater for an 8-inch Schmidt-Cassegrain
telescope and that the telescope will be powered from a 12-volt battery. Taking
3 watts as the desired heating value, the formula works out as

$$\text{Resistance} = \frac{(12 \text{ volts})^2}{3 \text{ watts}}$$
$$= 48 \text{ ohms}$$

This docs not mean that a single 48-ohm resistor should be used. To work
efficiently, the heat must be distributed evenly around the circumference of
the optic (in this case, the corrector plate). An 8-inch corrector plate has a
circumference of about 25 inches. Therefore, it would make much more sense
to use either six resistors at 8 ohms each or eight resistors at 6 ohms each.
Looking at the former combination, each resistor will dissipate 0.5 watts of
heat (because 3 watts ÷ 6 resistors = 0.5 watts per resistor).

Solder the resistors in series (that is, end to end) forming a string. Loop
the resistor string around but do not solder the ends together. Instead, solder
the end leads to a polarized plug-in style DC connector, such as Radio Shack
#270-026. The plug's mating half can then be attached to the battery source for
quick connection and disconnection without risking a short circuit. You may
also want to add a switch.

To determine how long the heater will run before draining the battery, use the following version of Ohm's Law:

$$\text{current (in amps)} = \frac{\text{voltage (in volts)}}{\text{resistance (in ohms)}}$$

Continuing with the example above:

$$\text{current} = \frac{12 \text{ volts}}{48 \text{ ohms}}$$
$$= 0.25 \text{ amp}$$

This means that to operate the heater, the battery must deliver 0.25 amp. To use the heater for an extended period, check the battery's amp-hour rating. A 1–amp-hour battery will power this heater for four hours (0.25 amp × 4 hours = 1 amp-hour), a 2–amp-hour battery will run the heater for eight hours, and so on. For further information on batteries, review the discussion in Chapter 6.

The heating element may be affixed directly to the outermost edge of a lens, mirror, or corrector plate (as shown in Figure 7.9), or it may be glued to the inside of a dew cap. Using an adhesive like RTV™ by Loctite (available from any hardware store), carefully glue each resistor in place, then tack down the plug. Be sure to leave enough slack in the plug's wire to plug and unplug the

Figure 7.9   *Resistor ring dew eliminator.*

battery. To prevent the possibility of a short circuit, choose where to affix the resistors wisely. To prevent a short circuit, be certain that the bare resistor leads are isolated from any metal on the telescope. Shorting may be a problem, especially if your telescope has a metal tube, so wrap the bare leads with electrical tape or shrink sleeving before permanently affixing the resistors.

An Easy-to-Make Mounting for Binoculars

In *Touring the Universe through Binoculars* (John Wiley & Sons, 1990), I described several binoculars and binocular mounts constructed by amateur astronomers and amateur telescope makers. Though all generated a lot of reader interest, people seemed most fascinated with the mounting by Jerry Burns. Burns, a consulting engineer from Moorestown, New Jersey, based his design on the flexible-parallelogram binocular mounting system that is so popular today. I first saw his creation at the Stellafane amateur telescope makers' convention in Springfield, Vermont, about a decade ago and immediately became entranced.

My own version of the flexible-parallelogram binocular mount is shown in Figure 7.10. The design is strongly reminiscent of swing-arm desk lamps, which may be raised and lowered to any position while the light itself remains aimed at a constant angle. This scheme has been adapted into the Burns design and is one of its nicest features. Once aimed, the binoculars may be raised and lowered to permit comfortable viewing by observers of differing heights without ever losing their target. How is this possible? Like the desk lamps, the frame forms a hinged parallelogram. Even though the angle between adjacent sides may change, the sides opposite each other remain parallel. Furthermore, the shorter vertical sides always hold the same angle relative to the tripod on which they are mounted. This means that anything attached to one of the shorter legs of the parallelogram will stay aimed in one direction even when the frame is raised up or down. An itemized list of what you need to make the mounting is given in Table 7.6, while Figure 7.11 shows the dimensions I used to create my own version of the flexible-parallelogram binocular mount.

Burns machined all of the mount's components from aluminum in his home workshop. Readers should note, however, that this mount can be made out of a wide variety of rigid materials; hard woods are an especially good choice. For instance, a friend of mine, lacking access to metal-working tools, made a very nice replica using poplar and wood screws.

Begin by cutting or machining the top and bottom beams. For simplicity, make both beams the same size (although a difference in length will not adversely affect the mounting). Although the drawing here specifies the dimensions I used, you may choose to stretch them if you wish.

Drill (and tap, if using metal) three holes along the sides of the beams. Their spacing must be identical on both beams, or the mounting will bind. On the top beam, drill two holes to attach the semicircular brake pad, while on

Figure 7.10 *The author's flexible-parallelogram binocular mount, based on the Burns approach.*

the bottom beam, drill (and tap, if using metal) two holes to attach the counterweight block.

Lay the beams aside for now and begin construction of the two cross tines, the two cross supports, and the two triangular supports. Exact dimensions of the pieces are not critical; what is critical is that the spacing of the near-vertical holes be consistent. Cut the bottom of the cross tines at a 75° angle as noted on the drawing.

The binocular plate must be the same width as the top and bottom beams, and it is typically cut from the same stock. Drill a 0.25-inch hole about a quarter of the way in from one end of the plate. This will be used to attach a standard 90° binocular-to-tripod adapter when the mounting is finally completed. Drill two holes directly opposite each other into the sides of the plate to attach it

Table 7.6   **Binocular Mount—Parts List**

Description	Quantity
Top beam, 15″ long (0.5″ x 1″ aluminum)	1
Bottom beam, 15″ long (0.5″ x 1″)	1
Cross tine, left, 5.625″ long (0.75″ x 1.25″)	1
Cross tine, right, 5.625″ long (0.75″ x 1.25″)	1
Cross support, 3.5″ long (0.125″)	2
Triangular support (0.125″)	2
Mounting plate, 2″ long (0.75″ x 1.8125″)	1
Binocular plate, 4″ long (0.125″ x 1″)	1
Brake pad, 1.5″ radius (0.5″)	2
Thumbscrew, ¼-20 x 1″ long (nylon-tipped, if possible)	2
Counterweight block, 4″ long (1″ x 1″)	1
Counterweight arm, 0.5-inch rod (length to suit)	1
Counterweight (weight to suit; try about 5 pounds at first)	1
Screws (#10-24 machine screws)	As required
Split-ring lock washers (#10)	As required

between the triangular supports as shown. Lastly, drill two holes through the top of the binocular plate for attaching one of the brake pads noted below.

Even when the mounting is perfectly balanced with a counterweight, some additional frictional drag may be required to keep the binoculars from moving too freely. The brake pads themselves are two half-circles cut from either metal or wood that is slightly thinner than the width of the beams. Drill two holes along the flat edge of each brake pad to match the holes in the top beam and the binocular plate. Join the pads to their respective mates using either counterbored or countersunk screws.

For the brakes, use a pair of nylon-tipped screws, available from better hardware stores. Drill and tap corresponding holes in the cross tines and the triangular supports where shown for the screws. When the mount is finally assembled, these screws will press against the brake pads to prevent the mount from moving too freely. Braking pressure may be adjusted by turning one or both of the screws by hand.

The counterweight arm and counterweight block may be combined into one if desired. Though I used a 0.5-inch diameter threaded rod for the counterweight shaft, a straight rod or even a hardwood dowel may be substituted successfully. Just how much counterweighting is required depends on the binoculars and the mounting. Though it can be calculated easily using simple mechanics, it is probably easiest to size the amount of counterweighting required for your binoculars by trial and error after the mount is completed.

When assembling the mount, be sure to insert nylon washers between all pivoting surfaces to prevent rubbing and possible binding. These include the 12 beam-to-support and beam-to-tine connections, as well as the two tine–to–binocular-plate connections. Use the appropriate size of screws and lock wash-

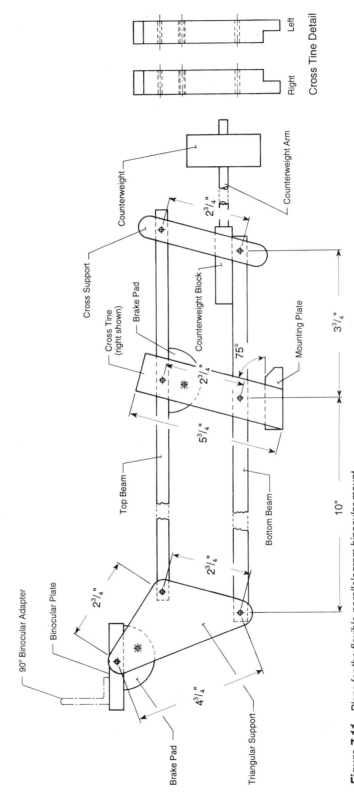

Figure 7.11 Plans for the flexible-parallelogram binocular mount.

ers to put the mounting together. Avoid tightening the screws too much, as the nylon washers may be damaged.

Finally, after the rest of the mounting is completed, measure the inside distance between the cross tines. Cut a piece of wood or metal to this size for the tripod plate. Drill and tap a central ¼-20 hole through the tripod plate for attaching the mount to a camera tripod or other support. (If you are working with wood, drill an oversized hole to accept a ¼-20 tee nut. Be sure to hammer the tee nut into the *top* of the tripod plate, opposite where the tripod will fasten.) Drill four holes (two on either side) through the bottom of the cross tines, and drill and tap corresponding holes into the tripod plate. Complete the assembly by securing the tripod plate to the cross tines.

The flexible-parallelogram binocular mount is elegant in its simplicity, practicality, and execution. Once you try it, you are sure to agree that it is a must for anyone touring the universe through binoculars!

A Simple Camera Tracking Device

Capturing the night sky on film is one of the most popular and potentially most expensive aspects of amateur astronomy. And although a detailed discussion of astrophotographic techniques is beyond the scope of this book (see Chapter 10 for an introduction), I thought it appropriate to include plans for a simple device that greatly reduces the money and equipment needed to take guided exposures of the stars.

Rather than simply snapping a picture, as one might do of a daytime scene, the stars demand exposures many minutes in length. During that time span, the stars appear to move slowly across the sky, a result of the Earth turning on its axis. If a camera is placed on a stationary tripod, the stars will be recorded on the film as trails rather than points. The only way to prevent the stars from trailing is to have the camera move with the stars.

Many expensive equatorial-mounted telescopes come equipped with elaborate clock drive mechanisms to help compensate for the motion of the sky. Does this mean that astrophotography is only for the rich and famous? Happily, the answer is no. Using simple hand tools and about $20 worth of materials, anyone can assemble the camera-tracking device shown in Figure 7.12. Called the *Scotch Mount*, this clever little gadget was first introduced by George Haig of Glasgow, Scotland, (hence the name) in the April 1975 issue of *Sky & Telescope*. The original Scotch mount design permits accurately guided exposures of up to about 30 minutes using any camera with a manually adjustable shutter and lens aperture, without an electric motor. Instead, *you* become the motor! How does the Scotch mount work? Let's begin by considering two facts. First, the Earth turns 360° in 24 hours. That translates to 15° per hour, or 5° every 20 minutes. Next, one of the most common pieces of hardware is the ¼-20 screw (0.25 inch in diameter, with 20 threads per inch). If this screw is turned at a rate of one revolution per minute (1 rpm), then it will travel 1 inch

Figure 7.12 *Scotch mount.*

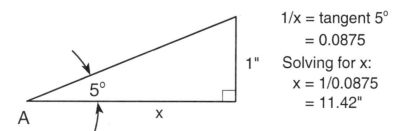

$1/x$ = tangent $5°$
= 0.0875
Solving for x:
x = 1/0.0875
= 11.42"

Figure 7.13 *Twenty-minute triangle.*

in 20 minutes. Putting both of these facts together creates what I call the 20-minute Triangle (Figure 7.13). Using a little trigonometry, it turns out that if a ¼-20 screw is placed exactly 11.42 inches from a pivot (e.g., a hinge) located at **A** and turned at 1 rpm, the triangle's hypotenuse will move at approximately the same rate as the sky. Anything mounted on the hypotenuse (such as a camera) will move right along with the stars.

Occasionally, a company will come along offering for sale a variation of the Scotch mount, either complete or in kit form. Keep in mind, though, that part of the beauty of this little contraption is its ease of construction. Anyone can build one in an afternoon, quite possibly from scrap material found in the garage or basement. Although a Scotch mount can be made out of many different materials ranging from aluminum to plastic, most builders prefer using 0.75-inch plywood. Table 7.7 is a suggested parts list.

Referring to Figure 7.14, begin the project by cutting two 6-by-12-inch pieces of plywood. Next, attach a 6-inch-long piano hinge to one of the short sides of each board. (No, you don't have to buy a piano first; piano hinges are available from most hardware stores.) What you end up with looks like a wood-bound book cover.

From the center of the hinge's pin, accurately measure 11.42 inches (that's 290 mm to us metric-minded individuals, or a hair's breadth under $11^{27}/_{64}$ inches) along the middle of the bottom board and make a mark. Drill a 0.3125-inch hole through the board and hammer in a ¼-20 tee nut (again, available from a local hardware store). Insert it so that the tee nut's flat flange lies flush with the bottom board's top surface. For the mount's all-important tangent screw, thread a 4-inch-long ¼-20 threaded rod through the tee nut. Attach a knob to the end of the tangent screw to act as the clockwheel and add a pointer to the bottom board.

Next, drill another 0.3125-inch hole through the bottom board and hammer in a second ¼-20 tee nut so that its flange is flush with the board's top surface. This will later serve as a tripod mounting socket to secure the finished Scotch mount to a photographic tripod.

Screw two #6 screw eyes into the top and bottom boards on the sides opposite the hinge. When all is assembled and ready to go, wrap a rubber band

Table 7.7   *Scotch Mount: Parts List*

Description	Quantity
Top board, 6″ x 12″ (0.75-inch)	1
Bottom board, 6″ x 12″ (0.75-inch)	1
Piano hinge, 6″ long (must match boards' short sides)	1
Tangent screw, ¼-20 x 4″-long round head screw	1
Tee nut, ¼-20 internal thread	2
Clock wheel (either from an old television or make your own)	1
Screw eyes, #6	2
Rubber band	1
Tripod head (get the smallest, lightest one you can find)	1

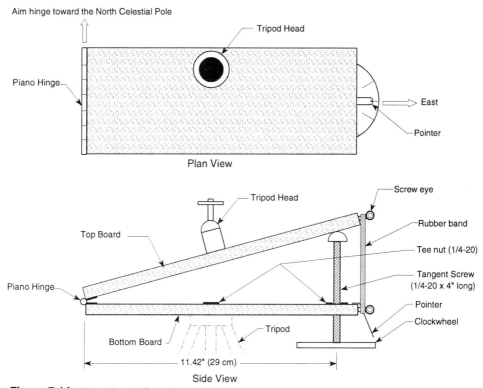

Figure 7.14 Plans for the Scotch mount.

between these two screw eyes to keep the boards in compression. This will prevent the boards from opening suddenly under the weight of the camera.

Finally, drill a 0.25-inch hole through the top board. This last hole is used to attach a miniature adjustable tripod head for mounting the camera itself to the mount. These are available from better-stocked camera stores.

The Scotch mount is as simple to use as it is to make. To begin your evening of photographic escapades, align the mount's polar axis (the hinge pin) to the celestial pole. Although precise polar alignment will be detailed in Chapter 9, the Scotch mount may be adequately aligned by sighting along the hinge toward the North Star. Note that although polar alignment does not have to be perfect, it should be done as carefully as possible. As an aid, I prefer to sight through a 6x30 finder held against the hinge. Make certain that the hinge lies on the left, or western, side of the mounting, or the mount will track in reverse.

Once set, aim the camera toward the chosen area of sky and open the shutter. Be very careful to move only the camera and not the entire mount, or polar alignment will be lost. Turn on the motor by rotating the clockwheel counterclockwise (as seen from the top) in time with the second hand of a watch. Thankfully, the clockwheel does not have to be turned continuously. Rather, the wheel may be turned in segments, which vary according to lens focal length. Table 7.8 shows how the turning frequency changes with different lenses.

Table 7.8   **Scotch Mount—Turn Frequency Compared to Focal Length of Lens**

Lens Focal Length	Turn Frequency	Maximum Exposure[1]
Wide angle (i.e., 35 mm or less)	180° every 30 seconds	35 min.
Normal (i.e., 40 mm to 65 mm)	90° every 15 seconds	20 min.
Telephoto[2] (i.e., 70 mm to 200 mm)	30° every 5 seconds	10 min.

Notes:

1. *Maximum practical exposure assuming perfect polar alignment.*
2. *In general, 200 mm is the longest practical focal length that can be used successfully with the original Scotch mount.*

As you can see from the table, the longer the focal length, the more precise the tracking required to ensure against any trailing of the stars. For example, suppose you wanted to photograph the sky with a 50-mm lens. After the camera was aimed toward the desired field, the clockwheel would have to be turned one-quarter of the way around every 15 seconds over the entire length of exposure. If a 28-mm lens was used instead, then the clockwheel would only have to be turned halfway around every 30 seconds, while a 135-mm telephoto requires the wheel be rotated one-twelfth of a turn every 5 seconds.

The original Haig design reaches a limit in its accuracy after a while. The top board (that is, the board on which the camera is attached) moves in a circle around the hinge, but the tangent screw travels in a straight line. Hence, while the mount holds true for short exposures, accuracy will eventually fall off due to *tangent error*. Maximum exposure is also listed in Table 7.8.

Tangent error can be reduced in many different ways. Some amateurs have successfully incorporated a pivoting tangent screw designed to tilt as the mount opens. Others have used a gear segment with a radius of 11.42 inches instead of a straight bolt, while still others have experimented with a bi-fold top board. This latter approach is well detailed in *Sky & Telescope*'s April 1989 (p. 436) and July 1989 (p. 102) issues. A few amateurs have even incorporated stepper motors, but to me all this would seem to detract from the mount's attractive simplicity. Regardless of the design you choose, the accuracy of the mount is strongly dependent on the precise placement of the tangent screw, length of exposure, and polar alignment. In fact, practice shows that polar alignment plays a much bigger role in tracking error than tangent error does.

Few devices are as simple or as effective as the Scotch mount. For those of us who like to take astrophotos but are forced to live within a limited budget or simply lack the room for a lot of elaborate equipment, the Scotch mount will quickly become an invaluable accessory.

A Theater of the Sun

Gerry Atkinson has enjoyed amateur astronomy for more than 15 years. But as a resident of West Springfield, Massachusetts, he quickly realized that his dark-sky observing time would be limited to weekends only, when he could escape the lights of the city for the dark countryside. Not wanting to be a weekend-only astronomer, Atkinson became interested in observing the brighter members of our solar system, an activity that can be enjoyed equally by urban and rural astronomers. To pursue his interest, Atkinson purchased a small refracting telescope, and he soon became fascinated with monitoring the ever-changing face of the Sun.

Although he tried full-aperture solar filters, Atkinson found them lacking for his daily program of recording and drawing sunspots. Instead, he decided to make an accessory popular among many solar observers: an enclosed projection box, or what I call a Theater of the Sun. Using his 3.1-inch f/12 refractor and an 18-mm Kellner eyepiece, Atkinson's Solar Theater, seen in Figure 7.15, projects an image of the Sun's disk 9 inches across. To view the image, the observer looks through the 3.5-by-8.5-inch peep hole on the side of the theater. Table 7.9 offers some suggested materials for constructing your own.

Figure 7.15   *Gerry Atkinson's solar theater.*

Figure 7.16 shows plans for Atkinson's version of a solar projection box, but unless you are using the same size of telescope and eyepiece, the dimensions of your solar theater probably will be different. To find out just how different, set up your telescope on the next sunny day and project the Sun through it onto a piece of paper. (If you are unfamiliar with how to do this safely, read the section on observing the Sun in Chapter 10 before trying!) With the Sun shining through the telescope, measure the distance behind the eyepiece required to produce an image of at least 6 inches across (preferably

Table 7.9 *Solar Theater*

Description	Quantity
Foam board, 0.5″ thick 21″ wide (cut exact length to suit)	5
Balsa wood, trailing-edge, 1″ wide x 0.25″-thick, approx. 15″ long	12
Wooden block, 1.375″ thick, 4.25″ x 4.25″	1
Plywood, 0.125″ thick, 6″ x 7.25″	1
Right-angle bracket	6
Hinge	1
Felt, adhesive-backed	As required
White glue	As required
Clear dope (paint)	As required
Miscellaneous hardware	As required

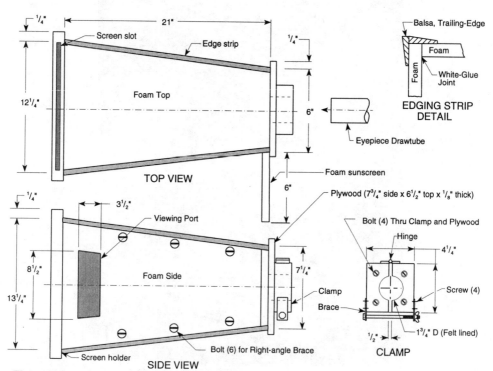

Figure 7.16 *Plans for the Atkinson solar theater.*

larger). Keep the projection distance as short as possible by using the highest-power eyepiece that will cast a full-disk image of the Sun. This is an important design consideration, because added length means added weight and the potential for balancing problems. (Atkinson's finished box weighs just 1.5 pounds.)

For the walls of the box, use 0.5-inch-thick foam board, readily available from hobby and craft stores. The four walls may be cut to size using a razor-style modeling knife. Before assembly, apply adhesive-backed, black felt to the inner face of each wall. The felt decreases stray light from bouncing off the walls and lowering image contrast.

Glue the four walls together using a white glue like Franklin Titebond.™ Do not use acetone-based adhesives such as modeling cement as they will dissolve the foam. If in doubt, test the glue on a scrap piece of foam first. To help reinforce the long sides of the box, Atkinson suggests internally mounted aluminum right-angle brackets, three on each side, painted flat black.

Strips of 1-inch-wide-by-0.25-inch-thick balsa wood along each outside edge reinforce the edges of the box. Atkinson selected model-airplane tapered trailing-edge stock from a local hobby shop for a more sculpted look. Cover these long balsa strips with model-airplane silk and brush on clear dope (paint) finish to fill the weave and strengthen the balsa.

Pieces of balsa also frame the screen end of the box. A slot on the top of the box frame allows the screen to be inserted, while grooves on the inside of the other three frame strips ensure that the screen will slide down evenly and be held into place. The screen itself is made from heavyweight Bristol board, available from art-supply stores. Be sure to get the smooth or plate, surface, not the matte.

Cut a piece of 0.125-inch plywood 6.5 by 7.75 inches and drill a hole in its center at least a half-inch larger than the outer diameter of your telescope's focusing tube (again, your numbers may vary). Attach the plywood to the telescope end of the foam box using white glue.

A second piece of wood, preferably a hardwood such as 1.375-inch-thick oak or poplar, serves to clamp the theater onto the telescope. (Atkinson used the base of an old golf trophy for his!) Cut the hardwood to about 4.25 inches square. Measure the diameter of your focusing mount's drawtube and drill an equal-sized hole in the center of the wooden block. Split the block in half as shown. Drill a hole through one side of the block to accept a clamping bolt and hinge the other side. Line the inside of the hole with adhesive-backed felt to prevent the clamp from marring the telescope's finish. Finally, attach the clamp to the 0.125-inch plywood using some wood glue and four screws.

To mount the theater onto your telescope, slip the clamp over the draw-tube and tighten the bolt. Select and slide an eyepiece into place and you are ready to go. Atkinson emphasizes the importance of attaching a side-mounted sunscreen to the telescope end of the box to shadow the observer's head when viewing through the peep hole. The screen may be easily made from leftover foam board and attached to the box using Velcro™-style hook-and-loop material. If the telescope is too greatly out of balance when the theater is attached, either slide the telescope forward in its mounting rings (if so equipped) or add some counterweight to the objective end of the tube. By loosening the wooden clamp, the box is removed for storage or nighttime stargazing.

A Radio-Controlled Eyepiece Focuser

A sharp focus is crucial for both high-magnification observations and through-the-telescope astrophotography. One of the latest innovations to help make this exacting task easier is the electric focusing device, outlined in the previous chapter. Although beneficial, these gadgets have wires that are easily entangled, especially at night, as well as prices that many amateurs find prohibitive.

One amateur who decided to address both obstacles by coming up with his own solution is Carl Lancaster of Riverside, Connecticut. His result: the wireless radio-controlled focusing mount seen in Figure 7.17. The design is based around a reversible 1 rpm DC motor and an inexpensive radio-controlled car, both available from many hobby or electronic-supply shops. Although Lancaster designed his mechanism for a Meade 2-inch focuser on his 10-inch Newtonian, it is versatile enough to be adapted to just about any kind of telescope with a little ingenuity. Table 7.10 lists the ingredients needed for the radio-controlled focuser. One of each is required.

For the electronics, Lancaster used the Racer 27 radio-controlled car from Radio Shack, catalogue number 60-4092. This particular car was chosen for

Figure 7.17 *Carl Lancaster's radio-controlled focuser.*

Table 7.10 **Radio-Controlled Eyepiece Focuser—Parts List**

Description	Part Number
Radio-controlled car	Radio Shack #60-4092, ''Racer 27''
Motor	Edmund Scientific #Y41,327
Light-emitting diode, red	Radio Shack #276-068
Potentiometer, 5k-ohm	Radio Shack #271-1714
Switch (S1), DPDT	Radio Shack #275-614
Resistor, 1 k-ohm, 0.25-watt	Radio Shack #271-1321
Resistor, 470-ohm, 0.5-watt	Radio Shack #271-019
Battery holder, four AA	Radio Shack #270-391
Battery holder, one 9-volt	Radio Shack #270-325
Battery clip, 9-volt	Radio Shack #270-326
Box, plastic	Radio Shack #270-627
Wire, antenna (22 gauge)	Radio Shack #278-1224
Shaft collar, clamp type	Small Parts, Inc., part number K-SCX-8
Shaft, aluminum	Small Parts, Inc., part number K-ZRA-8-6
Foam rubber	As needed
Teflon tape	As needed
Miscellaneous hardware	As needed

its low price, its battery supply (four commonly available AA 1.5-volt batteries and two 9-volt batteries) and its hand-control unit. Unlike some radio-controlled cars, whose motors run on forever once activated, this particular model stops when the hand-operated speed control is centered. (Note: Any radio-controlled car with separate forward and reverse switches can be used.) Take the body of the car off and remove the receiver circuit mounted in the chassis. Be sure to note and label the battery terminals and antenna connection.

To house the receiver circuit, the 9-volt battery and the four 1.5-volt AA batteries required to power the motor, Lancaster selected a 2-by-6.25-inch plastic project box from Radio Shack. Mounted on top of the box are an on-off double-pole, double-throw (DPDT) toggle switch for power to the motor, a red LED to show when power is on, and a small potentiometer to help control the motor's speed. As Figure 7.18 shows, all of these items are electrically connected between the 9-volt battery and the receiver/motor. The 5k-ohm potentiometer must be connected in parallel with the 470-ohm resistor and together in series between the receiver circuit and the motor. Before closing the box, pack it with foam rubber to prevent its components from shifting and possibly shorting out. Solder a 12-inch piece of 22-gauge stranded wire to the receiver's antenna terminal. Drill a small hole in the bottom of the box and thread the wire through. To work properly, the wire antenna should be stretched out as straight as possible. (Brave souls may choose to follow Lancaster's method of drilling a hole through both the box and the telescope tube and running the wire along the inside of the instrument.) Finally, attach the box to the tube near the focusing mount using either screws or hook-and-loop material.

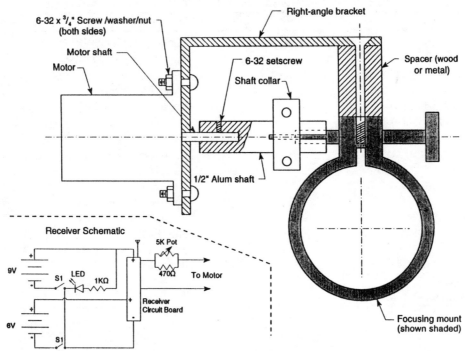

Figure 7.18   *Plans and schematic for the Lancaster radio-controlled focuser.*

The most difficult part of the project is adapting the motor to the focusing mount. For this, Lancaster had to remove one of the focuser's finger knobs. On most eyepiece mounts, the knob can be unscrewed easily. A homemade slip clutch connects between the motor's shaft and the focusing mount. This is made from a small piece of aluminum shaft and a clamp-type shaft collar. At one end of the aluminum shaft, drill a centralized 0.125-inch hole to accept the motor's axle; at the opposite end drill an oversized hole to fit your focusing mount's spindle [in the case of the Meade focuser, the spindle is 21/64 inch (0.328 inch) in diameter, so Lancaster used a 0.375-inch drill.] Using a hacksaw or a band saw, cut two perpendicular slits across the diameter of the focuser end of the aluminum shaft. Make the slits the same length as the depth of the hole just drilled. Next, drill and tap a small hole on the motor end of the aluminum shaft for a setscrew (used to secure the motor's axle). Before attaching the shaft to the eyepiece focuser, wrap a thin piece of Teflon™ tape around the focuser's spindle to create a slip clutch. This will permit manual adjustment of the focuser, if ever the need arises. Teflon™ tape is available from most hardware and plumbing suppliers.

Finally, make a bracket to attach the motor to the telescope. You may use one similar to that shown in Figure 7.18 or create your own. Lancaster used a 90° bracket, available from any hardware store, and a wooden block as a spacer. The two are held together with a couple of wood screws. To attach the block to the focuser, he removed the two short screws used to hold the rack-and-pinion gearing's pressure plate in place and substituted two longer screws. The new screws pass through the block and into the focuser's housing. Tighten the screws just enough to hold the whole assembly together; the slip clutch will provide enough friction to keep the focuser from spinning too freely.

(A postscript to the Lancaster radio-controlled focuser: Besides adding the radio-control mechanism, Lancaster also modified the focusing mount heavily, as you can see from the photograph. He points out that this is not necessary for the focuser to work. In addition, he also added a radio-controlled dual-axis drive corrector, but that will have to wait for another book!)

A Backyard Observatory

I've saved the biggest project for last. For most amateurs, owning their own observatory is only a dream. But for George Viscome of Lake Placid, New York, this dream became a reality after some well-thought-out planning, some hard work, and about $550 worth of materials.

Viscome's observatory (Figure 7.19) was built around his massive yoke-mounted 14.5-inch f/6 Newtonian reflector. The *yoke mount* is another variation on the equatorial mounting; although not often used by amateurs, probably due to its size and weight, the yoke mount is known for its stability. A detailed write-up of Viscome's telescope and mounting was featured in issue 42 of *Telescope Making* magazine.

After investigating different observatory designs including the popular roll-off roof and traditional dome approaches, he chose a unique system of hinging the roof in sections. The building is oriented with the roof sections sloping east and west and the peaks toward the north and south. Each section

Figure 7.19 *George Viscome's observatory. Photo courtesy of George Viscome.*

is hinged independently, allowing the versatility of opening all or only some of the roof panels at a time. For a full-sky view, all four roof panels plus the south gable end are folded down, blossoming like the petals of a flower. Or, if only a small part of the sky is to be observed, then only the panels in that direction need to be opened. In this latter case, the closed sections can serve as a block, useful in areas of light pollution or wind. In Viscome's design, the northern part of the sky below the North Celestial Pole is blocked by the rigid north gable of the observatory. He opted for this minor inconvenience in favor of a full-height entrance door in the north wall. If desired, the north gable can be hinged like the south, but the door into the observatory will have to be down-sized to only about 3 feet high, accordingly.

Where does a project like this begin? Before the first nail can be driven, follow the real-estate industry's Rule of the Three Ls: location, location, location! Find a spot where the telescope will have the best view in all directions. Then take a trip to Town Hall to check your municipality's building code. Are there any restrictions on building a garden shed on private property? Fill out the required paperwork and submit it to the proper office.

Next, draw up plans for your observatory similar to those shown in Figure 7.20. Decide just how large you want the observatory to be. The dimensions

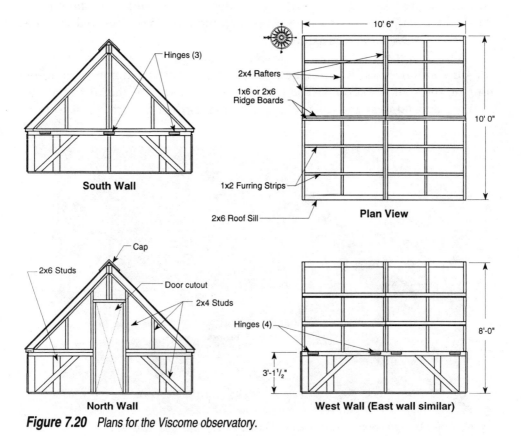

Figure 7.20 *Plans for the Viscome observatory.*

shown here are suggestions only and are by no means set in stone. If your telescope does not require that much room, then shrink the numbers as you see fit (but not too much). If more room is required, let the observatory grow as needed. Should you foresee the possibility of someday upgrading your scope to something larger, take this into consideration when sizing up the building. The important thing is to make the observatory fit your needs, not the other way around.

Begin construction by staking out the area, making sure that everything remains square throughout the project. In the four corners, sink the footings that the building will be anchored to. Viscome chose 4-foot-long, 5-by-5-inch pressure-treated stockade fence posts, though you may prefer to use concrete piers.

Construct the 3-foot-high walls from 2-by-4-inch wood framing (use 4-by-4s in the four corners) and 0.5-inch plywood sheathing. Frame each wall separately, then add the sheathing as shown. As a final touch, attach a 2-by-4 along the top outer ridge of the east and west walls to help support the weight of the roof and provide an overhang (drip-edge). Throughout construction, Viscome recommends using long sheetrock screws, instead of common nails, driven by an electric screwdriver or drill. With the help of a friend, raise each wall into place, bolt or nail them together, and secure them to the sunken corner posts with metal strapping.

Next, build the north and south triangular gables from 2-by-4 framing and 0.25-inch plywood sheathing. Because the north pier of his telescope's yoke-style equatorial mount stands a bit over 7 feet tall, Viscome sloped the roof at a 45° angle. As an added benefit for those who live in colder climates, the roof's steep pitch also helps snow to slide off quickly.

Viscome chose to divide the roof into four sections, to lessen the weight of each section and thus make it easier to open the observatory for use. If your plans call for smaller dimensions, then you might be able to get away with only two roof sections.

Construct the roof panel framing using 2-by-4s, 2-by-6s, and 1-by-2-inch furring strips. Each section is secured to the observatory wall with two large hinges. Choose the largest hinges you can find in your local hardware store, but don't be surprised if the selection is rather limited and quite pricey. To get around this predicament, Viscome elected to make his own hinges. He began with two 4" × 8" × 0.25" steel plates and an 8-inch long, 0.5-inch diameter pipe for each hinge. At a local welding shop, the plates were laid on top of each other, and the pipe was placed on one side. The ends of the pipe were welded to the top plate, while the center was welded to the bottom, with none of the welds overlapping each other. Back home, he cut the pipe between the welds, inserted a tight-fitting threaded rod into the pipe, and tapped it through with a hammer. To keep the rod from sliding out, he had the rod tack-welded to the pipe at both ends. (An alternative to the tack-welding might be to use a longer threaded rod and two pair of hex nuts, one pair for each end. Insert the rod so that equal lengths protrude out either end of the pipe. Before screwing on the nuts, apply a little Loctite™ thread adhesive to the rod. Screw one nut onto either end and bolt it down quite tightly. Then screw on the second nut, tightening it against the first to act as a jam nut.) Voilà, a homemade hinge!

With two hinges attached to the bottom of each roof panel—and with the help of at least two other people—lift each roof frame into place on the observatory. Fasten the triangular gables first, followed by the rectangular panels. Be sure to use temporary bracing to hold each panel in position as the others are lifted into place. A chain on a hook, with one end attached to the east wall section, is used to hold the south gable up as the building is being opened or closed. It is unhooked when the gable is to be lowered.

The final step in the observatory's assembly is adding the roof itself. For this, Viscome chose corrugated aluminum roofing sheets. These prove to be much lighter than steel or wood and require practically zero maintenance. To ensure that the building will remain weatherproof, the southern aluminum panels overlap the northern by 4.5 inches. The roof ridge is attached to the western panels only and lifts off with them when the observatory is opened. Five large bolts and wing nuts slip through holes in the roof framing from the inside to lock the roof closed when not in use.

Large metal screw eyes are fastened into the frame near the top of each roof panel. Viscome uses a metal hook at the end of a 10-foot pole to hook the screw eye and pull (*carefully!*) each section open. Because the observatory is usually opened and closed after dark, Viscome found it helpful to tape a small penlight at the hook end of the pole to help light the way. As designed, a maximum of about 25 pounds of support or lift is required when opening or closing a roof section. Top off the observatory with a coat of paint and some decorative trim, and you are ready for the dedication ceremony.

8

Till Death Do You Part

If this were a book about how to run a business, then this chapter might be entitled "Standard Operating Procedures" because it details methods and strategies that, although not part of the business's primary product or service, help to enhance the company's operational efficiency. This chapter contains lots of little tidbits to help you get the most out of your telescope. It addresses a wide variety of topics, ranging from care and maintenance to traveling with a telescope.

Love Thy Telescope As Thyself

Unlike so many other products in our throw-away society, in which planned obsolescence seems the rule, telescopes are designed to outlast their owners. They require very little care and attention, cost nothing to keep, and eat very little. With a little common sense on the part of its master, a telescope will return a lifetime of fascination and adventure. But if neglected or abused, a telescope may not make it to the next New Moon. Are you a telescope abuser?

Storing Your Telescope

Nothing affects a telescope's life span more than how and where it is stored when *not* in use. *How* to store a telescope will be addressed farther along in this chapter, but first let's consider *where* the best places are to keep an idle instrument. The choice should be based on a number of different factors that, at first glance, might appear unrelated. A good storage place should be dry, dust-free, secure, and large enough to get the telescope in and out easily. Ideally, a telescope should always be kept at or near the outside ambient air temperature; doing so reduces the cool-down time required whenever the telescope is first set up at night. The quicker the cool-down time, the sooner the telescope will be ready to use.

Without a doubt, the best place to keep a telescope is in an observatory. It offers a controlled environment and easy access to the night sky. Of course, not everyone can afford to build a dedicated observatory, nor is an observatory always warranted. Clearly, an observatory is pointless if the nearest good observing site is an hour's drive away. In cases such as these, a few compromises must be struck.

If an observatory is not in the stars (so to speak), other good places to store telescopes include a vented, walled-off corner of an unheated garage or a wooden tool shed. I keep my telescopes in a corner of my garage that is completely walled off to protect the optics from the inevitable dust and dirt that accumulates in garages as well as from any automotive exhaust that could damage delicate optical surfaces. A pair of louvered vents were installed in the outside wall of the garage to let air move freely in and out of the telescope room, reducing the risk of mildew. Wooden tool or garden sheds share many of the advantages of observatories and garages for telescope storage, but again I recommend installing one or two louvered vents in the shed's walls for air circulation. Metal sheds are not as good, as they can build up a lot of interior heat on sunny days.

Many amateurs choose to store their equipment in basements, which are certainly secure enough and large enough to qualify. Furthermore, they offer easy access provided there is a door leading directly to the outside. Their cool temperatures also keep the optics closer to that of the outside air. While all of these considerations weigh in their favor, most basements fail when it comes to being dry and free of dust. If a basement is your only alternative, invest in a dehumidifier. Clothes closets, another favorite place to hide smaller telescopes, also fall short because clothes act as dust magnets. Remember, unless a spot meets all of the criteria, continue the search.

Regardless of where a telescope is kept, seal the optics from dust and other pollutants when it is not in use. Usually this is simply a matter of putting a dust cap over the front of the tube. Most manufacturers supply their telescopes with a custom-fit dust cap just for this purpose. Use it diligently. If the telescope did not come with a dust cap or if it has been lost over time, then a plastic shower cap makes a great substitute. If you are into a more sophisticated look, some of the companies listed in Appendix A sell dust caps made of rubber in a wide variety of sizes. Further, if the telescope or binoculars came with a case, use it. Not only will a case add a second seal against dust, but it also will protect the instrument against any accidental knocks or bumps.

A dark, damp telescope tube is the perfect breeding ground for mold and mildew. To avoid the risk of turning your telescope into an expensive petri dish, be sure that all of its parts are dry before sealing it up for the night. Tilt the tube horizontally to prevent water from puddling on the objective lens, primary mirror, or corrector plate. No matter how careful you are, optics are bound to become contaminated with dust eventually. A moderate amount of dust has, surprisingly, little effect on a telescope's performance. But if there's a great deal of it, or if the optics have become coated with a film or mildew, the observer sees dimmer, hazier views that lack clarity.

Havng a telescope that is a little dusty is cause not for panic, but for cautious action. "Clean a telescope?!" you ask. "Isn't that a job that should be left to professionals?" Not at all. While to the uninitiated it might seem a formidable task, it's actually not, as you are about to discover.

An optic should be cleaned only when dust or stains are apparent to the eye; otherwise, leave well enough alone. *Never* clean a telescope lens or mirror just for the sake of cleaning it, because every time an optic is touched, there is always the risk of damaging it. Remember this rule: If it ain't broke, don't fix it.

The methods described here are for cleaning *outer* optical surfaces only. Unless you really know what you are doing, I strongly urge against dismantling sealed telescopes (such as refractors and catadioptrics), binoculars, and eyepieces. Dirt and dust will never enter a sealed tube if it is properly stored and protected. Nevertheless, if an interior lens or mirror surface in a sealed telescope becomes tainted by film or mildew, it should be disassembled and cleaned only by a qualified professional. Contact the instrument's manufacturer for recommendations on how to do so. If you don't, you may discover that the telescope is much easier to take apart than it is to put back together! I know someone who once decided to take apart his 4-inch achromatic objective lens to give it a thorough cleaning. All was going well until it came time to put the whole thing back together. Seems he tightened a retaining ring in the lens holder just a little too much and . . . CRACK! The edges of the crown element fractured. Don't make the same mistake, y'hear?

Never start to clean a telescope if time is short. For instance, it is not time to decide that your telescope is absolutely filthy as the Sun is setting in a crystal-clear sky. Instead, check the optics well beforehand so there are no surprises. To help you along the way, I have divided the cleaning process into two parts, one for lenses and corrector plates and another for mirrors.

Cleaning Lenses and Corrector Plates

Begin the cleaning process by removing all abrasive particles that have found their way onto the lens or corrector plate. This does NOT mean blowing across the lens with your mouth; you'll only spit all over it. Instead, use either a soft camel's-hair brush (Figure 8.1a) or a can of compressed air, both available from photographic supply stores. Some brushes come with air bulbs to allow blowing and sweeping at the same time. If the brush is your choice, lightly whisk the surface of the lens in one direction only, flicking the brush free of any accumulated dust particles at the end of each stroke.

Many amateurs prefer to use a can of compressed air instead of a brush for dusting optical surfaces, as this way the lens is never physically touched.

Figure 8.1 *The proper way to clean a lens or corrector plate: (a) Lightly brush the surface with a soft camel's-hair brush, then (b) gently wipe the lens with lens-cleaning solution.*

Hold the can perfectly upright, with the nozzle away from the lens *at least* as far as recommended by the manufacturer. If the can is too close or tilted, there is a chance that some of the spray propellant will strike the glass surface and stain it. Also, it is best to use several short spurts of air instead of one long gust.

With the dust removed, use a gentle cleaning solution for fingerprints, skin oils, stains, and other residue. Don't use a window glass–cleaning spray or other household cleaners. They could damage the lens' delicate coatings beyond repair. Photographic lens-cleaning fluid can be used but can occasionally leave a filmy residue behind. One of the best lens-cleaning solutions can be brewed right at home. In a clean container, mix three cups of distilled water with a half cup of pure-grade isopropyl alcohol. Add two or three drops of a mild liquid dishwashing soap (you know, the kind that claims it will not chap hands after repeated use).

Dampen a piece of sterile surgical cotton (not artificial so-called cotton balls) or lens tissue with the solution. Do not use bathroom tissue or facial tissue; they are impregnated with perfumes and dyes. Squeeze the cotton or lens tissue until it is only damp, not dripping, and gently blot the lens (Figure 8.1b). Avoid the urge to use a little elbow grease to get out a stubborn stain. The only pressure should be from the weight of the cotton ball or lens tissue. Once done, use a dry piece of lens tissue or cotton to blot up any moisture.

The operation for cleaning the corrector plate of a catadioptric telescope is pretty much the same as just detailed. The only difference is in the blotting direction; in this case, begin with the damp cotton or tissue at the secondary mirror holder in the center of the corrector and move out toward the edge. Follow a spokelike pattern around the plate, using a new piece of cotton or tissue with each pass. As you stroke the glass, turn the cotton or tissue in a backward-rolling motion to carry any grit up and away from the surface before it has a chance to be rubbed against the optical surface. Overlap the strokes until the entire surface is clean. Again, gently blot dry.

Mirror Cleaning

Cleaning a mirror is much like cleaning a lens in that special care must be exercised to ensure against damaging the fine optical surface. In fact, a mirror is even more susceptible to scratches than a lens. The mirror's thin aluminized coating is extremely soft, especially when compared to abrasive dirt, and so is easily gouged. This is not meant to scare you out of cleaning your mirror if it really needs it but only to heighten your awareness.

The operation for cleaning a telescope's primary or secondary mirror requires it to be removed from the telescope and the cell that holds it in place. Consult your owner's manual for more details on mirror removal. With the naked mirror lying on a table, blow compressed air across its surface to rid it of any large dust and dirt particles. (Remember, keep the can of compressed air vertical.) Do not use a brush for this step, as even the softest bristles can damage a mirror's coating.

Next, inspect the mirror's coating for pinholes and scratches. A good aluminum coating should last at least ten years, even longer if the mirror has been well cared for. To check its condition, hold the mirror, its reflective side toward you, in front of a bright light. It is not unusual to see a faint bluish image of the light source through the mirror if the source is especially bright, but its image should appear the same across the entire mirror. If not, there may be thin, uneven spots in the coating. Any scratches or pinholes in the coating also will become immediately obvious as well. A few small scratches or pinholes, while not desirable, can be lived with. But if scratches or pinholes abound or if an uneven coating is detected, then the mirror should be sent out for re-aluminizing. Appendix A lists several companies that re-aluminize mirrors; consult any or all of them for prices and shipping details. It is also a good idea to have the secondary mirror re-aluminized at the same time, because it probably suffers from the same problems as the primary. In fact, most re-aluminizing companies will work on the secondary at no additional cost.

If the coating is acceptable, bring the mirror to a sink. Be sure to clean the sink first and lay a folded towel in the sink as a cushion, just in case—OOPS—the mirror slips. Run lukewarm tap water across the mirror's reflective surface. This should lift off any stubborn dirt particles that refused to dislodge themselves under the compressed air. End with a rinse of distilled water. Tilt the mirror on its side next to the sink on a soft, dry towel and let the water drain off the surface. Examine the mirror carefully. Is it clean? If so, quit.

If you want to go further, thoroughly clean the sink to remove any gritty particles that may not have washed down the drain. Next, fill the sink with enough tepid tap water to immerse the mirror fully and add to it a few drops of gentle liquid dish soap (the same as used in the lens-cleaning solution). As shown in Figure 8.2a, carefully lower the mirror into the soapy water and let it sit for a minute or two. With a big, clean wad of surgical cotton, sweep across the mirror's surface ever so gently with the same backward-rolling motion, being careful not to bear down. Now is not the time to act macho. After you've rolled the cotton a half-turn backward, discard it and use a new piece. If stains still exist after this step is completed, let the mirror soak in the water for five to ten minutes and repeat the sweeping with more new cotton.

With the surface cleaned to your satisfaction, drain the sink once again. Run tepid tap water on the mirror for a while to rinse away all soap. Then turn off the tap and pour room-temperature distilled water across the surface for a final rinse.

Finally, rest the mirror on edge on a towel where it may be left to air dry in safety. I usually rest it against a pillow on my bed (Figure 8.2b). Tilt the mirror at a fairly steep angle (greater than 45°), its edge resting on the soft towel, to let any remaining water droplets roll off without leaving spots. Close the door behind you to prevent any nonastronomers from touching the mirror. When the mirror is completely dry, reassemble the telescope. Recollimate the optics using the procedure described later in this chapter, and you're done!

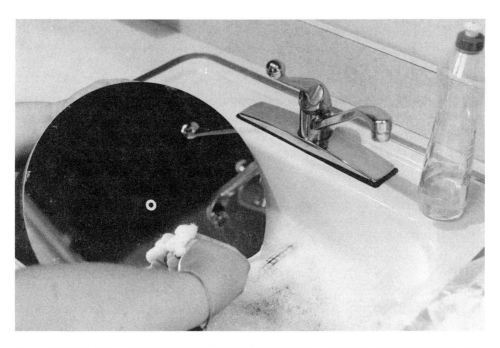

Figure 8.2 *The proper way to clean a mirror: (a) After using compressed air or a camel's-hair brush to remove any loose contaminants, wash the mirror in the sink using a gentle liquid dishwashing detergent. (b) After rinsing the mirror, tilt the mirror nearly vertically to prevent water from puddling.*

Other Tips

Other telescope parts require occasional attention as well. For instance, rack-and-pinion focusing mounts will sometimes bind if not lubricated occasionally. To prevent this from happening, spray a little silicone-based lubricant (such as WD-40™) on the mount's small driving gear. To do this, remove the screws (typically two) that hold the small plate onto the side of the focuser housing, taking care not to lose anything along the way. With the plate out of the way, take a look inside at the small pinion gear that meshes with and drives the focusing mount's tube up and down. Squirt a little (*very* little) lubricant on the pinion teeth, reaffix the cover plate, and wipe off any drips as required.

If your telescope has a cardboard tube, make sure its ends do not fray and begin to unravel. Tilt the tube horizontally, remove or otherwise protect the optics, and brush on a thin layer of varnish to seal the ends. Checking the tube's condition once every now and then can add years to its life.

If a metal telescope mounting begins to move roughly or starts to bind, put a drop or two of lubricating oil on the axes' bearing points. This will keep the telescope moving freely and evenly, rather than binding and grabbing. Some manufacturers recommend this be done at specific intervals, while others make no mention of it at all. If nothing is said in the owner's manual, then do it once a year or so.

The typical wood-Formica™-Teflon™ construction of Dobsonian mounts requires little in the way of maintenance. However, if your Dobsonian does not move freely in altitude or azimuth, take the mount apart and spray a little furniture polish on the contact surfaces. Buff the polish as you would a coffee table until it shines with a luster. Put the mounting back together, and take it for a test spin. The difference should be immediately noticeable.

Some clock drives also need an occasional check to keep them happy. Carefully remove the drive's protective cover plate or housing and put a little thin grease between the two meshing gears. While the drive is open, put a drop or two of thin oil on the motor's shaft as well. Finally, reassemble the drive and turn it on. Listen for any noises. Most clock drives hum as they slowly turn. If unusually loud, angry, grinding noises are coming from it, turn the drive off immediately and contact the manufacturer for recommendations.

Get It Straight!

Have you ever gotten as frustrated with your telescope as the character depicted in Figure 8.3? (Hopefully you stopped short of whacking it with an axe.) There is nothing more vexing or disappointing to an amateur astronomer than to own a telescope that doesn't work as expected. "That lousy company," you think to yourself. "I spend over a thousand dollars on a telescope, and what do I get? A big, expensive, piece of junk!" All too often we are quick to blame the poor images produced by our telescopes on faulty optics and poor workmanship. Yet this is not necessarily the case. Although some telescopes are truly

Figure 8.3 *From Russell W. Porter* by Berton Willard (Bond Wheelwright Company, 1976). Reprinted with kind permission of the author.

lacking in quality, most work acceptably *if they are in proper tune*. Even with the finest optics, a telescope will show nothing but blurry, ill-formed images if those optics are not in proper alignment. Correct optical alignment, or collimation, is a must if we expect our telescopes to work to perfection.

What exactly is meant by *collimation*? A telescope is said to be collimated if all of its optics are properly aligned to one another. Refractors, for instance, are collimated when the objective lens and eyepiece are perpendicular to a line connecting their centers (of course, this is assuming the lack of a star diagonal).

A Newtonian reflector is collimated when the optical axis of the primary mirror passes through the centers of the diagonal mirror and the eyepiece. Furthermore, the eyepiece focusing mount must be at a right angle to the optical axis. For our purposes here, all Cassegrain reflectors and all Cassegrain-based catadioptric telescopes will be lumped together. To be precisely collimated, their primary and secondary mirrors must be parallel to one another, with the center of the secondary lying on the optical axis of the primary. The eyepiece holder must be perpendicular to the mirrors, with the telescope's optical axis passing through its center. Refer to Figure 2.4 for clarification if these descriptions are unfamiliar.

The optics in some types of telescopes are apt to go out of alignment more easily than others. For example, it is rare to find a refractor that is not properly collimated unless its tube is bent or warped, usually the result of abuse and mishandling. On the other hand, most Schmidt-Cassegrain telescopes will experience some collimation difficulties in their lives, whereas Newtonian and Cassegrain reflectors are notoriously easy to knock out of alignment.

It's easy to tell if a telescope is collimated properly. On the next clear night, take your telescope outside and center it on a bright star using high magnification. Place the star slightly out of focus and take note of the diffraction rings that result. If the telescope is properly collimated, then the rings should appear concentric around the star's center. If, however, the rings appear oval or lopsided—and remain oriented in the same direction when you turn the focuser both inside and outside of best focus—then the instrument is in need of adjustment.

Collimating a Newtonian

In general, the faster the telescope (that is, the lower the f-number), the more critical its collimation. Though slower Newtonians may be collimated adequately by eye, I have chosen to lump them together with RFT scopes in the interest of brevity. Throughout the discussion, I will make it clear when a step is required for both and when it is for RFTs only.

Begin by purchasing or making a *sight tube*, which is essentially a long, empty tube with a pair of crosshairs at one end and a precisely centered pinhole at the other. The outer diameter of the sight tube must match the inside diameter of the telescope's eyepiece holder (typically, 1.25 inches). Both Tectron and AstroSystems make excellent sight tubes. Or you can make one yourself.

If you choose to make your own, take a trip down to the local plumbing supply house and pick up a 1.25-inch brass drainpipe used in bathroom sinks. The drainpipe necks down from a 1.25-inch inside diameter to a 1.25-inch outside diameter. Check several samples, selecting the roundest and straightest of the lot. Once back home, check to see that the tube fits snugly into the eyepiece holder. If it is too loose, wrap some masking tape around it, but make sure that the tape does not overlap. To position the crosshairs, place a ruler exactly over the center of the smaller end and make two corresponding marks on the tube's outer wall. Flip the ruler 90° and repeat the procedure. Now, using a thin hacksaw blade across the tube, cut four 0.0625-inch slots at the marks, creating an X pattern. Lay a short piece of white thread across one pair of slots. Place a spot of epoxy or other household cement on one slot and tape the thread to hold it in place. Stretch the thread across the tube's center and glue the other end into its slot. Once again, use tape to hold the thread taut until the glue dries. Repeat the procedure for the second crosshair. When the glue is dry, remove the tape and trim away excess thread. The pinhole is easily made using the plastic lid from a Kodak 35-mm film can. Drill a tiny hole (no larger than 0.0625-inch across) into the lid's central dimple. Put the lid over the large end of the tube and glue it in place. That's it!

As an aid for collimation, many Newtonians (especially those that are f/6 or less) come with small, black dots right in the middle of their primary mirrors. (Though at first it might appear that these dots will impede mirror performance, in reality they have no effect because they lie in the shadow of the diagonal mirror.) If your telescope mirror does not have a centered dot, now is the time to put one there. Remove the mirror from the telescope and place it on a large piece of tissue paper. Trace its diameter with a pencil and then move the mirror to a safe place.

Now to find the *exact* center of the mirror tracing: Set a drawing compass to half of the mirror's diameter. Place the point of the compass along the tracing's circumference and strike an arc in the center of the tracing. Move the compass to any other point along the circumference and strike another arc. Where the arcs cross is the tracing's center. (If your mirror is too large or your compass too small, substitute a string with a sharp pencil tied to one end and a thumbtack tied to the other.)

Return with the mirror and align (gently, please) the tissue paper so that the tracing matches the mirror's edge. Using the tracing as a template, make a tiny mark at the center of the mirror with a permanent felt-tip marker. Carefully check the spot's accuracy with the ruler; if acceptable, then enlarge the spot to about 0.125 inch (0.25 inch for mirrors of 12 inches and larger). Finally, stick an adhesive-backed hole reinforcer (the kind used to strengthen loose-leaf paper) on the mirror around the black dot.

Aim the telescope toward a bright scene (the daytime sky works well). Rack the focusing mount in as far as it will go, insert the sight tube, and take a look. The scene will probably look like one of those shown in Figure 8.4. Ideally, you should see the diagonal mirror centered in the sight tube. If it is, skip to the next step; if not, then the diagonal must be adjusted. Move to the front of the telescope tube and look at the back of the diagonal mirror holder. Most Newtonians use a four-vane spider mount to hold and position the diagonal in place. Spider mounts typically grasp the diagonal in a holder supported on a central bolt. Loosen the nut(s) holding the bolt in place and move the diagonal in and out along the optical axis until its outer diameter is centered in the sight tube. Before tightening the nut, check to make sure that the diagonal is not turned away from the eyepiece, because it can also rotate as it is moved. When done with this step, the view through the sight tube should look like Figure 8.4b.

Some less expensive telescopes use single rods attached to the eyepiece mounts to hold their diagonals in place. These are much more prone to being knocked out of adjustment and are unfortunately much harder to aim accurately. Loosen the setscrew that holds the rod in place, being certain to hold on to the diagonal or it will drop into the tube. By rotating the rod or moving it up and down, recenter the diagonal in the sight tube. Be careful not to bend the rod in the process. (Unless the rod is bent, the primary should appear centered in the diagonal at the end of this step. If so, skip the next step; if not, then repeat this process.)

With the diagonal centered in the sight tube, look through the tube at the reflection of the primary mirror. You ought to see at least part of the primary and the far end of the telescope tube. For the diagonal to be adjusted correctly,

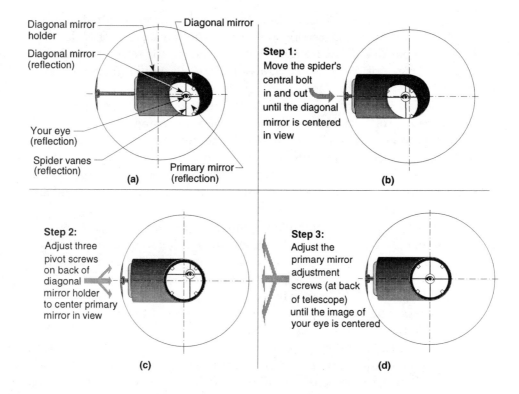

Figure 8.4 *Collimating a Newtonian reflector is as easy as 1-2-3-4. (a) The view through an uncollimated telescope. (b) Adjust the diagonal mirror's central post until the diagonal is centered under the eyepiece tube. (c) Turn the diagonal mirror's three adjustment screws behind the mirror until the reflection of the primary mirror is centered. (d) Finally, adjust the primary mirror's cell until the mirror image of your eye is centered in view.*

the primary must appear centered. To do this, most spider mounts have three equally spaced screws that, when turned, pivot the diagonal's angle. By alternately loosening and tightening the screws, move the reflection of the primary until it is centered in the sight tube. When properly aligned, the view after this step should look like Figure 8.4c.

Aiming the primary mirror will go much faster if you have an assistant. First, look at the back of the primary mirror's cell. You should see three (sometimes six) screws facing out. These adjust the tilt of the primary. Turn one or more of the adjustment screws until the diagonal's silhouette is centered in the sight tube.

With your helper at the adjustment screws, look through the sight tube. You should see the reflection of the primary centered in the crosshairs. In bright light, the primary's central spot should also be seen. Following your instructions, have your assistant turn one or more of the screws until the reflected image of your eye in the diagonal mirror appears centered as shown in Figure 8.4d. [Primary mirror mounts with six screws follow pretty much the same procedure. Three of the screws (usually the inner three) adjust the mirror, while the others prevent the mirror from rocking. If your telescope uses this type of mirror mount, the three outer screws must be loosened slightly before an adjustment can be made.]

If you own an NFT Newtonian, the mirrors should now be collimated adequately. If, however, you own an RFT, then some fine-tuning will be needed. For this, it is strongly recommended that you use a Cheshire eyepiece. Here's how it works. Shine a flashlight beam into the eyepiece's side opening and look through the peephole. Centered in the dark silhouette of the diagonal will be a bright donut of light—the reflection of the eyepiece's mirrored surface. The dark center is actually the hole in that surface. Adjust the primary mirror until its black reference spot is centered in the Cheshire eyepiece's donut. That's it.

The tell-all test is the *star test*. Take the telescope outside and let the optics fully adjust to the outside air temperature. Aim toward a moderately bright star. Unless your telescope is polar-aligned and the clock drive is turned on, use Polaris for the star test. Unlike all other stars, Polaris has the distinct advantage of not moving (at least, not much) in the sky, which makes the test a little easier to perform. Using a medium-power eyepiece, place the star in the center of the field. Move it slightly out of focus, transforming it from a point into a tiny disk surrounded by bright and dark rings. If the mirrors are properly collimated, then the rings should be concentric, like a bull's-eye. If not, one or both of the mirrors may need a little fine-tweaking. Be sure to recenter the star in the field every time an adjustment is made. (Defects in the mirror's curve can also cause the rings to appear irregular, but more about this later.)

Collimating a Schmidt-Cassegrain

Unlike the Newtonian, in which both the primary and secondary mirrors can be readily accessed for collimation, commercially made Schmidt-Cassegrain telescopes have their primary mirrors set and sealed at the factory. As such, an owner cannot adjust the primary if misalignment ever occurs. Fortunately, SCTs are rugged enough to put up with the minor bumps that might occur during set-up without affecting collimation.

This leaves only the secondary mirror to adjust, as shown in Figure 8.5. Take a look at the secondary mirror mount centered in the front corrector plate. There, you will see the heads of three adjustment screws spaced 120° apart. (Note that on some models, a plastic disk covers the screw heads. If so, it must be removed—carefully—to expose the adjustment screws.) In addition, some telescopes have a fourth, large screw or nut in the center of the secondary holder. If so, **DO NOT TOUCH IT!** Loosening that central screw will release the secondary from its cell and drop it into the tube. Talk about a quick way to ruin your day!

Remove the star diagonal and insert a sight tube into the eyepiece holder. Take a look. Ideally, you should see your eye centered in the secondary mirror. If the view is more like Figure 8.5a, turn one or more of the adjusting screws until everything lines up as in Figure 8.5b.

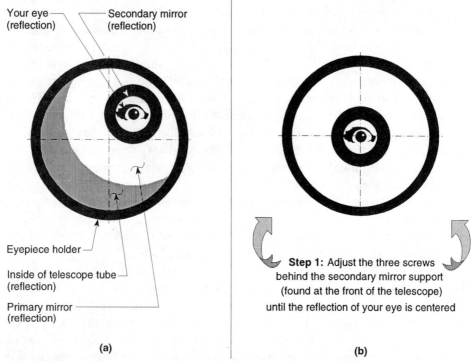

Your eye (reflection)
Secondary mirror (reflection)
Eyepiece holder
Inside of telescope tube (reflection)
Primary mirror (reflection)

(a)

Step 1: Adjust the three screws behind the secondary mirror support (found at the front of the telescope) until the reflection of your eye is centered

(b)

Figure 8.5 *Collimating a Schmidt-Cassegrain involves adjusting the secondary mirror (a) until its reflection appears centered in the primary (b).*

To check your success, move outside on a clear night. Set the telescope up as you would for an observing session, giving it adequate time to cool to the night air. With the telescope acclimated to the outdoor temperature, remove the star diagonal, if so equipped, and insert a medium-power eyepiece into the telescope. Center the instrument on a very bright star. Turn the focusing knob until the star moves out of focus and its disk fills about a third of the eyepiece field. Take a look at the dark spot on the out-of-focus disk. That's the dark silhouette of the secondary mirror. For the telescope to be properly collimated, the secondary's image must be centered on the star, creating a donut-like illusion. If the donut is asymmetric, then the diagonal must be adjusted.

Make a mental note of which direction the silhouette favors, go to the front of the telescope, and turn the adjustment screw that most closely coincides to that direction *ever so slightly*. Now, return to the eyepiece, recenter the star in view, and look at the dark spot again. Is it better or worse? If it's worse, turn the same screw the opposite way; if it's off in a different direction, turn one of the other screws and see what happens. Continue going back and forth between eyepiece and adjustment screws until the dark spot is perfectly centered in the star blob.

To double-check the adjustment, switch to a high-power eyepiece. Place the star back in the center of the field and defocus its image only slightly. If the secondary's dark outline still appears uniformly centered, then collimation was a success, and the telescope is set to perform at its best. If not, repeat the procedure, but with finer adjustments.

If the image is still not correctly aligned even after repeated attempts, then there is a distinct possibility that the primary is not square to the secondary. Focus on a rich star field. If any coma (ellipticity) is evident around the stars at the center of view, then chances are good that the primary is angled incorrectly. In this case, your only alternative is to contact either the dealer from which the telescope was purchased or the manufacturer.

To learn more about the fine art of telescope collimation, there are a couple of excellent treatments you should review. The first is *Perspectives on Collimation*, a booklet written by Vic Menard and Tippy D'Auria and available from Tectron Telescopes. The second is a pair of articles from the March and April 1988 issues of *Sky & Telescope* magazine entitled "Collimating Your Telescope" by optician extraordinaire Paul Valleli. Both are strongly recommended if you would like to learn even more about the art of telescope collimation.

Test Your Telescope

There is no way to guarantee that every telescope made, even by the finest manufacturers, will work equally well. Though companies have quality-control measures in place to weed out the bad from the good, a lemon is bound to slip through every now and then. That is why it is important to check an instrument right after it is purchased. Examine the telescope for any overt signs of damage. Next, look at the optics. They should be free of obvious dirt and scratches. The mounting should be solidly put together and move smoothly. If any problems are detected, contact the outlet from which the telescope was purchased immediately so that the problem can be rectified.

Though it is easy to tell clean, scratch-free optics from those that are not, a poor-quality lens or mirror may not be immediately obvious. Fortunately, an elaborate, fully equipped optical laboratory is not required to check the accuracy of a telescope's optics. Instead, all that you need are a well-collimated telescope, a high-power eyepiece, a piece of cardboard as large as the telescope's aperture, and a clear sky. With these four ingredients, any amateur astronomer can perform one of the most sensitive and telling optical tests available: the *star test*. A fifth ingredient (this book) will help you interpret the test's results.

If your telescope has a clock drive and is polar-aligned, aim the instrument toward a star of third magnitude or brighter. If your scope does not have a clock drive or, an equatorial mount or is not polar-aligned, it is best to aim toward a bright star near the celestial pole, because these move more slowly in the sky than, say, stars near the celestial equator. Switch to a high-power eyepiece, focus the image precisely, and examine the star closely. Recall from Chapter 1 that at high magnification a star will look like a bull's-eye; that is, a bright central disk (the Airy disk) surrounded by a couple of faint concentric rings (diffraction rings). By definition, any telescope that claims to have diffraction-limited optics must show this pattern. Is that what you see?

The answer is probably not, or at least not at first glance. Many factors affect the visibility of diffraction rings, such as telescope collimation, warm air currents inside the telescope tube, the steadiness of the atmosphere, and the telescope. Poor atmospheric steadiness (or *seeing*, as it is commonly referred to) causes star images to seem to boil, making it impossible to detect fine detail. Diffraction rings can only be seen under steady seeing conditions (see the section in Chapter 9 entitled "Evaluating Sky Conditions" for more on this). The aperture of the telescope also plays a big role in seeing diffraction rings. The larger the aperture, the smaller the diffraction pattern and thus the more difficult it is to see.

Slowly rack the image out of focus. The starlight should enlarge evenly in all directions, like ripples expanding after a small stone is tossed into a calm body of water. First, move the eyepiece slightly inside focus. Examine the out-of-focus star image. It should look like one of the patterns illustrated in Figure 8.6. Now, reverse the focusing knob, bringing the star slightly outside focus. Continue until its diameter matches the inside-focus image. Examine the ring pattern again. If the telescope has first-rate optics (that is, substantially better than the merely diffraction-limited variety), both extra-focal images should appear identical.

What if they don't? Compare the exact shapes of the inside-focus and outside-focus patterns with those shown in Figure 8.7. Are the patterns at least circular? No? What geometric pattern do they resemble? Oval? If the oval shapes look the same on both the inside and outside of best focus, the optics are doubtlessly out of collimation. If the images are dancing wildly, then either the telescope optics are not yet acclimated to the outside air temperature or the atmosphere is too turbulent to perform the test. What if the images are either triangular or hexagonal? If the patterns have one or more sharp corners, then the optics are probably being *pinched*, or distorted, by their mounts. Pinched optics are especially common in Newtonian reflectors when the clips holding the primary in the mirror cell are too tight.

"My telescope never focuses stars sharply. I just did the star test, but the out-of-focus patterns appear oval. When the eyepiece is racked from one side of focus to the other, the ovals flip orientation 90°." This telescope is suffering from *astigmatism*, which may be caused by poorly figured optics, pinched optics, or by the cooling-down process after the scope is taken outdoors. In Newtonian reflectors, a slightly convex or concave secondary mirror is also a common cause of astigmatism. If the axis of the star-oval is parallel to the telescope tube, suspect the secondary mirror.

The most common optical defect found in amateur telescopes is *spherical aberration*. Spherical aberration becomes evident when a mirror or lens has not been ground and polished to its required curvature. As a result, light from around the edge of the optic comes to a focus at a different distance than light

Here's how a star should look . . .

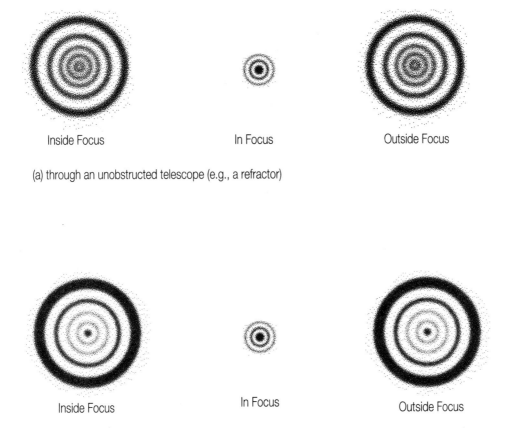

(a) through an unobstructed telescope (e.g., a refractor)

(b) through an obstructed telescope (e.g., a reflector or catadioptric)

Figure 8.6 *What should a star look like through the perfect telescope? Ideally all light should come to a common focus, causing a star to appear as a tiny disk (the Airy disk) surrounded by concentric diffraction rings. The star should expand to look like one of the illustrations here when seen out of focus, the image being identical on either side of focus: (a) through a refractor (unobstructed telescope) and (b) through a reflector or catadioptric (obstructed telescope).*

from the center. Spherical aberration comes in two varieties: one caused by undercorrected optics and one caused by overcorrected optics. Both produce similar effects; on one side of focus, the outermost part of the bull's-eye pattern is brighter or sharper than on the other side of focus. In the case of a Newtonian or Cassegrain design, the shadow of the secondary mirror in the center of the star disk will look larger on one side of best focus and smaller on the other.

What if, even when a star is brought out of focus, rings cannot be seen at all and instead all that can be seen is a round, mottled blob? This condition, more common in reflectors and catadioptrics than in refractors, indicates a *rough optical surface.* Mirror makers have an especially appropriate nickname for this: *dog biscuit.*

Some Common Problems . . .

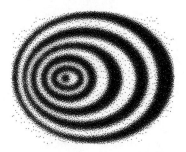

a. Misaligned (out-of-collimation) Optics

b. Atmospheric Turbulence

c. Pinched Optics (just one possibility)

Figure 8.7 *Something is amiss if the star test yields any of these results: (a) optics are out of collimation; (b) turbulent atmospheric conditions are present; (c) optics are being pinched or squeezed by their mounting; (d) optics have not acclimated to the outside air temperature; optics suffer from (e) astigmatism, (f) spherical aberration, and (g) rough optical surface.*

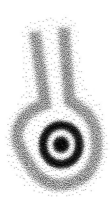

d. Optics Not At Thermal Equilibrium
(just one possibility)

e. Astigmatism

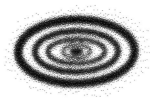

One Side of Focus   The Other Side of Focus

f. Spherical
Aberration

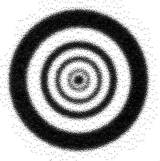

Inside Focus   Outside Focus

g. Rough Optical Surface
(Dog Biscuit)

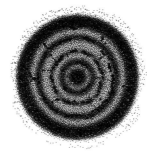

Figure 8.7 Continued

A final note on all this. Just how perfect must a telescope's optics be to produce good-quality images? That was the question raised by *Sky & Telescope* magazine in the March 1992 issue. To find the answer, they commissioned master optician Peter Ceravolo to make four 6-inch f/8 Newtonian reflectors. Each telescope was identical except for one very important factor: their primary mirrors, which varied in the precision of their ground-in parabolic curves, ranging from 1 to ½, ¼, and ¹⁄₁₀ wavefront error. The purpose of the experiment was to set up the instruments side by side and see if observers could judge them apart. The principal referees were Terence Dickinson and Douglas George, two experienced observers from Canada.

After careful testing and deliberation, both Dickinson and George were able to rank the telescopes in proper order. They found the telescope with 1-wavefront optics to be, in Dickinson's words, "a real turkey" characterized by ill-focused star blobs at anything but very low power. The ½-wavefront mirror performed better but was still judged unacceptable by the observers. The images it produced did not "snap into focus," as George put it. The difference between the ¼- and ¹⁄₁₀-wavefront telescopes was harder to distinguish. Both instruments showed clear detail on all sky objects tested, the difference only becoming noticeable after careful examination under excellent seeing conditions.

Later, *Sky & Telescope* editors carried the three best telescopes to the Stellafane amateur telescope makers convention in Vermont. After each telescope was set up at night and aimed at Polaris, conventioneers were invited to judge the scopes for themselves. From 8 P.M. to 2 A.M., 103 people examined the telescopes' performance and were interviewed afterward by *Sky & Telescope* editor Alan MacRobert. Virtually everybody could identify the telescope with the ½-wave wavefront error as clearly the worst. Ranking the other two was harder. But even so, a clear majority could detect the higher quality of the ¹⁄₁₀-wave wavefront telescope, especially during spells of steady seeing. (A ¹⁄₁₀-wave wavefront error equals ¹⁄₂₀-wave error on the mirror's surface, because errors are doubled by reflection. In this book, all wavefront errors are measured peak-to-valley.)

The moral to all this? There is no substitute for good optics. Diffraction-limited optics (those with ¼-wave wavefront error) are very good but detectably not the best. And anything less than diffraction limited is not acceptable. Pay close attention to manufacturers' claims and challenge all those that appear too good to be true. If you are considering a telescope made by the XYZ Telescope Company, write or call to inquire about their optics. Who makes them? What tests are performed to evaluate optical quality? And finally, what kind of guarantee is offered? Remember, if you want absolute perfection, you will probably have to pay a premium for it.

Have Telescope, Will Travel

Not too long ago, the most traveling a telescope would ever do was the trip from the house to the backyard, but not anymore. With the increased popularity of large regional or national star parties, as well as the ever-worsening problem of light pollution, many amateur telescopes routinely travel tens or hundreds of miles in search of dark skies.

Whenever a telescope is picked up and moved, the risk of damage is present. That is why some hobbyists tend to shy away from driving around with such delicate instruments. With a little forethought and common sense, though, this risk can be minimized.

Let's first examine traveling with a telescope by car. Begin by examining the situation at hand. If you own, say, a subcompact sedan and an 18-inch Newtonian, you have a big problem with only one solution: Buy another car. (That is exactly what a friend of mine did after he purchased an NGT-18 from JMI a few years ago! He traded in his sports car for a minivan.)

If your telescope and automobile are already compatibly sized, then it is just a matter of storing the instrument safely for transport. Take apart as little of the telescope as necessary. If there is room in the car to lay down the entire instrument, mounting and all, then by all means do so. Take advantage of anything that can minimize the time required to set up and tear down.

If the telescope tube must be separated from its mounting, it is usually best to place the tube into the car first. Be sure to seal the optics from possible dust and dirt contamination, which usually means simply leaving the dust caps in place, just as when the telescope is stored at home. If the telescope does not come with a storage case, protect the tube from bumps by wrapping it in a clean blanket, quilt, or sleeping bag. Strategically placed pillows and pieces of foam rubber can also help minimize screw-loosening vibrations. If possible, strap the telescope in place using the car's seatbelt.

Next comes the mounting. Carefully place it into the car, making sure that it does not rub against anything that may damage it or that it may damage. Wrap everything with a clean blanket for added protection. Again, use either the car's seatbelts or elastic shock cords (bungee cords) to keep things from moving around during a sharp turn. Be sure to secure counterweights, which can become dangerous airborne projectiles if the car's brakes are hit hard.

Transporting telescopes by air presents many additional problems. Their large dimensions usually make it impossible to carry them on a plane and store them under the seat! Therefore, we must be especially careful when packing these delicate instruments.

Some owners of 8-inch and smaller Schmidt-Cassegrain telescopes prefer to wrap their instruments in foam rubber and place them in large canvas duffle bags and carry them onto the plane, placing the scopes into the overhead compartments. Although this method usually works (some overhead compartments are just not large enough), the chance of damaging the telescope is great. This practice, therefore, is *not* recommended. Neither should an SCT be shipped in its standard footlocker-type carrying case. They are not designed for that much (potential) abuse.

Meade recommends a home-constructed wooden crate made of 2-by-2-inch wooden framing and 0.5-inch plywood side panels. The interior dimensions should be the same as the manufacturer's case to permit using the same precut foam padding. Complete the crate with a pair of strong hinges and a lock. Celestron tells us that their *hard plastic* cases, such as those that come with Ultima models, are strong enough to take the jostling that air cargo can go through. These cases can also be used with other catadioptric instruments, although they may not provide a custom fit.

Even if you will be using one of these high-impact cases, take a few additional precautions before sending your baby off onto the loading ramp. Begin by completely enclosing the telescope in protective bubble wrap, available in larger post offices and stationery stores. Bubble wrap is a two-layer plastic sheet impregnated with air to form a multitude of cushioning bubbles. The bubbles are available in many different sizes—select the larger variety. Another measure of protection is to place the telescope (case and all) into a heavy-duty multiwalled cardboard box. These boxes may be purchased from independent companies that specialize in wrapping packages for shipment.

Owners of other types of telescopes face even greater challenges. First and foremost, the optics must be removed and placed into a suitable case to be carried on the plane. The empty tube may then be bubble-wrapped, surrounded by styrofoam pellets (both inside and outside the tube) and placed into a strong wooden crate. Make certain that the shipping carton can be used again on the return trip, and bring along a roll of packing tape or duct tape just in case an emergency repair is needed.

Due to their weight, tripods and mountings pose special problems. I have traveled by air with a large tripod by first wrapping it in two thick sleeping bags and then packing everything in a large tent carrying case. Heavy equatorial mountings, on the other hand, must be packed professionally. Once again, seek a local crating company for help.

There is no industry wide policy regarding the transport of telescopes by air. Some airlines permit telescopes to be checked as luggage provided they do not exceed size and weight restrictions. (The purchase of optional luggage insurance is strongly recommended.) Other airlines will not accept telescopes as check-in items at all. In these instances, you must ship the instrument separately ahead of time via an air cargo carrier. Due to the amount of paperwork involved, especially on international shipments, air cargo services usually require that advance arrangements be made. Contact your airline well ahead of departure to find out exact details and damage insurance options.

Make sure that each piece of luggage has both a destination ticket and an identification tag and that both are clearly visible on the outside. Information on the ID tag should include your name, complete address, and telephone number. Although permanent plastic-faced identification tags are preferred, most check-in points provide paper tags that may be filled out on the spot. I always make it a habit to include a second identification label inside my luggage as well, just in case the outside tag is torn off.

Whatever you do, don't forget to bring along all the tools needed to reassemble the telescope once you arrive. It is best to keep the tools in a piece of checked baggage. While returning from Mexico after the July 1991 solar eclipse, a friend was stopped from boarding his flight because he was carrying a screwdriver. He ultimately had to check the screwdriver as a separate piece of luggage because all of his bags had already been boarded. It was quite a sight at the baggage claim area when his screwdriver came down the ramp among all these suitcases!

Finally, compile a thorough inventory of all equipment that you plan to bring. Include a complete description of each item, such as its dimensions, color, serial number, manufacturer, and approximate value. U.S. Customs requires owners to register cameras and accessories with them on a "Certificate of Registration for Personal Effects Taken Abroad" form before departure. Contact your nearest Customs office for further information. Keep a copy of the list with you at all times while traveling, just in case any item is lost or stolen. Carriers will be able to find the missing piece quicker if they know what to look for.

9

A Few Tricks of the Trade

A telescope alone does not an astronomer make. Sure, telescopes, binoculars, eyepieces, and other assorted contraptions are all important ingredients for the successful amateur astronomer, but there is a lot more to it than that. If a stargazer lacks the knowledge and skills to use this equipment, then it is doomed to spend more time indoors gathering dust than outdoors gathering starlight. Here is a look at some techniques and tricks used by amateur astronomers when viewing the night sky.

Evaluating Sky Conditions

Clearly, nothing affects our viewing pleasure more than the clarity of the night sky. Just because the weather forecast calls for clear skies does not necessarily mean that it's time to get out the telescope. As amateur astronomers are quick to discover, "clear" is in the eye of the beholder. To most people a clear sky simply means an absence of obvious clouds, but to a stargazer it is much more.

To an astronomer, sky conditions may be broken down into two separate categories: *transparency* and *seeing*. Transparency is the measure of how clear the sky is, or in other words, how faint a star can be seen. Many different factors, such as clouds, haze, and humidity, contribute to the sky's transparency. The presence of air pollutants, both natural and otherwise, also adversely affect sky transparency. Artificial pollutants include smog and other particulate exhaust, whereas volcanic aerosols and smoke from large fires (for example, forest fires) are forms of natural air pollution. Still, the greatest threat to sky transparency comes not from nature but from ourselves. We are the enemy, and the weapon is uncontrolled, badly designed nighttime lighting.

Today's amateur astronomers live in a paradoxical world. On one hand, we are truly fortunate to live in a time when modern technology makes it possible for hobbyists to own advanced equipment once in the realm of the professional only. On the other, we hardly find ourselves in an astronomical Garden of Eden. Though technology continues to serve the astronomer, it is also proving to be a powerful adversary. The night sky is under attack by a force so powerful that unless drastic action is taken soon, our children may never know the joy and beauty of a supremely dark sky. No matter where you look, lights are everywhere: buildings, gigantic billboards, highway signs, roadways, parking lots, houses, shopping centers, and malls. Most of these lights are supposed to cast their light downward to illuminate their earthly surroundings. Unfortunately, many fixtures are so poorly designed that much of their light is directed horizontally and skyward. The result: light pollution, the bane of the modern astronomer.

Have you ever driven down a dark country road toward a big city? Long before you get to the city line, a distinctive glow emerges from over the horizon. Growing brighter and brighter with each passing mile, this monster slowly but surely devours the stars; first the faint ones surrender, but eventually nearly all succumb. Several miles from the city itself, the sky has metamorphosized from a jewel-bespangled wonderland to a milky, orange-gray barren desert. While I know of no astronomer advocating the total and complete annihilation of all nighttime illumination, we must take a critical look at how it can be made less obtrusive.

Responsible lighting must take the place of haphazard lighting. But let's face it—not many people will be interested in light conservation if the point is debated from an astronomical perspective only. To win them over, they must be convinced that more efficient lighting is good for *them*. The public must be educated on how a well-designed fixture can provide the same amount of illumination over the target area as a poorly designed one, but without extraneous light scattered toward the sky and with a lower operating cost. That latter phrase, *lower operating cost*, is the key to the argument. The cost of operating the light will be lower because all of its potential is specifically directed where it will do the most good. The wattage of the bulb may now be lowered for the same effect, resulting in a lower cost. Everybody wins—taxpayers, consumers, and, yes, even the astronomers!

Can one person effect a change? That was the dream of David Crawford, the person behind the International Dark-Sky Association. The non-profit IDA has successfully spearheaded anti–light-pollution campaigns in Tucson, San Diego, and many other towns and cities. It provides essential facts, strategies, and resources to light-pollution activists worldwide. For more information on how you can join the fight against light pollution, contact the IDA at 3545 North Stewart, Tucson, Arizona 85716.

In midnorthern latitudes, the clearest nights usually take place immediately after the crossing of an arctic cold front. After the front rushes through, cool, dry air usually dominates the weather for 24 to 48 hours, wiping the atmosphere clean of smog, haze, and pollutants. Such nights are characterized by crisp temperatures, high barometric pressure, and low relative humidity.

To help judge exactly how clear the night sky actually is, many amateurs living in the northern hemisphere use the stars of Ursa Minor (the Little Dipper) as a reference because they are visible every hour of every clear night in the year. Figure 9.1 shows the major stars of Ursa Minor with their corresponding visual magnitudes. Note that in each case the decimal point has been eliminated to avoid confusing it with another star. Therefore, the 21 next to Polaris indicates it to be magnitude 2.1, and so on. You may find that all of the Dipper stars are visible only on nights of good clarity, while other readers may be able to see them on nearly every night. Still others, observing from light-polluted environs, may never see them all.

The night sky is also judged in terms of seeing, which refers not to the clarity but rather the sharpness and steadiness of telescopic images. Frequently, on clear, dark nights of exceptional transparency, the twinkling of the stars almost seems to make the sky come alive in dance. To many, twinkling adds a certain romantic feeling to the heavens, but to astronomers, it only detracts from the resolving power of telescopes and binoculars.

The twinkling effect, called *scintillation*, is caused by turbulence in our atmosphere. Density differences between warm and cold layers refract or bend the light passing through, causing the stars to flicker. Ironically, the air seems steadiest when a slight haze is present. Although any cloudiness may make faint objects invisible, the presence of thin clouds can actually enhance subtle details in brighter celestial sights such as double stars and the planets.

While the Earth's atmosphere greatly influences seeing conditions, image steadiness also can be adversely affected by conditions inside and immediately surrounding the telescope. If you are like most amateur astronomers, you probably store your telescope somewhere inside your house or apartment . . . your *warm* house or apartment. Moving the telescope from a heated room out into the cool night air immediately sets up swirling heat currents as the instrument and its optics begin the cooling process toward thermal equilibrium. Peering through the eyepiece of a warm telescope on a cool night is like looking through a kaleidoscope, with the stars writhing in strange ritualistic dances. As the telescope tube and its optics reach equilibrium with the outside air, the images will begin to settle down. The night's observing may then begin in earnest.

How long it takes for these heat currents to subside depends on the telescope's size and type as well as the local weather. Newtonian and Cassegrain reflectors seem to acclimate themselves the fastest. Still, these can require at least one hour in the spring, summer, and fall, and up to two hours in the winter (the greater the temperature change, the longer it will take). Refractors and catadioptric instruments, because of their sealed tubes, need up to twice as long!

Several steps can be taken to minimize the time required for instrument cooling. For openers, find a cool and dry place to store the telescope when not in use. If a dedicated observatory is impractical, then good alternatives include a vented wooden garden shed or a sealed-off corner of an unheated garage. This way, the telescope's temperature will always be close to the outside tem-

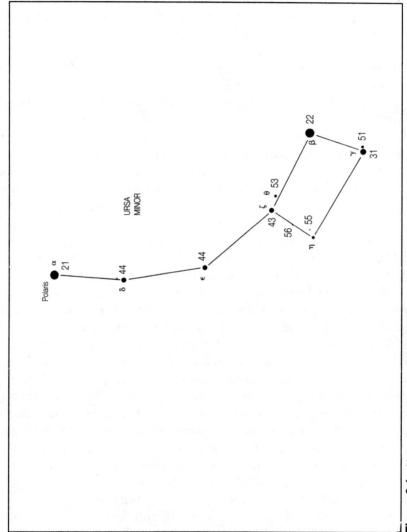

Figure 9.1 *How clear is the sky tonight? Many amateurs use the visibility of the Little Dipper as a gauge.*

perature even before it is set up. If you must drive with your telescope to an observing site, try to travel without using the heater on the way there.

Tilting the open end of a reflector vertically like a smokestack can help speed air-temperature equalization because warm air rises. In the case of sealed tubes, where a lens or corrector prevents the smokestack effect, it is best to turn the instrument broadside to the wind. The cooling breeze will help wick away heat as it is radiated by the telescope tube.

If the telescope is of your own making, select a material other than metal for the tube. Wood, cardboard, and fiberglass have a lower heat capacity than metal and therefore will have less heat to radiate once set outside. Unfortunately, nonmetal tubes can also retain the warm air inside the tube, but in practice the benefit outweighs the disadvantage.

A variation on this theme is not to use a solid tube at all but rather an open-truss framework to support the optics. Such an idea has been popular among observatory-installed telescopes for a century or more, as it speeds thermal equalization while cutting down the instrument's weight. Many Dobsonian-style reflectors utilize this same principal for the same reasons, as well as the resulting ability to break down a large instrument into a relatively small package for transport. Unfortunately, while the speed of temperature adaptation is increased, so are the chances of stray light intruding into the final image. Open tubes are also more susceptible to interference from the observer's body heat, turbulence caused by cross winds, and possible dewing of the optics. A sleeve made from black cloth and wrapped around the truss acts to slow these interferences but will not eliminate them.

As noted in Chapter 4, some manufacturers build cooling vents into their instruments' designs—a good idea. Some, such as JMI's NGT instruments, Obsession and Starsplitter Newtonians, and Takahashi's Schmidt-Cassegrains, even offer small, flat, so-called muffin fans to help rid their telescopes of trapped pockets of warm air. Amateur telescope makers would do well to consider including some type of venting system in their instruments as well, for they go a long way in speeding up a telescope's reaching temperature equilibrium. Just be sure to seal the vents against dust infiltration when the telescope is not in use.

A telescope is not the only piece of equipment that needs to adjust to the change in temperature after being brought outdoors for the first time at night; the Earth must as well. The ground you place the instrument on, having been exposed to the comparatively warm temperatures of daytime, also has to adapt to the cool of night. Different surfaces absorb heat better than others. Concrete and blacktop are the worst offenders because they readily absorb and retain heat. Grass, though also requiring a cool-down period, is better because it does not retain as much heat.

Your Observing Site

This brings up another hot topic among amateur astronomers today: where to view the sky. Choosing a good observing site is becoming increasingly difficult.

The ideal location should be far from all sources of light pollution and civilization in general. In addition, it should be as high above sea level as possible to avoid low-lying haze and fog, it must be safe from social ne'er-do-wells and other possibly harmful trespassers, and it should allow for an obstacle-free view of the horizon in all directions. Wouldn't it be nice if this was a description of your backyard? While we can all dream of finding such a Shangri-la, a few compromises usually must be made.

Over the years I have used several different observing sites with varying degrees of success. Many national, state, county, and local parks and beaches offer excellent areas, but their accessibility may be limited to daytime hours only or by residency. Ask your local park office if special access is available. The local authority that oversees the state parks near my home offers a stargazing permit that allows after-hours access to more than half a dozen parks for a small annual fee. The parks not only have much better horizons than most observing sites but also offer the added benefit of round-the-clock security patrols, an important consideration in these times. (Unfortunately, the patrol cars are outfitted with more lights than a small city, but I guess you have to take the good with the bad.)

Other good alternatives include both private and public golf courses. They have wide open expanses but may also suffer from restrictions and excessive security lighting around the clubhouse. If the owner of the course is apprehensive at first, why not offer to run a free observing session for club members in return for nighttime access? Though they may not be bona fide amateur astronomers, most people jump at the chance to see celestial wonders such as the rings of Saturn and the Moon. Flat farmland can also provide a secluded view, but unless the land is your own, be sure to secure permission from the owner beforehand. The last thing you want is to be chased by a gun-wielding farmer at two in the morning!

Where are the best observing sites? Here's my top-ten list:

10. A beach (watch out for fog and salt spray!)
 9. A flat rooftop (given a while to cool down after sunset)
 8. A town, county, state, or national park
 7. An open field or farmland
 6. A club observatory
 5. Your yard
 4. A *daytime-only* airport or landing strip
 3. The desert
 2. A golf course
 1. A hill or mountain

No matter where it is, a good observing site must be easy to reach. I know many urban and suburban amateurs who never see starlight during the week; instead, they restrict their observing time to weekends only, because the closest dark-sky site is more than an hour's drive away. Isn't that a pity? First, these amateurs may spend more time commuting to the stars than they do actually looking at them. Second, odds are they are forsaking many clear nights each month just because they believe the sky conditions closer to home are unusable. You really have to ask yourself if the local sky conditions are truly that bad. Remember, a telescope will show the Moon, Sun, and the five naked-eye planets as well from the center of a large city as it will from the darkest spots on Earth. Hundreds, if not thousands, of double and variable stars are also observable through even the most dismal of sky conditions. True, there is something extraspecial about observing under a star-filled sky, but never forgo a clear night just because the ambience is less than ideal. As the old saying goes, where there's a will, there's a way.

Star Parties and Astronomy Conventions

Just as they prove useful when investigating telescopes and accessories, local astronomy clubs are a tremendous resource for finding suitable observing sites near your home. In all likelihood, their members have already researched the area and have come up with some reasonable observing sites. Many of the larger societies have built observatories on their own land for use by members on clear nights. These same sites also permit members to bring along their own telescopes, setting them up side by side for a night under the stars. Not only is observing with a group safer, but it's also a lot more fun when you have someone with whom to share the excitement.

Perhaps as a reflection of the increasing interference of light pollution, the past few decades have seen a tremendous growth in regional and national star parties and astronomy conventions. At these, hundreds, even thousands of amateur astronomers travel to remote spots to set up their telescopes and enjoy dark skies far superior to those back home.

The oldest such gathering in North America is the annual Stellafane amateur telescope makers' convention, held each summer in Springfield, Vermont (Figure 9.2). Many of the 1,500 to 2,000 annual attendees think of Stellafane (a contraction of *stellar fane*, meaning "shrine to the stars") as sort of an astronomical mecca and would never think of missing the yearly pilgrimage up Breezy Hill outside of town. (I've missed only one year since 1969.) Some bring finely crafted homemade telescopes to be entered in friendly competition for optical and mechanical excellence. Others attend for the bargain-laden swap table or to hear talks given by some of the best-known amateur and professional astronomers and opticians. Still others come simply to meet old acquaintances and make new friends. All are there to witness the beauty of the Vermont night sky—which unfortunately is being eroded by the lights of Springfield. For information on the next convention, write to Stellafane, P.O. Box 50, Belmont, Massachusetts 02178 after June 1. Keep in mind that local hotel and motel accommodations grow scarce during the convention, so early reservations are a must. On-site camping is also available, though primitive by some standards. Personally, I think of it this way: There is the weekend of

Figure 9.2 *The granddaddy of all amateur astronomy conventions, Stellafane hosts thousands of hobbyists each summer atop Breezy Hill in Springfield, Vermont.*

Stellafane, and there are 51 others (give or take) to live through until the next Stellafane.

For West Coast amateur astronomers, there is the Riverside Telescope Makers' Conference held at Big Bear Lake outside of Los Angeles each Memorial Day weekend. Sometimes jokingly called "Stellafane West," Riverside attracts up to 2,000 amateurs each year for observing, judging of homemade telescopes, a swap table, and a huge display of the finest commercial telescopic equipment that money can buy. Talks are also presented by prominent amateur and professional astronomers, with topics covering just about every aspect of the science. Though both Riverside and Stellafane are geared toward the amateur telescope maker, anyone can come and join in on the fun. Contact RTMC, 9045 Haven Avenue #109, Rancho Cucamonga, California 91730 for more information on the next Riverside conference.

If you long to observe under some of the finest conditions to be found anywhere on the continent, then the Texas Star Party is the place to be. Held each Spring at the Prude Ranch near Fort Davis, the Texas Star Party offers nearly a week of day and night nonstop astronomy. (If Fort Davis sounds familiar, it might be because the McDonald Observatory is also near there.) Talks during the day and clear skies at night (it *never* rains during one of these conventions, does it?) leave precious little time for sleep at the TSP. On-site accommodations include group bunkhouses, campsites, and motel rooms, but as is the case with Stellafane, they are usually booked far in advance. Write to TSP Registrar, P.O. Box 386, Wylie, Texas 75098 for further information.

Several other conventions and star parties of note are held throughout the year. Of special interest is Autumn's Okie-Tex Star Party. Also held at the Prude Ranch, Okie-Tex gives observers the chance to see the sights of autumn and winter without the winter's cold. Contact the Oklahoma City Astronomy Club, P.O. Box 21221, Oklahoma City, Oklahoma 73156 for more information.

Speaking of winter, the Winter Star Party is one of the fastest-growing events on the astronomical social calendar. Sponsored by Miami's Southern Cross Astronomical Society, the Winter Star Party is held at a scout camp in the Florida Keys each February. The southerly vantage point (about 25° north latitude) affords absolutely breathtaking views of such wonders as Omega Centauri, the Eta Carinae Nebula, the Southern Cross, and Alpha Centauri, along with the advantage of warm temperatures. It's a great way to beat the winter blahs. For more information contact Bob and Sharon Grant, 5401 SW 110th Avenue, Miami, Florida 33165, but be forewarned that due to the WSP's incredible popularity and the small size of the scout camp, advance reservations are required *just to get in*!

Midwesterners may choose from either Astrofest or Hidden Hollow, both held in September. Astrofest is held at a 4-H camp in Kankakee, Illinois, and is sponsored by the Chicago Astronomical Society, whereas Hidden Hollow convenes in Mansfield, Ohio. Both attract a bevy of well-known speakers and a wide variety of homemade and commercial telescopes, and both feature dark-sky observing. For information, write to either Astrofest, P.O. Box 596, Tinley Park, Illinois 60477, or Hidden Hollow, P.O. Box 1118, Mansfield, Ohio 44901.

Canadian amateurs have three national astronomy conventions from which to choose. All are usually held near the August New Moon. For those in the east, there is Nova East in New Brunswick's Fundy National Park. Contact the Halifax Centre of the Royal Astronomical Society of Canada, 1747 Summer Street, Halifax, Nova Scotia, B3H 3A6, Canada. The North York Astronomical Association hosts Starfest each year at the Riverplace Campground in Mount Forest, Ontario. Details may be obtained by writing Starfest, 26 Chryessa Avenue, Toronto, Ontario, M6N 4T5, Canada. Finally, amateurs in western Canada and the northwestern United States gather each year for the Mount Kobau Star Party held on a rugged mountain near Osoyoos, British Columbia. For further information, write to 4100 25th Avenue, Vernon, British Columbia, V1T 1P4, Canada.

Dozens of smaller conventions are held across the country and around the world throughout the year. To help spread the word about when and where they will occur, *Sky & Telescope*, *Astronomy*, and *Astronomy Now* magazines contain monthly listings giving information and addresses. Take at look at a current issue to see if there is an upcoming event near you. If so, by all means try to attend. Astronomy conventions and star parties are great ways to meet new friends, learn a lot about your hobby and science, and get a great view of the night sky.

Finding Your Way

Once a site is selected, it is time to depart on your personal tour of the sky. Although most novice enthusiasts begin with the Moon and brighter planets, most objects of interest in the sky are not visible to the unaided eye or even through a side-mounted finderscope. How can a telescope be aimed their way if the observer cannot see the target in the first place? That is where observing technique comes into play. To locate these heavenly bodies, one of two different methods must be used. But before any of these systems can be discussed, it is best to become fluent in the way astronomers specify the location of objects in the sky.

Celestial Coordinates

Like the Earth's spherical surface, the celestial sphere has been divided up by a coordinate system. On Earth, the location of every spot can be pinpointed by its unique longitude and latitude coordinates. Likewise, the position of every star in the sky may be defined by *right ascension* and *declination* coordinates.

Let's look at declination first. Just as latitude is the measure of angular distance north or south of the Earth's equator, declination (abbreviated *Dec.*) specifies the angular distance north or south of the celestial equator. The celestial equator is the projection of the Earth's equator up into the sky. If we were positioned at 0° latitude on Earth, we would see 0° declination pass directly through the zenith, while 90° north declination (the North Celestial Pole) would be overhead from the Earth's North Pole. From our South Pole, 90° south declination (the South Celestial Pole) is at the zenith.

As with any angular measurement, the accuracy of a star's declination position may be increased by expressing it to within a small fraction of a degree. We know there are 360° in a circle. Each of those degrees may be broken into 60 equal parts called *minutes of arc*. Further, every minute of arc may be broken up to 60 equal *seconds of arc*. In other words:

$$\text{one degree } (1°) = 60 \text{ minutes of arc } (60')$$
$$= 3{,}600 \text{ seconds of arc } (3{,}600'')$$

When minutes of arc and seconds of arc are spoken of, an angular measurement is being referred to, not the passage of time.

Right ascension (abbreviated *R.A.*) is the sky's equivalent of longitude. The big difference is that, while longitude is expressed in degrees, right ascension divides the sky into 24 equal east-west slices called *hours*. Quite arbitrarily, astronomers chose as the beginning or zero-mark of right ascension the point in the sky where the Sun crosses the celestial equator on the first day of the Northern Hemisphere's spring. A line drawn from the North Celestial Pole through this point (the Vernal Equinox) to the South Celestial Pole represents 0 hours right ascension. Therefore, any star that falls exactly on that line has a right ascension coordinate of 0 hours. Values of right ascension increase toward the east by one hour for every 15° of sky crossed at the celestial equator.

To increase precision, each hour of right ascension may be subdivided into 60 minutes, and each minute into 60 seconds. A second equality statement summarizes this:

$$\text{one hour R.A. (1 h)} = 60 \text{ minutes R.A. (60 m)}$$
$$= 3{,}600 \text{ seconds R.A. (3,600 s)}$$

Unlike declination, where a minute of arc does not equal a minute of time, a minute of R.A. does.

The stars' coordinates do not remain fixed. Due to a 26,000-year wobble of the Earth's axis called *precession*, the celestial poles actually trace circles on the sky. Right now, the North Celestial Pole happens to be aimed almost exactly at Polaris, the North Star. But in 13,000 years it will have shifted away from Polaris and will instead point toward Vega, in the constellation Lyra. The passage of another 13,000 years will find the pole aligned with Polaris once again.

Throughout the cycle, the entire sky shifts behind the celestial coordinate grid. While this shifting is insignificant from one year to the next, astronomers find it necessary to update the stars' positions every 50 years or so. That is why you will notice that the right ascension and declination coordinates are referred to as "epoch 2000.0" in this book and most other contemporary volumes. These indicate their exact locations at the beginning of the year 2000 but are accurate enough for most purposes for several decades on either side.

Star-Hopping

The simplest method for finding faint sky objects, and also the one preferred by most amateur astronomers, is called *star-hopping*. Star-hopping is a great way to learn your way around the sky while developing your skills as an observer. Before a telescope can be used to star-hop to a desired target, its finderscope must be aligned with the main instrument. Take a look at the telescope's finder. Chances are it is held by six thumb screws in a set of mounting rings. Begin the process by aiming the telescope toward a distant identifiable object. Though the Moon, a bright star, or a bright planet may be used, I suggest using instead a terrestrial object such as a distant light pole or mailbox. The reason for this is quite simple: They don't move. Celestial objects appear to move because of the Earth's rotation, making constant realignment of the telescope necessary just to keep up. Center the target in the telescope's field and lock the mounting's axes. To alter the finder's aim, adjust the front three thumb screws until the finder's tube is centered in the front mounting ring. Now loosen the back three screws. Move the finder by hand until the target is centered in the finder's crosshairs. Once set, tighten all of the adjustment screws and check to see that the finder or telescope did not shift in the process.

Some finderscopes are held by only three adjustment screws and are adjusted in much the same way (though they are much more prone to misalignment). Aim at a terrestrial target as previously suggested and lock the telescope's axes. Look through the finder to see which direction is out of alignment.

Loosen the two opposite screws and move the finder by hand until the target is centered. Tighten all thumb screws, making a final check to see that the finder is aimed correctly.

With the finder correctly aligned to the telescope, the fun can begin. After deciding on which object you want to find, pinpoint its position on a detailed star atlas, such as one of the charts found in the next chapter. Scan the atlas page for a star near the target that is bright enough to be seen with the naked eye. Once a suitable star is located on the atlas, turn your telescope (or binoculars) toward it in the night sky. Looking back at the atlas, try to find little geometric patterns among the fainter suns that lie between the naked-eye star and the target. You might see a small triangle, an arc, or perhaps a parallelogram. Move the telescope to this pattern, center it in the finderscope, and return to the atlas. By switching back and forth between the finderscope and the atlas, hop from one star (or star pattern) to the next across the gap toward the intended target. Repeat this process as many times as it takes to get to the area of your destination. Finally, make a geometric pattern among the stars and the object itself. For instance, you might say to yourself, "My object lies halfway between and just south of a line connecting star A and star B." Then locate star A and star B in the finder, shift the view to the point between and south of those two stars, and your target should be in (or at least near) the telescope's field of view. Don't worry if you get lost along the way; breathe a deep sigh and return to the starting point.

The charts found in Chapter 10 are drawn with the star-hopper in mind. Each has at least one naked-eye star shown, and many display large portions of prominent constellations. In addition, the written description of each featured object includes a short passage describing how to star-hop to its position.

Let's imagine that we want to find M31, the Andromeda Galaxy. (M31 is this galaxy's catalogue number in the famous Messier listing of deep-sky objects. If you don't know what the Messier catalogue is, or for that matter, what a deep-sky object is, read the next chapter.) Begin by finding it in Figure 9.3. M31's celestial address is right ascension 0 hours 42.7 minutes (written 00^h $42^m.7$), declination $+41° 16'$. Notice that it is located northwest of the naked-eye stars Beta (β), Mu (μ), and Nu (ν) Andromedae. Find these stars in the sky and center your telescope's aim on them. Hop from Beta to Mu, then on to Nu. M31 lies a little over a degree to the west-northwest of this last star and should be visible through the finderscope (indeed, it is visible to the naked eye under moderately dark skies).

Easy, right? Now let's tackle something a little more challenging. Many observers consider M33 (seen back in Figure 5.3), the Great Spiral Galaxy in Triangulum, to be one of the most difficult objects in the sky to find. Most references note it as magnitude 5.9, which means that if the galaxy could somehow be squeezed down to a point, that would be its magnitude. However, M33 measures a full degree across, so its *surface brightness*, or brightness per unit area, is very low.

M33 and its home constellation are also plotted on Figure 9.3. How can we star-hop to this point? One way is to locate the naked-eye stars Alpha Trian-

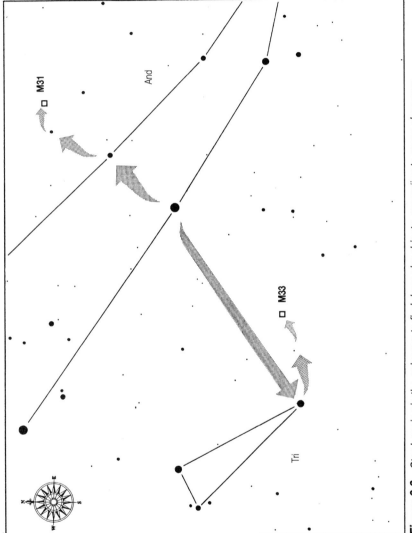

Figure 9.3 Star-hopping is the only way to find deep-sky objects, according to some observers. Here are suggested plans of attack for locating M31, the Andromeda Galaxy, and M33, the Great Spiral in Triangulum. Both are in the autumn sky.

guli and our old friend Beta Andromedae. Aim at the former with the finder-scope and move slightly toward the latter. About a quarter of the way between Alpha Trianguli and Beta Andromedae, and a little to the south, lies a 6th-magnitude star. Spot it in your finderscope. Connect an imaginary line between it and Alpha and then extend that line an equal and opposite distance from the 6th-magnitude star. There, you should see the galaxy's large, hazy glow. Nevertheless, when you peer through your telescope at that point, nothing is there. Recheck the map. No, that's not it; the telescope is aimed in the right direction. It's got to be there! What could be wrong?

At frustrating times such as this, it is best to pull back from your search and take a breather. Then go back, but this time use the averted-vision technique mentioned later in the section of this chapter entitled "The Eyes Have It." Look a little to one side of where the galaxy should be. By glancing at it with peripheral vision, its feeble light will fall on a more sensitive area of the eye's retina. Suddenly, there it is! You will wonder how you ever missed it before.

Setting Circles

This brings us to the second method that amateur astronomers use to find faint objects in the sky. It sounds easiest in theory, because you don't have to know the sky. But in fact it's no shortcut—as the next few pages of (complicated) instructions should make clear.

Most equatorial telescope mounts come outfitted with a pair of round, graduated scales, one on each axis, known as *setting circles* (Figure 9.4). The circle on the polar axis is divided into 24 equal segments, with each segment equal to one hour of right ascension. The setting circle attached to the declination axis is divided into degrees of declination, from 90° North to 90° South. With an equatorial mounting accurately aligned to the celestial pole (or any mounting with the properly encoded alt-azimuth digital setting circles), a sky object may be located by dialing in its pair of celestial coordinates.

Is that all there is to it? Unfortunately, no. First, to use setting circles, the mounting's polar axis must be aligned to the celestial pole. Begin by leveling the telescope. This may be done by adjusting the length of each tripod leg (if so equipped) or by placing some sort of a block (a piece of wood, a brick, etc.) under one or more corner footpads of the mount. Don't spend a lot of time trying to make the mount perfectly level; close is good enough. In fact, to be completely correct, it is not necessary to level the mount at all. The only thing that matters is that the polar axis be aimed at the celestial pole. In practice, however, it is easier to polar-align a mount that is level than one that is not, so take the time to do so. As an aid, many instruments come outfitted with bubble levels on their mountings.

Next, check to make sure that your finderscope is aligned to the telescope and that the entire instrument is parallel to its polar axis. This latter step is usually accomplished simply by swinging the telescope around until the dec-

Figure 9.4 *Setting circles allow an observer to take aim at a sky object by knowing its celestial coordinates.*

lination circle reads 90°. Most declination circles are preset at the factory, though some are adjustable; others may have slipped over the years. If you believe that the telescope is not parallel to the polar axis when the declination circle reads 90°, consult your telescope manual for advice on correcting the reading. If your manual says nothing, or worse yet you cannot find it (it must be here somewhere, you think), try this test. Align the telescope to the polar axis as best you can by eye and lock the declination axis. Using only the horizontal and vertical motion (azimuth and altitude, respectively) of the mounting, center some distant object, such as a treetop or a star, in your finderscope's view. Now, rotate the telescope about the polar axis *only*. If the telescope/finderscope combination is parallel to the polar axis, then the object will remain fixed in the center of the view; in fact, the entire field will appear to pivot around it.

If, however, the object moved, then the finder and the polar axis are not parallel to one another. Try it again, but this time pay close attention to the direction in which the object shifted. If it moved side to side, shift the entire mounting in azimuth (horizontally) exactly half the *horizontal* distance it moved. If it moved up and down, then the mounting's altitude (vertical) pivot is not set at the correct angle. Loosen the pivot and move the entire instrument one-half the *vertical* distance that the object moved. Because, in all likelihood, it shifted diagonally, this will turn into a two-step procedure. Take it one step at a time, first eliminating its horizontal motion, then the vertical.

With the polar axis and telescope now parallel, it is time to set the polar axis parallel to the Earth's axis. Some equatorial mounts come with a polar-alignment finderscope built right into the polar axis. These come with special clear reticles surrounding the celestial pole. Consult the telescope's manual for specific instructions.

Polar-alignment finderscopes are certainly handy, but for those of us without such luxuries, the following method should work quite well. With the telescope level and the declination axis locked at $+90°$, point the right ascension axis by eye approximately toward Polaris. Release the locks on both axes and swing the instrument toward a star near the celestial equator. Once aimed at this star, spin the right ascension circle (taking care not to touch the declination circle) until it reads the star's right ascension. Table 9.1 suggests several suitable stars for this activity. Note that the stars' positions are given at five-year intervals. Choose the pair closest to your actual date. (Although this slight shift is not of much concern when aligning a telescope to use setting circles, it is of great consequence for long-exposure through-the-telescope astrophotography.)

Swing the telescope back toward the celestial pole, stopping when the setting circles read the position of Polaris (also given in Table 9.1). If the telescope is properly aligned with the pole, then Polaris should be centered in view. If not, lock the axes and shift the entire mounting horizontally and vertically until Polaris is in view. Repeat the procedure again, until Polaris is in view when the circles are set at its coordinates. Due to the coarse scale of most circles supplied on amateur telescopes, a polar alignment within roughly 0.5 to 1° of the celestial pole is usually the best you can get. (Digital setting circles are much more accurate, but still the alignment need not be overly precise to be useful.)

Once the mounting is adjusted to the pole and the right ascension circle (also known as the *hour circle*) is calibrated with the coordinates of a known star, the mount's clock drive must be turned on. Many equatorial mounts have a direct link that turns the hour circle in time with the telescope. This way, as the sky and its coordinate system shift relative to the horizon, the hour circle will move along with them. Some less-sophisticated equatorial mounts, however, do not have driven hour circles. In these instances, the hour circle must be recalibrated not just once a night but before each use—inconvenient, to say the least. The easiest way to do this is to reset it on a reference star immediately before swinging the telescope toward a new target.

Incidentally, though the previous paragraphs describe the procedure for aligning to the North Celestial Pole, the actions are the same for observers in the Southern Hemisphere, except the mounting must be aligned to the South Celestial Pole. Everywhere you see "Polaris" or "North Star," substitute in "Sigma Octantis" or "South Star." Sigma Octantis, a 5.5-magnitude sun, is located about 1° from the South Celestial Pole. Its celestial coordinates are also given in Table 9.1. While not as obvious as its northern counterpart, Sigma works well for the task at hand.

Table 9.1   **Suitable Stars for Setting-Circle Calibration**

Star	Epoch	Right Ascension		Declination	
		h	m	°	'
Alpheratz (Alpha Andromedae)	(1995)	00	08.2	+29	04
	(2000)	00	08.5	+29	05
	(2005)	00	08.7	+29	07
	(2010)	00	09.0	+29	09
Hamal (Alpha Arietis)	(1995)	02	07.2	+25	26
	(2000)	02	07.5	+25	28
	(2005)	02	07.9	+25	29
	(2010)	02	08.0	+25	31
Aldebaran (Alpha Tauri)	(1995)	04	35.6	+16	30
	(2000)	04	35.9	+16	30
	(2005)	04	36.2	+16	31
	(2010)	04	36.4	+16	31
Procyon (Alpha Canis Minoris)	(1995)	07	39.1	+05	14
	(2000)	07	39.3	+05	14
	(2005)	07	39.6	+05	13
	(2010)	07	39.9	+05	12
Regulus (Alpha Leonis)	(1995)	10	08.1	+11	59
	(2000)	10	08.3	+11	58
	(2005)	10	08.6	+11	57
	(2010)	10	08.9	+11	55
Arcturus (Alpha Boötis)	(1995)	14	15.4	+19	13
	(2000)	14	15.7	+19	11
	(2005)	14	15.9	+19	10
	(2010)	14	16.1	+19	08
Altair (Alpha Aquilae)	(1995)	19	50.5	+08	52
	(2000)	19	50.8	+08	53
	(2005)	19	51.0	+08	53
	(2010)	19	51.3	+08	54
Polaris (Alpha Ursae Minoris)	(1995)	02	26.6	+89	15
	(2000)	02	31.8	+89	16
	(2005)	02	37.6	+89	17
	(2010)	02	43.7	+89	18
South Pole Star (Sigma Octantis)	(1995)	21	04.3	−88	59
	(2000)	21	08.6	−88	57
	(2005)	21	13.1	−88	56
	(2010)	21	17.2	−88	55

With the mounting polar-aligned, the setting circles calibrated, and the clock drive switched on, it is now a relatively simple matter to swing the telescope around until the circles read the coordinates of the desired target. If all was done correctly beforehand, the target should appear in, or at least very near, the eyepiece's field of view. Be sure to use your eyepiece with the widest field when first looking for a target's field before moving up to a higher-power ocular.

What if it's not there? Don't give up the ship immediately. Instead, try recalibrating the setting circles on a known star near the intended target's location. This technique, called *offsetting*, is a lot more accurate than trusting the setting circles to give the correct reading by themselves.

Now that the technique for using setting circles is familiar, here is why they should *not* be used, especially by beginners who are unfamiliar with the sky. To my way of thinking (and you are free to disagree), using setting circles or a computer to help aim a telescope reduces the observer to little more than a couch-potato sports spectator flipping television channels between football games on a Sunday afternoon. Where is the challenge in that? Observational astronomy is not meant to be a spectator's sport; it is an activity that is best appreciated by doing. There is something very satisfying in knowing the sky well enough to be able to pick out an object such as a faint galaxy or an attractive double star using just a telescope, a finder, and a star chart. Even if you get your setting circles perfectly aligned and master their use, you will be missing out on the thrill of the hunt. Stalking an elusive sky object is much like searching for buried treasure; you never know what else you are going to uncover along the way. As you set your attention toward one target, you might turn up other objects that you have never seen before. Of course, there is also that slight possibility of striking astronomical pay dirt by stumbling onto an undiscovered comet. Imagine that thrill!

Alright, so maybe I have been hard on setting circles, maybe even a bit unjustifiably. Make no mistake; setting circles are very useful tools *when used for the right reasons*. There is no question that they are required to aim the large, cumbersome telescopes in professional observatories where viewing time is at a premium. They also serve a very real purpose for advanced amateurs who are involved in sophisticated research programs, such as searching for supernovae in distant galaxies or estimating the brightness of variable stars. Both of these activities involve rapid, repetitive checking of a specific list of objects. Setting circles can also come in very handy when light pollution makes star-hopping difficult. However, they should never be used as a crutch. Most amateur astronomers will do much better by looking up and learning how to read the night sky rather than looking down at the setting circles.

The Eyes Have It

Though we may think of our binoculars and telescopes as wonderful optical instruments, there is no optical device as marvelous or versatile as the human eye. Experts estimate that 90% of the information processed by our brains is received by our eyes. There is no denying it; we live in a visual cosmos. To better understand how we perceive our surroundings, both earthly and heavenly, let us pause a moment to ponder the workings of our eyes.

The human eye (Figure 9.5) measures about an inch in diameter and is surrounded by a two-part protective layer: the transparent, colorless *cornea* and the white, opaque *sclera*. (Remember, don't shoot until you see the whites of their scleras.) The cornea acts as a window to the eye and lies in front of a pocket of clear fluid called the *aqueous humor* and the *iris*. Besides giving the eye its characteristic color (be it blue, brown, or any of a wide variety of other hues) the iris regulates the amount of light entering the eye and, more importantly, varies its focal ratio. Under low-light conditions, the iris relaxes, dilating the *pupil* (the circular opening in the center of the iris) to about 7 mm. Under the brightest lighting, the iris contracts the pupil to about 2.5 mm across, increasing the focal ratio and masking lens aberrations to produce sharper views. From the pupil, light passes through the eye's *lens* and across the eyeball's interior, the latter being filled with fluid called the *vitreous humor*. Both the lens and cornea act to focus the image onto the *retina*. The retina is composed of ten layers of nerve cells, including photosensitive receptors called *rods* and *cones*. Cones are concerned with brightly lit scenes, color vision, and resolu-

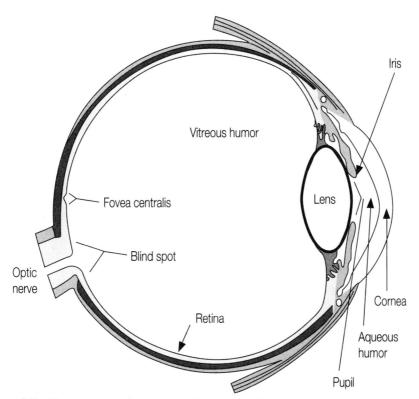

Figure 9.5 *The human eye, the astronomer's favorite tool.*

tion; rods are low-level light receptors that cannot distinguish color. There are more cones toward the *fovea centralis* (the center of the retina and our perceived view), while rods are more numerous toward the edges. There are neither rods nor cones at the junction with the optic nerve (the eye's *blind spot*).

From an astronomical point of view, we are most interested in the eye's performance under dimly lit conditions, our so-called night vision. Try this test on the next clear, moonless night. Go outdoors from a well-lit room and look toward the sky. Chances are you will be able to pick out a few of the brightest stars shining in what looks like an ink-black background. Face away from the sky, wait three or four minutes, and then look up again. You will immediately notice that many more stars seem to have appeared in that short time, an indication that the eye has begun to adapt to its new, darker environment.

Complete the exercise by turning away from the sky a second time, making certain to block your eyes from any stray lights. Wait another 15 to 20 minutes, then look skyward once again. This time there will be even more stars than before. Over the ensuing time, the eye has become fully dark-adapted. Not only has the pupil dilated, but a shift in the eye's chemical balance has also occurred. The build-up of a chemical substance called *rhodopsin* (also known as *visual purple*) has increased the sensitivity of the rods. Most people's eyes become adjusted to the dark in 20 to 30 minutes, though some require as little as 10 minutes or as long as one hour. This is enough to begin a night's observing, but *complete* dark adaptation takes up to another hour to occur.

While the eye's sensitivity to dim lighting increases dramatically during the dark-adaptation process, it loses most of its sensitivity to color. As a result, visual observers can never hope to see the wide range of hues and tints in sky objects that appear so vivid in astronomical photographs. Nebulae are good examples of this. Unlike stars, which shine across a broad spectrum, nebulae are only visible at specific, narrow wavelengths. Different types of nebulae shine at different wavelengths. For instance, emission nebulae, those clouds that are excited into fluorescence by the energy from young stars buried deep within them, shine with a characteristic reddish color. Red, toward the long end of the visible spectrum, is all but invisible to the human eye under dim illumination. As a result, emission nebulae are among the toughest objects for the visual astronomer to spy.

The eye is best at perceiving color among the brighter planets, such as Mars, Jupiter, and Saturn, as well as some double stars where the color contrast between the suns can be striking. Most extended deep-sky objects display little color, apart from the greenish and bluish tints of some brighter planetary nebulae and the golden and reddish-orange tinges of some star clusters.

While the eye's blind spot does not adversely affect our night vision, the fact that the fovea centralis is populated only by cones does. What this means, quite simply, is that the center of our view is *not* the most sensitive area of the eye to dim light, especially when it comes to diffuse objects such as comets and most deep-sky objects. Instead, to aid in the detection of targets at the threshold of visibility, astronomers use a technique called *averted vision*. Rather than staring directly toward a faint object, look a little to one side or the other of where the object lies. By averting your vision in this manner, you direct the target's dim light away from the cone-rich fovea centralis and onto the peripheral area of the retina, where the light-sensitive rods stand the best chance of revealing the faint target.

Another way to detect difficult objects is to tap the side of the telescope tube very lightly. Your peripheral vision is very sensitive to motion, so a slight back-and-forth motion to the field of view will frequently cause faint objects to reveal themselves. I know it sounds a bit strange, but try it. It works, but be gentle.

Here's an aside for all readers who wear eyeglasses. If you suffer from either nearsightedness or farsightedness, it is best to remove your glasses before looking through the eyepiece of a telescope or binoculars, refocusing the image until everything is sharp and clear. On the other hand, if you suffer from astigmatism or require eyeglasses with thick, curved lenses, then it is best to leave the glasses on. Of course, if you use contact lenses, leave them in place when observing as you would doing any other activity.

Frequently, localized light pollution will also mask a faint object. The distraction caused by glare seen out of the corners of your eyes from nearby porch lights, streetlights, and so on can be enough to cause a faint celestial object to be missed. To help shield our eyes from extraneous light, many eyepieces and binoculars sold today come with built-in rubber eyecups. While they prove adequate under most conditions, eyecups may not block out all peripheral light. Here are a couple of tricks to try if eyecups alone prove inadequate.

This first idea was already mentioned in *Touring the Universe through Binoculars*, but I think it merits repeating here. Buy a cheap pair of ordinary rubber underwater goggles, the kind you can find at just about any toy or sporting-goods store. Cut out half of the goggles' front window. If you prefer to use one eye over the other for looking through your telescope, make certain to cut out the correct side. Of course, both windows would need to be cut out if used with binoculars. Spray the goggles with flat black paint, and they're done. To test your creation, put the goggles on (after the paint has dried, please!) and go out under the stars. The blackened goggles should provide enough added baffling to keep stray light from creeping around the eyepiece's edge and into your eyes.

Here's another approach to the same situation that works well for viewing through telescopes but is not really applicable to binocular use. This involves wearing a dark turtleneck shirt or sweater, but not in the way you are used to. Instead of slipping it on from bottom to top, stick your head into the shirt through the neck opening. Let the shirt rest on your shoulders. Whenever you look through the eyepiece, simply pull the shirt up and over your head to act as a cloak against surrounding lights. This idea was first used by photographers a century ago, and it still works today. The only drawback, apart from looking a little strange to civilians, is that the eyepiece may dew over more quickly because of trapped body heat. Still, I can sometimes see up to a half-magnitude fainter just by using this cloaking device. Give it a try.

Record Keeping and Sketching

One of the best habits an amateur astronomer can develop is keeping a logbook of everything he or she sees in the sky. Recording observations serves the dual purpose of both chronicling what you have seen as well as how you have developed as an observer. It's also a great way to relive past triumphs on cloudy nights.

Though you are free to develop your own system, I prefer to record each object on a separate sheet of paper, including a few descriptive notes and a drawing. See Figure 9.6 for an example of a generic observation form. Most of the entries should be self-explanatory. *Transparency* rates sky clarity on a scale of 0, indicating complete overcast, to 10, which is perfect. *Seeing* refers to the steadiness of the atmosphere, from 0 (rampant scintillation) to 10 (very steady, with no twinkling even at high power). Because both of these are subjective judgments, as previously noted, I like to include the magnitude of the faintest star in Ursa Minor visible to the naked eye. This helps put the other two values in perspective. (The only exception to this is if the sky near the object under observation is noticeably different than that near Ursa Minor. In this case, record the faintest star visible near the object itself.)

It is difficult, if not impossible, to convey the visual impact of subtle heavenly sights with words alone, which is why including a drawing with all written observations is so important. The drawing should convey the perceived image as accurately as possible. Right now, some readers might be thinking, "I can't even draw a straight line." That's alright, because few objects in space are straight. Actually, sketching celestial objects is not as difficult as might be thought initially. It just takes a little practice.

Although astrophotographers require elaborate, expensive equipment to practice their trade, the astro-artist can enjoy his or her craft with minimal apparati. Besides a telescope or tripod-mounted binoculars, all you need to begin is a clipboard, a pad of paper, and a few pencils. All of the supplies are available from almost any art or stationery supply store for about $20. Here are a few specific things to look for. The delicate textures of celestial bodies are best rendered using an artist's pencil with soft lead such as H or HB or with sketching charcoal. (Just be careful with the charcoal as it tends to get away from those unfamiliar with it.) As for surface media, most astro-artists prefer smooth white paper as opposed to rag bond typing paper. The grain of rag paper tends to overwhelm the fine shading of most astronomical sketches. Most sketching pads have a fine surface grain ideal for the activity, but even computer printer paper will suffice. Lastly, the paper must be kept from blowing away while the artist moves between eyepiece and sketch. The simplest approach is a clipboard with a dim red flashlight clipped on or otherwise held in place. An even better idea is to make a backlit clipboard such as the one highlighted in Chapter 7.

Sketching a sky object is a multistep process, one that requires patience and close attention to detail. First, examine the target at a wide range of mag-

Observation Record

Object: _____ Constellation: _____

Date: _____ Time: _____

Observing Site: _____

Sky Trans'y: _____ Naked-eye Limit: _____ Seeing: _____

Telescope: _____

Eyepiece: _____ Filter: _____

Notes: _____

_____

_____

_____

_____

Figure 9.6 *Suggested observation form for recording celestial sights.*

nifications. Select the one that gives the best overall view as the basis for the sketch. Begin the drawing by lightly marking the positions of the brighter field stars surrounding the target object. Next, go back to each star and change its appearance to match its perceived brightness. Convention has it that brighter stars are drawn larger than fainter ones. Once the field is drawn accurately, lightly depict the location and shape of the target itself. The last step is to shade in the target to match its visual impression. Using your finger, a smudging tool, or a soft eraser to smudge the lead, recreate the delicate shadows and brighter regions. Remember, the drawing is, in effect, a negative image of the object. As such, the brightest areas should appear the darkest on the sketch, and vice versa. Finally, examine the target again with several different eyepieces, penciling in any detail that previously went unseen.

Here are some final tips for beginning astro-artists. First, never try to hurry a drawing, even if it is bitter cold and your hands are turning blue. Better to put the pencil down, take a five- or ten-minute warm-up break, and then go back. Second, do not try to create a Rembrandt at the telescope. Instead, make only a rough (but accurate) sketch at the telescope itself; the final drawing may be made later indoors at your leisure. Last, avoid the urge to add a little stylistic license, such as putting spikes around the brighter stars or drawing in detail that isn't quite visible but that you think is there from looking at photographs. Remember, a good astro-artist is an impartial reporter.

Once filled out, the observation record may be filed in a large loose-leaf notebook. It is handiest to separate observations by category. Individual headings might include the Moon, planets, variable stars, deep-sky objects, and so on. Interstellar objects may be further broken down first by type and then by increasing right ascension beginning at 0 hours. Members of the Solar System might be separated by object and then filed chronologically.

The eye of the experienced observer can detect much more tenuous detail in sky objects than a beginner can spot. Does this mean that the veteran has better eyesight? Probably not. Like most things in life, talent is not inborn. It has to be nurtured and developed with time. That's why it is important to keep notes. You will be amazed at how far your observing skills have come when you look back at your early entries a few years later.

Observing versus Peeking (A Commentary)

Are you an astronomical observer or an astronomical peeker? There is a difference . . . a big one! A peeker flits from one object to the next, barely looking at each before . . . *swish* . . . it's off to another. He or she never writes down what was seen, let alone makes a simple drawing. Whenever asked what he or she saw during an observing session, the peeker will only say, "Oh, I don't know, just some stuff." The fact of the matter is that peekers usually cannot remember what they have seen and what they have not.

An observer takes a slow, methodical approach to the study of the night sky. Most compile long lists of objects they want to see before going outside in an effort to use each moment under the stars as effectively as possible. Unlike our friend the peeker, the true observer is not out to break any land-speed records. He or she prefers to take it a little more slowly, savoring each photon that reaches the eye.

Perhaps it is a sign of the hectic times in which we live, but more and more amateur astronomers seem to be peekers. If you are one of them, I have to ask you this: What's your rush? Take a deep breath and relax. Resist the urge to race impulsively across the sky. By taking a slower, more deliberate tour, you will see the heavens in a new and exciting light. Become an astronomical observer, a connoisseur of the universe.

10

It's Time to Solo!

Although this book is intended as a guide to equipment and not as a guide to observing the sky, it would be remiss not to devote some space to what can be seen with the telescopes, binoculars, and accessories of which we speak. With that in mind, this chapter will take you on a quick trip across the heavens. If this is your maiden voyage, you are about to witness sights that few people are aware even exist. Many will defy description. If you are a seasoned traveler, then you know of what I speak. Regardless of how many years you look toward the sky, there will always be something new and wondrous to observe.

The Solar System

The Moon

The Moon is a fascinating place to visit through any telescope, regardless of aperture, magnification, or degree of sophistication. Even the smallest instruments (yes, including binoculars) will display the stark lunar terrain in all its "magnificent desolation," as Apollo 11's Edwin Aldrin put it. The surface of the Moon today appears much as it did in the past. Rugged mountain ranges, expansive flat plains, deep valleys, and innumerable craters all dating back to the beginnings of our Solar System await the amateur explorer.

Each lunar phase holds something exciting for the sightseeing astronomer. Figures 10.1 and 10.2 show the Moon at its first-quarter and last-quarter phases, respectively. To help guide your way across the barren lunar surface, each identifies several of the most interesting and spectacular features the Moon has to offer. The crescent phases after New Moon through first quarter

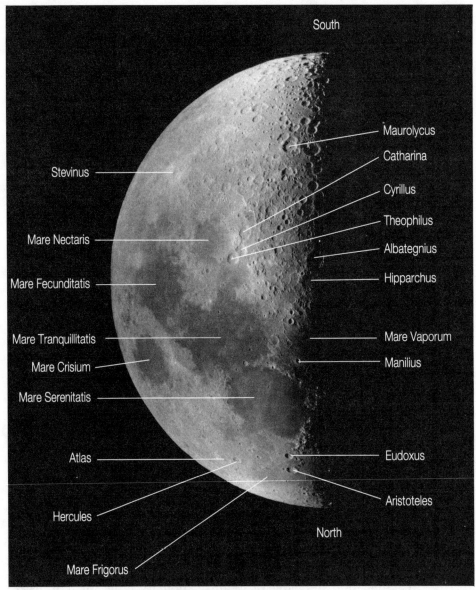

Figure 10.1 *First-quarter Moon. Photograph by Richard Sanderson (6-inch f/12 Astrophysics refractor, T-Max 100 film, ⅟₃₀ second at f/12). South is up.*

display a tremendous variety of lunar terrain. Dominating the equatorial zone are the vast expanses of the lunar seas: (or *maria*; the singular form is *mare*) Mare Crisium, Mare Fecunditatis, Mare Tranquilitatis, and Mare Serenitatis. To their north are many scattered large craters, while the south polar region is awe-inspiring in its coarse beauty. Of special interest are the craters Clavius and Tycho. Both premier the night after first quarter.

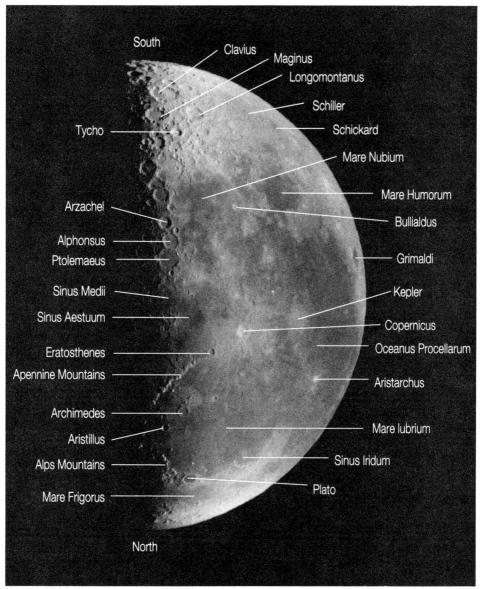

Figure 10.2 *Last-quarter Moon. Photograph by Richard Sanderson (6-inch f/12 Astrophysics refractor, T-Max 100 film, 1/30 second at f/12). South is up.*

After first quarter, sunlight slowly pours into Oceanus Procellarum, the Ocean of Storms, which is the largest of all lunar maria. The Ocean of Storms holds many wondrous sights, including the craters Copernicus and Kepler. To the north, the crater remnant Sinus Iridum (the Bay of Rainbows) and the unusual dark-floored crater Plato receive first light. To the south, Tycho seems ablaze with fire as its magnificent system of bright rays scatters nearly to the Moon's equator. Once Full Moon passes, the sequence reverses, with shadowing and lighting effects adding a different perspective to our neighbor.

Many amateurs, especially those with larger instruments, shy away from looking at the Moon because of its often-overwhelming glare. But this is no excuse, because the Moon's brightness can be easily controlled. For about $15, a Moon filter will reduce the intensity to a comfortable level. Though there can be benefits to viewing the Moon through color filters (green is especially useful for increasing subtle contrasts), most people prefer either neutral-density or variable-polarizing filters to maintain proper color balance. Another way to cut down on the glare is to stop down your telescope with a mask. Stopping down a telescope to about 3 inches will make the moonlight more manageable, though it will bring a loss of detail at high power.

For example, a 4-inch f/7 refractor (focal length = 28 inches) can be turned into an f/10 instrument by reducing its aperture to 2.8 inches. Simply cut a circle 2.8 inches across in the cardboard and secure the mask centrally in front of the objective lens.

A telescope with a central obstruction (that is, a secondary mirror) requires a slightly different approach. Measure the distance between the outer edge of that obstruction and the inside edge of the telescope tube. Cut a circle of this diameter in the cardboard. Make sure to offset the opening in the mask from the telescope's optical axis to ensure that no part of the secondary-mirror support blocks any portion of the view when the mask is attached to the telescope. For instance, a 14-inch f/5 Newtonian (focal length = 70 inches) requires an offset 7-inch mask to change the focal ratio to f/10. But because the distance between the diagonal mirror and the tube will be less than this, a smaller circle must be cut in the cardboard accordingly.

There is no one magnification that is best for viewing the Moon. Instead, try experimenting. Scan the surface with a low power to take the whole thing in. Then, when you spot an interesting area, switch to a medium or high power for detailed study. Under steady seeing conditions, it is often possible to exceed the 60-power-per-inch rule when studying the Moon. (This is best attempted with a clock-driven telescope; viewing at extremely high power without one is like trying to play tag with a speeding train! Use the lunar drive rate, if your telescope is so equipped.)

Often as it orbits Earth, the Moon will pass in front of, or *occult* stars and, on infrequent occasions, even a planet. Watching as the Moon slowly crosses in front of another body can be a lot of fun. For advanced amateurs, precisely timing the occurrence of an occultation is also a useful scientific activity. By knowing the exact moment of disappearance (called *ingress*) or reappearance (*egress*) of a star or planet, scientists may detect slight fluctuations in the Moon's orbital distance and speed. It is also possible to determine if the occulted star is single or an unresolved double.

Occultation predictions are computed and published annually by the International Lunar Occultation Center (ILOC) in Japan. ILOC predictions for the occultations of brighter stars are also available in the annual Royal Astronomical Society of Canada's *Observer's Handbook*, as well as in both *Sky & Telescope* and *Astronomy* magazines.

Also listed in the same sources are predictions for total, partial, and penumbral lunar eclipses. A lunar eclipse occurs when the Full Moon passes through Earth's shadow. Because the Moon's orbit about Earth is not quite parallel to Earth's path around the Sun, but rather tilted 5°, eclipses do not occur every Full Moon. Instead, the Moon misses Earth's shadow, passing either above or below. No more than three times a year, the Moon will cross Earth's orbit at Full Moon (at a point called a *node*), causing a lunar eclipse. Appendix C lists upcoming lunar eclipses through the year 2000.

The Planets

Each of the nine planets that circle our Sun has a unique appearance through amateur telescopes. Saturn has its rings, Jupiter has its belted atmosphere and four Galilean moons, Mars displays its polar caps and mottled orange surface, and Venus drifts through different Moon-like phases. Each of these traits, and many more, can be seen through even modest telescopes.

Because detail on the planets is usually quite subtle, observers want as much contrast as possible. Three factors affect the visibility and clarity of planetary features: optical quality, telescope design, and seeing conditions. All other things being equal, the degree of contrast visible depends on the size of a telescope's central obstruction (for example, from a secondary mirror), if any—the smaller the central obstruction, the greater the contrast. Refractors are ideal, because they do not have any central obstruction; that is, they have a clear aperture. Long-focus Newtonian reflectors have a very small central obstruction (the longer the focal length, the smaller the secondary mirror) and, therefore, good contrast. Short-focal-length Newtonians, Cassegrains, and Schmidt-Cassegrains have large central obstructions, ergo lower contrast for a given aperture.

If seeing conditions (or the optics) are poor, try stopping down its aperture to about 3 or 4 inches by using a cardboard mask. This is, at best, a compromise, for as aperture is decreased, so is resolution. With this in mind, many amateurs prefer using color filters to heighten contrast between different-colored areas on planets.

From the discussion in the previous chapter, you'll recall that *seeing* refers to the steadiness of the night sky. A turbulent night (that is, a night when the stars appear to twinkle with great fervor) is a poor night for planet-watching. For the best views, choose instead a night when the stars shine steadily (often when a slight haze or cloudiness is present). Also, the brightness of the naked-eye planets may dazzle the eye when viewed through larger instruments. Filters can help lessen the glare. To learn more about the art of observing the other members of our Solar System, consider joining the Association of Lunar and Planetary Observers (P.O. Box 143, Heber Springs, Arkansas 72543).

Appendix D lists the locations for the four easily seen naked-eye planets—Venus, Mars, Jupiter, and Saturn—through the year 2000. For further details, as well as the locations of Mercury, Uranus, Neptune, and Pluto, consult one of the monthly astronomy magazines listed in Chapter 6.

Mercury. Timing is everything when it comes to finding Mercury. Because of its proximity to the Sun, it is only seen near the horizon in very heavy twilight, no more than an hour after sunset or before sunrise. Therefore, you must have a clear view in the direction of sunset or sunrise, free of any low-lying fog, haze, or clouds. Good places for Mercury-spotting include a beach; a large, flat field; or the top of a hill.

Certain times of the year are better for spotting Mercury than others. The best chance for observers in the Northern Hemisphere to see Mercury in the evening is when the planet is at greatest eastern elongation in the spring or greatest western elongation in the early predawn morning hours of autumn. It is during these seasons that the ecliptic makes its largest angle with the horizon, carrying Mercury higher above the horizon than at any other time of the year.

To find Mercury, scan the horizon slowly and carefully near where the Sun just set or is about to rise for a bright starlike object. Once spotted, Mercury will show only a tiny disk. The only distinctive characteristic of Mercury is that because it is located between Earth and Sun, it goes through phases like the Moon. Trying to determine the phase can be difficult because of glare from the bright surrounding sky. To help improve contrast and reduce the effect of atmospheric turbulence, try using a Wratten No. 21 (orange), 23A (red), or 25 (deep red) color filter. A light blue Wratten No. 80A filter may help bring out some hazy surface mottlings, as may a polarizing filter.

Though Mercury will usually pass above or below the Sun as seen from Earth, every now and then it will appear to slide directly across the solar disk. Such an event is called a *transit*. By using at least $50\times$ and following proper safety precautions (outlined later in this chapter), astronomers can watch as tiny Mercury slowly moves across the Sun's disk.

The last transit of Mercury took place in November 1993, although it was not visible from North America. The next transits are set for November 15, 1999; May 6–7, 2003; and November 8, 2006. The 1999 transit will be visible only from Antarctica and southernmost Australia. The 2003 event will be visible from Europe, Asia, Africa, and Australia, whereas the 2006 transit will be seen from North and South America and the Pacific. As the dates draw closer, consult the astronomical periodicals and annuals for exact times and circumstances of the transits.

Venus. Like Mercury, Venus's orbit lies inside of Earth's, therefore holding it hostage to either the western sky after sunset or the eastern sky before sunrise. However, though Venus is visible for no more than about three hours after sunset or before sunrise, the planet's outstanding brightness makes it an outstanding sight. At times reaching a brilliance greater than magnitude -4, Venus outshines all other objects in the sky except the Sun and Moon. For all its eminence, however, Venus hides its secrets well. The planet's atmosphere proves impenetrable even with the largest telescopes on Earth. Though its surface may be hidden from view, Venus can be observed going through a series of phases similar to the Moon's with even the smallest telescopes. Watching the planet's phases over the course of time can be great fun to keep track of and record.

By knowing the planet's current phase, it is also possible to know where Venus is in its orbit. Venus displays a large, thin crescent phase (Figure 10.3) when it is closest to Earth and about to pass between us and the Sun (a point in its orbit called *inferior conjunction*). When Venus is farther from Earth, its globe will look noticeably smaller, even though more of its disk will appear lit by the Sun. At these times, Venus looks more like a quarter or gibbous Moon. As it rounds the far side of the Sun (referred to as *superior conjunction*), nearly the entire lighted side of the planet will face Earth.

On rare occasions, Venus may be seen to transit the face of the sun at inferior conjunction. Transits of Venus occur in pairs separated by eight years, with over 100 years between successive pairs. The last transits of Venus were seen in 1874 and 1882. The next will take place on June 8, 2004, and June 6, 2012. At least part of the 2004 event will be visible from Europe, Africa, Asia, and Australia. The 2012 transit will be seen from western North America, the Pacific region, eastern Asia and Australia. The Midnight Sun will allow the few astronomers north of about latitude 70° (inside the Arctic Circle and parallel to northern Alaska) to see both events in their entirety. Once again, check the periodicals and annuals for exact times.

Figure 10.3 *Crescent Venus. Photograph taken at high noon (!) by George Viscome (12.5-inch f/10 Cassegrain, Tech Pan 2415 film, with No. 25A [red] filter).*

Mars. No planet has quite the allure for the planetary observer as does Mars (Figure 10.4). Though the canals and tales of Martians of a century ago are gone, something still fascinates us about this distant world.

Part of that fascination might be the fact that Mars is well-placed for observation only once every 26 months, when our two planets are closest. This is referred to as *opposition*. At opposition, Mars lies directly opposite the Sun in our sky, rising in the East at sunset and remaining visible all night long.

Some oppositions are better for observing Mars than others. At a poor opposition, Mars will grow to only about 14 seconds of arc (abbreviated 14″) across. Even at its best, Mars will never appear larger than 25 arc-seconds (25″). This is slightly larger than a ringless Saturn, but much smaller than either Venus or Jupiter. Therefore, at least a high-powered 3-inch refractor or 6-inch reflector and steady seeing are required to see any of Mars' famed features, such as the polar caps or larger dark markings. Be sure to use your best-quality eyepieces when searching for elusive Martian detail.

Mars also benefits greatly from the use of color filters. Orange (Wratten No. 21) and red (Nos. 23A or 25) filters increase the contrast of the dark surface markings against the bright orange desert regions. Blue (Nos. 38A or 80A) and green (No. 58) filters are best for showing the polar caps as well as any haze, fog, or clouds in the planet's thin atmosphere.

Regardless of telescope, eyepiece, or filters, Mars is not an easy planet to study. Even with the planet near opposition, the best instrumentation will reveal little if the observer doesn't possess two ingredients that money can't buy: determination and patience. Long periods may lapse when Mars appears as little more than a quivering mass swimming in the turbulence of our atmo-

Figure 10.4 *Mars. Photograph by Gregory Terrance (16-inch f/5 Newtonian reflector, Lynxx PC CCD camera, exposure 1 second at f/45).*

sphere. But on the rare occasion of atmospheric tranquility, the real Mars will come through to reward the observer.

From Earth, the most prominent features of Mars are its two polar caps. During an opposition, small telescopes reveal the caps as small but unmistakably white. During the oppositions when Mars comes comparatively close to Earth (next occurring in 2003), the southern cap will be tilted toward Earth. At distant oppositions (such as those in the mid- and late-1990s), the northern cap is exposed toward Earth.

Scattered across the Martian surface are *albedo features*, the famed dark markings on Mars. Although most of the Martian surface appears bright orange, these regions appear noticeably darker by contrast. Albedo features are permanent markings, but they do appear to change slightly in size and shape. This metamorphosis is believed to be caused by the swirling Martian winds. Though the planet's atmosphere is much thinner than Earth's, its winds will occasionally reach hurricanelike force, shifting sands that alternately cover and uncover different portions of the rocky surface.

The most easily spotted albedo feature on Mars is Syrtis Major, a triangular wedge extending from north to south across the planet's equator. Almost equally dark but quite a bit smaller is Meridiani Sinus, a dark patch along the equator about 90° to the west of Syrtis Major.

Jupiter. No doubt about it, Jupiter is one of the most impressive members of the Solar System to observe. What makes it so special? For openers, Jupiter is one of the brightest objects in the sky, surpassed only by the Sun, Moon, Venus, and rarely Mars. Jupiter is also huge; 88,000 miles in diameter at the equator. Even at more than 391 million miles from Earth, this translates to an average apparent equatorial diameter of 47 arc-seconds, larger than any other planet except Venus when it is near inferior conjunction. All this adds up to a spectacular view.

Even with the smallest astronomical telescopes, observers can spot the distinctive equatorial bulge of Jupiter. While Jupiter measures about 47 arc-seconds across its equator, it spans only about 44 arc-seconds from pole to pole. This *polar flattening* is the result of Jupiter's rapid rotation—a day on Jupiter is approximately 9 hours and 55 minutes long.

Telescopes both large and small can also divide the planet's impenetrable clouds into bright zones and dark belts, as seen in Figure 10.5. Especially prominent are Jupiter's broad, bright Equatorial Zone and the North and South Equatorial Belts. Larger telescopes and medium to high powers will show swirls, ovals, and festoons within the turbulent atmosphere. The Great Red Spot, a huge oval cyclonic storm found in Jupiter's South Tropical Zone, may also be seen. Don't expect to see the Red Spot immediately. There's only a 50-50 chance of it being on the earthward side of Jupiter in the first place, and even then its coloration will be very subtle. Rarely a bright red, the Red Spot varies in color from a pale pink or orange to very pale tan-white.

Figure 10.5   *Jupiter. Photograph by Gregory Terrance (16-inch f/5 Newtonian reflector, Lynxx PC CCD camera, exposure ⅓ second at f/30).*

Filters can help highlight details in the Jovian cloud bands. Try using a yellow-green (Wratten No. 11) or orange (No. 21) filter to pick out subtle features in the darker belts. A green (Nos. 56 or 58) or blue (Nos. 38A or 80A) filter helps accentuate the Red Spot's visibility, though it still may be difficult to pick out.

Jupiter owns at least 16 moons by modern reckoning, though only the four brightest are visible in most amateur telescopes. These are, moving in order from the planet outward, Io, Europa, Ganymede, and Callisto. All were discovered by Galileo when he turned his first crude telescope toward Jupiter, and so today they are called the *Galilean satellites*. All slowly but constantly change their positions relative to the planet and each other. Io moves the quickest, circling Jupiter in a little less than two days, while distant Callisto takes about two weeks. Watch and record their changing positions and see if you can tell which moon is which.

In addition to appearing dutifully by the side of Jupiter, the Jovian satellites also perform great slight-of-hand tricks with their master world. Frequently, one or more of the satellites will be eclipsed or occulted behind Jupiter, while at other times, they will transit in front of the planet.

Transits are two-part events. As the satellite crosses in front of Jupiter, its tiny, round shadow may also be seen projected onto the top of the Jovian atmosphere. The entire affair may last several hours.

The location of the shadow relative to the satellite depends on where Jupiter is in our sky relative to the Sun. Before Jupiter reaches opposition, the satellite's shadow precedes the satellite itself, allowing both to be seen prior to the transit's beginning. After opposition, the reverse is true, with the shadow now following the satellite. At opposition (and for several nights either side), the shadow is cast almost exactly behind the satellite.

While the shadow of a satellite is easy to spot during a transit, seeing the satellite itself is not. Usually, the tiny whitish disk of the satellite is overpowered by the brighter clouds of Jupiter, rendering it invisible. Your best chance at spying the satellite in front of Jupiter will come by monitoring the scene before the event starts and then tracking the satellite as it begins its trans-Jovian trek.

An eclipse and occultation both occur every time a Jovian satellite hides behind Jupiter. The only difference between these two events is in what's doing the hiding. When the satellite moves into Jupiter's shadow, it is said to be in eclipse, while an occultation takes place when a satellite disappears behind Jupiter itself.

Saturn. Who can forget the rush of excitement at his or her first view of the planet Saturn? There, before the observer's eyes, stands a golden globe encircled by a bright, white ring. It's a heavenly sight surpassed by few others.

Saturn (Figure 10.6) is surrounded by a thick atmosphere that is similar in nature and composition to that of Jupiter. However, Jupiter's atmosphere is rich in detail; Saturn's is usually dull and nearly featureless. Most outstanding is the planet's whitish equatorial region and bordering northern and southern temperate zones. Try a deep yellow (Wratten No. 15) or orange (No. 21) to highlight some of the subtle features in the Saturnian atmosphere.

Every now and then, Saturn will surprise observers, such as back in September 1990. A bright white region grew to encircle fully one-quarter of the planet's equatorial zone. The so-called white spot of Saturn then slowly faded from view, disappearing almost two months after it first formed. Similar spots occurred in 1903, 1933, and 1960, approximately the same span of time it takes Saturn to complete one 29.5-year orbit of the Sun.

Figure 10.6 *Saturn. Photograph by Gregory Terrance (16-inch f/5 Newtonian reflector, Lynxx PC CCD camera, exposure 1 second at f/30).*

At first glance, Saturn might appear encircled by a single, solid ring. Closer inspection, however, shows that the ring is actually divided into separate zones, in much the same way as the selections on a phonograph record (remember them?) are divided by blank spaces. In the case of Saturn's rings, most amateur telescopes will reveal two major divisions. The outermost part of the ring is called the *A ring* and appears a bright gray. The broader, inner ring is called the *B ring* and is colored pure white. Separating the two is a pencil-thin dark gap called *Cassini's Division*. Under exceptional seeing conditions, all three may be distinguished through telescopes as small as 2 inches in aperture. Larger instruments will also show a fourth component to the ring system. The *C (or Crepe) ring* lies inside the B and appears dark gray. Due to its darkness, the C ring is most easily seen against the planet's bright disk. Trained observers can see great detail in the rings through large telescopes in perfect seeing, whereas the eyes of passing spacecraft Voyagers 1 and 2 back in the 1980s revealed hundreds of intricate subdivisions within the Saturnian family of rings.

The amount of detail visible in the rings depends strongly on their tilt to our line of sight. Because the rings are tilted approximately 27° to the ecliptic, they present different faces at different points in Saturn's 29-year solar orbit. For approximately 15 years we see one side of the rings, and then for the next 15 we see the other. In between, the rings are oriented edge-on, and they disappear from view. The rings of Saturn will next appear edge-on in 1995, while maximum presentation of their southern face will occur in 2003. After that, the rings will again appear edge-on in 2010, with their northern face seen for the next 15 years thereafter.

Saturn has at least 18 satellites orbiting it. The brightest is named Titan and shines at 8th magnitude. With a sharp eye and a little luck, an observer using just a pair of 7× binoculars can spot Titan. Small telescopes will show Titan easily as it circles Saturn once every 16 days. At least five others—Enceladus, Tethys, Dione, Rhea, and Iapetus—are within range of 8- to 10-inch telescopes, although they may prove difficult to tell apart from faint background stars. Only when the Saturnian system is tilted edge-on to our view will the satellites perform disappearing acts, such as occultations, eclipses, and transits, like Jupiter's.

Uranus, Neptune, and Pluto. Uranus shines between magnitude 5.5 and 6, making it barely perceptible with the unaided eye under superb skies. From most of our observing sites, however, it will probably require a telescope or binoculars to be seen. Through small instruments, it looks like an ordinary star, standing out from the crowd only by its barest hint of pale green. A 6- or 8-inch telescope will display a tiny, featureless disk, while 10-inch and larger instruments can also reveal the brightest of its 15 satellites.

Neptune's challenge to amateur astronomers is simply in finding it. Again, its pale but peculiar blue-green color helps it to stand out from the crowd of background stars. Measuring a few seconds of arc in diameter, Neptune appears so small that it remains a point when viewed with less than about 100×. Higher magnifications will show its tiny disk, though there is little hope of seeing any detail in its cloudy atmosphere.

We finally arrive at distant, icy Pluto. Even when it lies at its minimum distance of 2.76 billion miles from the Sun (which last occurred in 1989), Pluto is still an extremely faint object. Appearing no brighter than 13th magnitude, it usually requires at least an 8-inch telescope to be seen (although Pluto has been spotted in telescopes as small as 6-inchers under exceptional conditions). The only way to tell Pluto apart from the myriad of other faint stars is by using a detailed finder chart. Once the chart's field is within the telescope's view, look very carefully for a faint star-like object. That will be Pluto. Plot its position on the chart, then revisit the spot over several nights to confirm your sighting by watching the planet creep slowly against the background stars.

Annual finder charts for Uranus, Neptune, and Pluto are published in the January issues of *Sky & Telescope* and *Astronomy* magazines. Maps are also found in the Royal Astronomical Society of Canada's annual *Observer's Handbook* and Guy Ottewell's yearly *Astronomical Calendar*. These star maps are absolutely essential when looking for these elusive planets.

Asteroids. Lying between the orbits of Mars and Jupiter are thousands of small, irregularly shaped chunks of rocky debris. These are the asteroids or minor planets, leftovers from the formation of the Solar System over 4.5 billion years ago. Many asteroids are bright enough to be seen through amateur telescopes, though not even the largest, 600-mile-wide Ceres, will show up as anything more than a point of light.

Whenever a minor planet of about 8th magnitude or brighter is due to be visible, astronomical periodicals will publish a finder chart with its path plotted against the stars. To find the asteroid, employ the same technique used for locating Pluto as outlined above. Once your telescope or binoculars is centered on the area of sky where the asteroid is located, glance back and forth between the eyepiece field and the asteroid's finder chart. The asteroid will appear as a point of light near the predicted position on the chart.

If asteroiding interests you, then consider subscribing to the *Minor Planet Observer*, a monthly periodical written and published by Brian Warner. Each issue of the *MPO* features several asteroids that will be visible in the evening sky for the month of publication. To inquire about subscriptions, write BDW Publishing, Box 818, Florissant, Colorado 80816.

Comets. These are the nomads of our Solar System. Traveling in highly elliptical orbits, these frozen chunks of dirty ice arrive from beyond the orbit of Neptune, swing close by the Sun, and then return to the farthest reaches of the Solar System from where they came. As a comet swings by the Sun, solar energy sublimates frozen gases and dust on its surface to create the comet's coma, or head. Radiation pressure from the Sun and the solar wind combine to press against the coma's gas and dust, thus creating the comet's long tail.

Each year, a dozen or more comets either are discovered (about half by amateur astronomers) or reappear on schedule. Most are quite faint, but several may eventually become bright enough to be seen in amateur telescopes and binoculars. Each appears as a round, amorphous glow (the coma) highlighted by a brighter center. Careful telescopic study may reveal intricate filamentary structures in the coma, the result of gaseous jets erupting from the comet's solid nucleus.

Brighter comets (that is, those that become visible to the naked eye) may also display ghostly tails extending behind the coma and opposite the Sun. The brightest comets have dual tails (one called the *ion tail*, the other called the *dust tail*) that can extend 30° or more in length. These, the so-called great comets, are among the most spectacular sights ever to grace the night sky. The most recent great comet was Comet West seen in 1976 (and in Figure 10.7). Unfortunately, astronomers cannot predict when the next great comet will appear in our sky, though the odds say we are long overdue. Maybe one will be discovered tonight, or maybe not for another 50 years or more!

It is always fascinating to observe and record how a comet changes in brightness and appearance as it moves across the sky and around the Sun. Use

Figure 10.7 *Comet West. Photograph by author (400-mm f/6.3 lens, Tri-X film, 15-minute exposure).*

a low-power, wide-field eyepiece to view the comet at first, then increase magnification to check areas of interest. Be especially watchful of any fine detail in the comet's coma. Note its overall magnitude by comparing the coma's in-focus image to the out-of-focus glows of nearby stars of known brightness. A star atlas such as the *AAVSO Star Atlas* will be a great help in locating stars for comparison.

Next, estimate the apparent diameter of the coma by comparing its size to the true field of the eyepiece. Also check for any coloration in the coma or tail. Brighter comets frequently exhibit a bluish tinge. Lastly, estimate the length and shape of the comet's tail. Try using a Wratten No. 80A (light blue) filter to increase the contrast of the comet's tail against the background sky.

The most readily available source of information on any comets currently visible may be heard on a recorded telephone message called "Skyline," sponsored by *Sky & Telescope* magazine. Skyline is updated every Friday with news of the latest astronomical discoveries and events and may be heard by calling 1-617-497-4168 (toll rates to Cambridge, Massachusetts, apply).

The Sun. Before discussing *what* to look for, it is critically important to emphasize *how* to look at the Sun. Quite simply: **ALWAYS PRACTICE SAFE SUN!** The Sun's ultraviolet rays, the same rays that cause sunburn, will burn the retinas of your eyes almost instantaneously when concentrated by a telescope or binoculars. Moments later, the retina will be further cooked by the focused heat of visible and infrared light. Without taking proper safety precautions, permanent eye damage, even blindness, will result.

Happily, there are simple methods of looking at the Sun in complete safety. The safest is to use your telescope as a projector to cast an image of the Sun onto a white projection screen. The biggest drawback to solar projection is that without proper baffling, the Sun's image will be washed out by the bright surroundings. To overcome this, many dedicated solar amateurs have constructed projection boxes that attach to their instruments. For plans on how to make such a solar theater, review the design in Chapter 7.

If you wish to view the sun directly through a telescope or binoculars, special solar filters are required. DO NOT use photographic neutral-density filters, smoked glass, or overexposed photographic film, as suggested by some people. They can all lead tragically to blindness. The proper filters are commonly made of aluminized mylar film or glass coated with a nickel-chromium alloy and were discussed in detail in Chapter 6. Regardless of type, the filter must be securely mounted IN FRONT OF the telescope or binoculars before looking through the eyepiece. This way, the dangerously intense light and heat from the Sun are reduced to safe levels before entering the optics and your eyes.

Never place a sun filter between the eyepiece and your eyes (even though some inexpensive telescopes come with such filters). Telescopes not only magnify the Sun's light but also magnify its intense heat as well, which can quickly crack glass filters or burn through thin mylar film, allowing undiminished sunlight to burst into your unprotected eyes and cause permanent blindness.

Solar filters must be treated with great care, or they will quickly become damaged and unsafe to use. Regularly inspect the filters (especially the mylar type) for pinholes and irregularities in the coating by holding the filter up to a bright light. A small hole can be sealed with a tiny dot of flat black paint without causing the image to suffer. Dab just a bit of paint over the hole using a toothpick. If, however, more serious damage is detected, then the filters must be replaced immediately.

Whether you use a filter or projection, never look through the telescope's finderscope to aim the instrument at the Sun. (Be sure to keep the front dust cap on the finder; otherwise, the Sun's focused heat may melt the finder's crosshairs!) To aim at the Sun, take a look at the telescope's shadow cast on the ground. Move the telescope back and forth, up and down, until its shadow is at its shortest. The Sun should then be in the field of view.

One of the most interesting observations to make is to monitor the fluctuating number of sunspots across the Sun's visible surface, or *photosphere* (Figure 10.8). Sunspots appear at disturbed areas in the Sun's powerful, complicated magnetic field. Each consists of a black central portion, the *umbra*, and a surrounding grayish area called the *penumbra*. They may range in size from hundreds to thousands of miles in diameter.

Try to draw the Sun's disk every second or third day over a span of a month or more. With each successive sketch, sunspots will be seen to come and go, changing size and shape as they travel across the solar disk.

As described in Chapter 6, hydrogen-alpha solar filters display the Sun at one precise wavelength—656 nanometers. Viewing at that wavelength reveals a turbulence within our star that goes unsuspected through white-light sun filters. Huge, flamelike solar prominences, filaments, and other fascinating sights, normally visible only during a total solar eclipse, can be watched and monitored daily with a hydrogen-alpha filter.

As often as three times a year, the New Moon will pass in front of part or all of the Sun, eclipsing it as seen from somewhere on the Earth's surface. The occurence of a solar eclipse over a populated area always generates wide publicity in the news media and excites the curiosity of the general public.

Solar eclipses come in three varieties: partial, annular, and total. The same precautions must be exercised when viewing either a partial or annular (when a bright ring, or annulus, of sunlight remains even at maximum eclipse) solar eclipse as for viewing the uneclipsed Sun. Only when the Moon completely covers the Sun during the total phase of an eclipse is it safe to view without protection.

As this book is released in the spring of 1994, the astronomical world is on the brink of an outstanding annular eclipse that will pass across North America on May 10th. In case you miss this one, Appendix C lists upcoming total and annular eclipses from now until 2000.

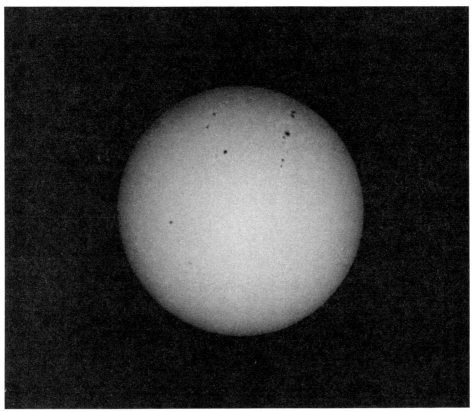

Figure 10.8 The Sun. Photograph by Brian Kennedy (8-inch f/10 Celestron 8 SCT, f/5 telecompressor, T-Max 100 film, $^{1}/_{250}$ second).

Deep-Sky Objects

Beyond the edge of our Solar System lies the rest of the universe, a most fascinating place to visit. Out there lie buried treasures that our eyes alone cannot see. But with the aid of a telescope or binoculars, exciting sights unequalled by any here on Earth await the amateur astronomer. Thousands upon thousands of stars, either single, multiple, or set in vast clusters, dot the night sky. Huge clouds of gas and dust—the nebulae—signal where new stars are being born or mark the graves of stars that once shone mightily. All of these lie within our galaxy, the Milky Way. Beyond it lie thousands of other galaxies, each a separate island in the endless ocean of the universe.

Be forewarned that most deep-sky objects appear disappointing at first. After being inundated with magnificent photographs of spectacular star clusters, colorful nebulae, and swirling galaxies, you get a telescope, aim it toward one of these distant masterpieces, and look through the eyepiece. What do you see? No color, no swirls, no magnificence; only a faint, gray smudge. Sad but true, the human eye cannot possibly perceive the level of detail or color that can be recorded on photographic film.

Disappointed? You'll probably be a little let down at first, but think about it. Here you are looking at sights hundreds, thousands, even millions of light years away with your own telescope. Also consider *what* you are looking at. As your friends watch reruns on television, you are seeing things that far less than 1% of humanity has ever seen before! There's a certain satisfaction in knowing that. Without even being aware of it, you might be looking toward unknown worlds scattered throughout gigantic systems of stars. Perhaps, on one of those distant worlds, another creature is looking through his or her (or its) equivalent of a telescope toward us, pondering the universe and your existence as well! Astronomy, it turns out, can be very cerebral. That can be its greatest attraction of all.

Here is an inventory of some of the wonders of the universe that await your inquisitive eye.

Double and Multiple Stars

Nearly half of the stars we see at night are double or multiple stars. Though not all doubles can be resolved visually through telescopes, of those that can, no two pairs appear exactly the same. Some are separated by great distances, and others seem to touch each other. Of course, there are also the impostors— two stars that just happen to lie along the same line of sight from Earth. These are referred to as *optical doubles*.

Many double stars shine pure white, while others flicker with distinctive colors. Most appear in pairs, though in some cases as many as six or seven stars belong to a single system.

Variable Stars

Most stars shine consistently at the same brightness, day after day, year after year. Others, called variable stars, appear to fluctuate in brightness. Some variables oscillate rhythmically over a predictable period of time; others burst unpredictably, flaring up to six magnitudes in brilliance in a matter of minutes or hours.

There are three major classes of variables: *eclipsing binaries*, *pulsating stars*, and *eruptive variables*. Eclipsing binaries are pairs of stars that are, strictly by chance, seen nearly edge-on from our earthbound vantage point. The stars alternately pass in front of each other, causing temporary diminishings of their combined light. Pulsating stars actually expand and contract in size, causing them to alternately fade and brighten. The most common type of pulsating variables are the long-period variables. These can take weeks or months to complete brightness cycles. Eruptive variables change in brightness very quickly.

Examples of all of these are visible through amateur telescopes and binoculars. To learn more about observing variable stars and how you can make a valuable contribution to their study, contact the American Association of Variable Star Observers at 25 Birch Street, Cambridge, Massachusetts 02178.

Star Clusters

These come in two varieties. Randomly shaped swarms of mostly young blue and white stars are called *open* or *galactic clusters*. Each may contain from a dozen to several hundred individual points of light. Some are loose groupings with little or no apparent central concentration, whereas others are squashed into stellar traffic jams. Most open clusters are found within the spiral arms of our galaxy.

The second type of star clusters, called *globular clusters*, encircle the Milky Way likes moths around a flame. Globulars are huge, spherical conglomerations of hundreds of thousands or even millions of stars. Unlike the young, brash stars in open clusters, the stars in globular clusters are generally among the oldest known. Even at the vast distances at which they lie, globulars appear surprisingly bright. Of the more than 100 known within our galaxy, most are visible in 3-inch telescopes. Typically, at least a 6-inch telescope is needed to resolve some of the stellar constituents into separate points of light.

Nebulae

Diffuse nebulae are enormous clouds of dust and gas—primarily hydrogen—and are found chiefly in the spiral arms of the Milky Way and in other spiral galaxies. Diffuse nebulae that shine brightly (relatively speaking) are called bright nebulae and come in two different varieties. If the gases in a nebula are excited into luminescence by the energy from young, hot stars within, the nebula itself will glow, much like the gas in a neon sign. These are called *emission nebulae*. Other nebulae are illuminated only by reflecting light from the stars they surround and are therefore called *reflection nebulae*. Emission nebulae appear reddish in photographs, while reflection nebulae appear bluish.

Dark nebulae are similar in constitution to their brighter cousins above but do not emit or reflect light. Instead, they obscure light from all objects behind, appearing as starless holes silhouetted against a star-filled field.

While diffuse nebulae mark stellar birth, *planetary nebulae* chronicle stellar death. Planetary nebulae are small, spherical shells of gas expelled by less massive stars during their death throes. These spheres are characterized by the glow of ionized oxygen, causing most to appear green or turquoise. The star that expelled the gaseous shell remains buried in the middle of the nebula and is referred to as the *central star*.

When the most massive stars go through their death throes, they go out in a big way! During its final moments, much of the star's mass is expelled in a tremendous supernova explosion. When the explosion clears, all that remains of the star is its dense core and an expanding cloud of gaseous debris called a *supernova remnant*.

Galaxies

All of the objects spoken of until now are within the confines of our galaxy, the Milky Way. Beyond lie other galaxies, each made up of tens of millions to hundreds of billions of individual stars.

Galaxies are divided into three major classifications. Those that resemble huge pinwheels, with two or more long, curved arms extending away from a central core, are called *spiral galaxies*. The Milky Way is considered to be a spiral galaxy. The second type, *elliptical galaxies*, show no hint of spiral structure. Instead, ellipticals appear as huge, oval spheres with no internal organization of any sort. Finally, if a galaxy has no distinctive shape, then it is categorized as an *irregular galaxy*. Some irregulars seem to exhibit hints of ill-defined spiral arms, though any evolutionary tie remains unknown.

A Celestial Inventory

To inventory the universe, astronomers have devised several different ways of labeling stars and other celestial objects. Only the brightest stars, such as Vega and Betelgeuse, bear names from antiquity that remain familiar today. While many others also have exotic names, few modern-day astronomers take the time to remember them.

Only a few short years before the invention of the telescope, astronomer Johannes Bayer created that era's most detailed atlas of the night sky. He chose to identify the brightest stars in each constellation by lowercase letters from the Greek alphabet. He usually labeled a constellation's brightest star *alpha* and then, working his way through the traditional constellation figure from head to toe, labeled succeeding stars *beta*, *gamma*, and so on. Once completed, he repeated the head-to-toe sequence for any fainter stars that remained, sometimes until all 24 letters of the Greek alphabet were used up. There are many exceptions to this pattern, but it holds true for the most part. Bayer's classification stuck and is still in use today. Table 10.1 lists the Greek alphabet by name and corresponding letter.

Therefore, the bright star Vega is called Alpha Lyrae, while nearby Albireo is Beta Cygni (*Lyrae* and *Cygni* being the genitive forms of the constellation names Lyra and Cygnus, respectively).

In order to extend the Greek alphabet system, British astronomer John Flamsteed assigned numbers to all stars of about fifth magnitude and brighter in each constellation. These Flamsteed numbers begin at 1 in each constellation and increase from west to east. Fainter stars have subsequently been cat-

Table 10.1	**The Greek Alphabet**	
	alpha	α
	beta	β
	gamma	γ
	delta	δ
	epsilon	ε
	zeta	ζ
	eta	η
	theta	θ
	iota	ι
	kappa	κ
	lambda	λ
	mu	μ
	nu	ν
	xi	ξ
	omicron	o
	pi	π
	rho	ρ
	sigma	σ
	tau	τ
	upsilon	υ
	phi	φ
	chi	χ
	psi	ψ
	omega	ω

alogued in many lists such as the SAO (Smithsonian Astrophysical Observatory), HD (Henry Draper) and other directories.

Nonstellar deep-sky objects are also catalogued individually. While some of the more spectacular examples have unofficial nicknames (such as the Orion Nebula or the Andromeda Galaxy), most do not; they do, however, have catalogue numbers assigned to them. For instance, in this and other astronomy books, the Orion Nebula is referred to as M42, and the Andromeda Galaxy is listed as M31. These are the entries assigned to them in the famous Messier catalogue of deep-sky objects. The Messier catalogue, listing 109 of the finest nonstellar objects in the sky, was largely created in the eighteenth century by Charles Messier, a French comet hunter. (The list actually goes up to M110, but it is now generally agreed that M102 was a mistaken repeat observation of M101.)

Another important catalogue of deep-sky objects is the New General Catalogue, or NGC, compiled at the end of the nineteenth century by an astronomer named John Dreyer. The NGC lists more than 7,800 clusters, nebulae, and galaxies, including all but a few of the Messier objects. For example, M42 is NGC 1976, while M31 is NGC 224. An extension of the NGC is the Index Catalogue (IC).

Selected Deep-Sky Targets

Searching for and observing deep-sky objects has become one of the most popular pastimes for today's amateur astronomer. It's a great way to learn the night sky and hone your observational talents all at the same time.

The following is a brief selection of my favorite deep-sky objects. Each is accompanied by a description of its visual appearance through a variety of telescopes, as well as a finder chart and a photograph. All (except open clusters) are also accompanied by a drawing made through either an 8-inch or 13.1-inch telescope, as noted. By relating the descriptions and drawings to the photos and then to what you see through your own telescope, subtle details will become more readily visible. Other data listed for each object include its constellation, epoch 2000 right ascension (R.A.) and declination (Dec.) coordinates, visual magnitude, apparent diameter, and tips on how to find it. Refer to the all-sky chart (Figure 10.9) to find out which of these celestial masterpieces will be visible tonight.

These showpiece objects are just a beginning. Once you have conquered these, move on to Appendix E, which lists the entire Messier catalogue as well as the finest non-Messier objects. There is a lot out there to see.

Spring

M44 (NGC 2632)   Open Cluster in Cancer
R.A. 08^{h}40^m.1   Dec. +19° 59'   Mag. = 3.1   Diameter = 95'

Where to look: Find the keystone-shaped body of Cancer the Crab about halfway between the twin stars Castor and Pollux of Gemini and Regulus in Leo. Under moderately dark skies, M44 is visible as a soft glow inside the Crab's body. (See Chart 10.10a.)

Description: Not all observers will immediately appreciate the grandeur of M44, the Beehive Cluster (Figure 10.10b). Due to its incredible 95' span, M44 is difficult to fit into the field of view of ordinary eyepieces above 30× to 40× magnification. As a result, the clustering effect becomes lost, and the observer may feel less than impressed. However, through binoculars, finders, and rich-field instruments, the bees in this celestial hive come alive in sparkling style! Close to six dozen points of light are seen across a field littered by the combined glow of another 130 fainter stars.

M81 (NGC 3031)   Spiral Galaxy in Ursa Major
R.A. 09^{h}55^m.6   Dec. +69° 04'   Mag. = 7.0   Diameter = 26'×14'
M82 (NGC 3034)   Irregular Galaxy in Ursa Major
R.A. 09^{h}55^m.8   Dec. +69° 41'   Mag. = 8.4   Diameter = 11'×5'

Where to look: Draw a diagonal line between Phecda (Gamma Ursae Majoris, the southeast star in the Big Dipper's bowl) and Dubhe (Alpha Ursae Majoris, the northwest Bowl star) and extend that line an equal distance toward

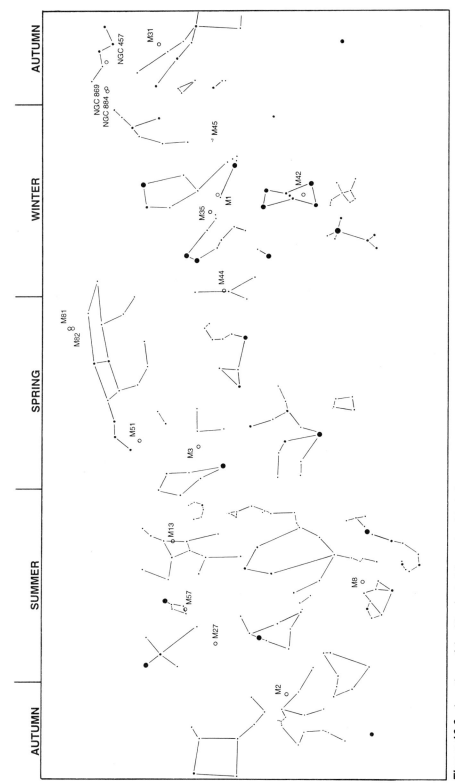

Figure 10.9 Locations of the 16 selected deep-sky objects to come.

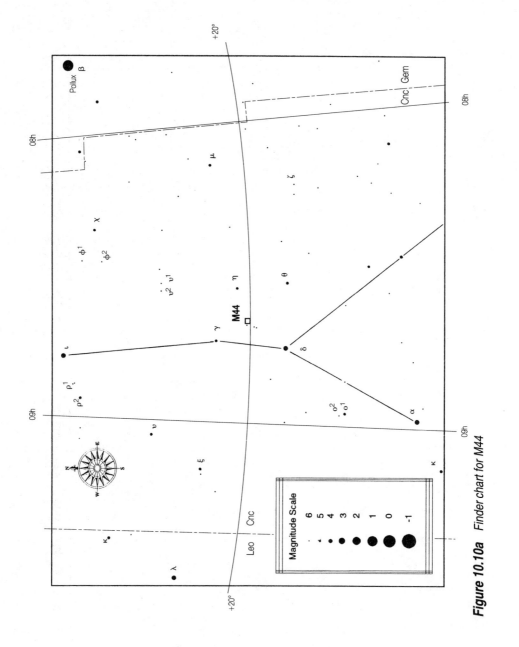

Figure 10.10a Finder chart for M44

Figure 10.10b *M44, the Beehive Cluster. Photograph by George Viscome (135-mm telephoto lens at f/3.5, hypered Tech Pan film, 20-minute exposure). South is up.*

the northwest. Just west of that point is an isosceles triangle made up of the 5th-magnitude stars Rho, Sigma[1] and Sigma[2] Ursae Majoris. Follow the base of the triangle from Sigma[1] to Sigma[2] and extend it about 3.5° to the northeast, where lies 5th-magnitude 24 Ursae Majoris and, just over a degree farther east, M81 and M82. (See Chart 10.11a.)

Description: M81 and M82 (Figures 10.11b and 10.11c) combine to create one of the finest close-set pair of galaxies found anywhere in the sky! M81 is an outstanding example of a type-Sb spiral galaxy. Given sharp eyes and mod-erately-dark skies, observers can spy its distinctive oval shape even in binocu-lars. Small- and medium-aperture amateur telescopes show the bright core surrounded by the fainter spiral halo, while 12-inch and larger scopes begin to suggest the galaxy's curled structure.

Irregular galaxy M82 reveals a long, slender form in all amateur telescopes. Six- to 8-inch instruments hint at the unusual dark lane that cuts through the galaxy's core, while 10-inch and larger scopes add many mottled bright and dark patches across the length of the object.

M51 (NGC 5194)   Spiral Galaxy in Canes Venatici
R.A. 13^h 29^m.9   Dec. +47° 12′   Mag. = 8.4   Diameter = 11′ × 8′

Where to look: Begin at Alkaid (Eta Ursae Majoris) at the end of the handle of the Big Dipper. Viewing through your finder, slide 2° to the west of Eta and pause at 6th-magnitude 24 Canum Venaticorum. This star forms a slender isosceles triangle with Eta and M51. Recreate that triangle in your finderscope

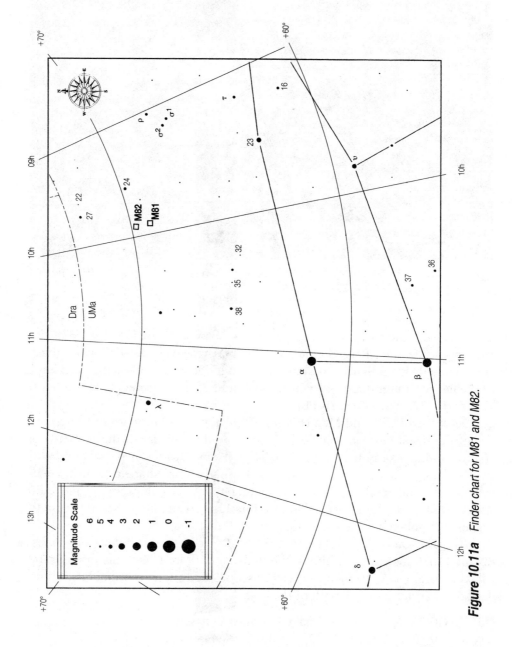

Figure 10.11a Finder chart for M81 and M82.

Figure 10.11b *M81 and M82. Photograph by George Viscome (14.5-inch f/6 Newtonian, hypered Tech Pan 2415 film, 20-minute exposure). South is up.*

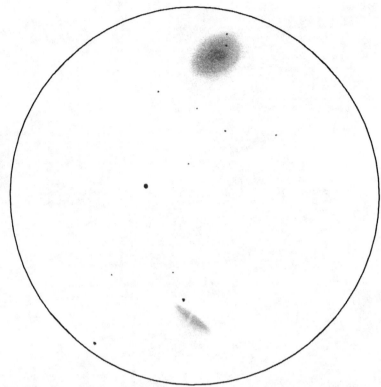

Figure 10.11c   *M81 and M82. Drawing by the author using an 8-inch f/7 Newtonian and 24-mm Tele Vue Wide Field eyepiece (59×).*

and then look through your telescope. You should be close enough to the object to find it with a low-power eyepiece. (Though not shown on Chart 10.12a, a small parallelogram of four 7th-magnitude stars surrounds M51. Hop from Eta to 24 Canum Venaticorum, then into this parallelogram. Once these four stars are spotted, look for M51 just inside the northeastern corner.)

Description: One of the finest examples of an Sc spiral galaxy found anywhere in our skies. M51 and its close companion galaxy NGC 5195 (Figures 10.12b and 10.12c) are both visible in 10×70 binoculars (maybe even smaller) as a pair of faint fuzzies situated in a nice star field. Their apparent sizes expand as telescope aperture grows, but not until the 8-inch aperture class is used will they take on a new and exciting appearance. In these scopes (and all larger), M51 begins to show some of its remarkable spiral structure. An 8-inch will hint at the spirality only on clear, dark nights, while my 13.1-inch f/4.5 Newtonian will show the arms regularly from suburban skies. And the view I had of M51 through a 24-inch f/5 Newtonian a few years ago took my breath away! It looked just like a photograph.

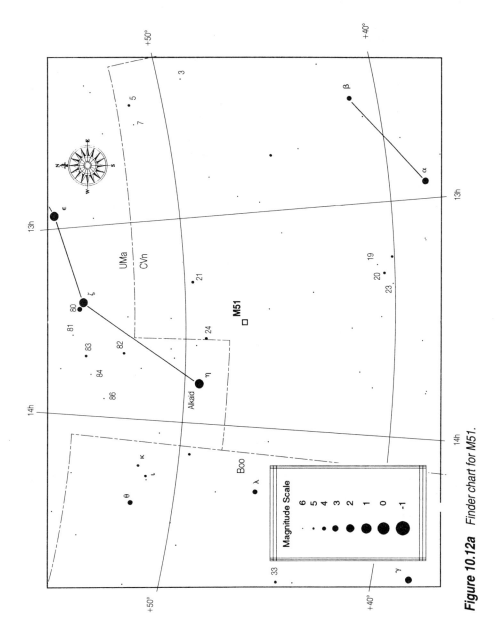

Figure 10.12a *Finder chart for M51.*

Figure 10.12b *M51, the Whirlpool Galaxy. Photograph by George Viscome (14.5-inch f/6 Newtonian, hypered Tech Pan 2415 film, 77-minute exposure). South is up.*

M3 (NGC 5272)   Globular Cluster in Canes Venatici
R.A. 13ʰ 42ᵐ.2   Dec. +28° 23′   Mag. = 6.4   Diameter = 16′
   Where to look: From Beta Comae Berenices, scan with your finderscope about 6° due east, keeping an eye out for a lone 6th-magnitude star. Right next door, and about equally bright, will be M3. (Expanding the view beyond what

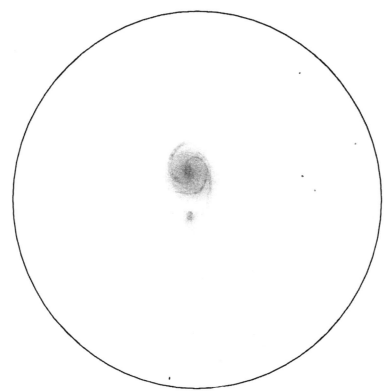

Figure 10.12c *M51. Drawing by the author using a 13.1-inch f/4.5 Newtonian and a 12-mm Tele Vue Nagler eyepiece (125×).*

is shown in Chart 10.13a, M3 lies a little less than halfway between Arcturus in Boötes to Cor Caroli in neighboring Canes Venatici.)

Description: M3 (Figures 10.13b and 10.13c) is a beautiful globular cluster for all amateur telescopes. Although small instruments will show only a nebulous patch of interstellar cotton, 6-inch and larger scopes begin to resolve the group into a myriad of faint suns. Try the highest power that optics and sky conditions permit for the best view.

Summer

M13 (NGC 6205)   Globular Cluster in Hercules
R.A. 16ʰ 41ᵐ.7   Dec. +36° 28′   Mag. = 5.9   Diameter = 16′

Where to look: Begin at the Hercules Keystone quadrangle, found about two-thirds of the way from the bright star Arcturus to brilliant Vega. Aim at Eta Herculis, the Keystone's northwestern star. Move your telescope along the

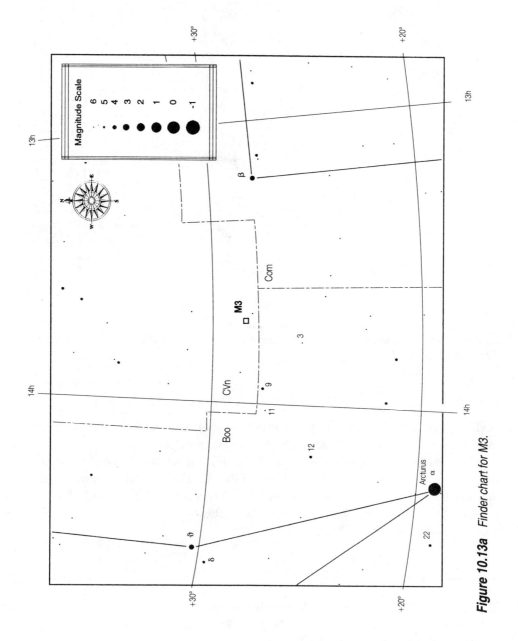

Figure 10.13a Finder chart for M3.

Figure 10.13b *M3. Photograph by George Viscome (14.5-inch f/6 Newtonian, Tri-X film, 20-minute exposure). South is up.*

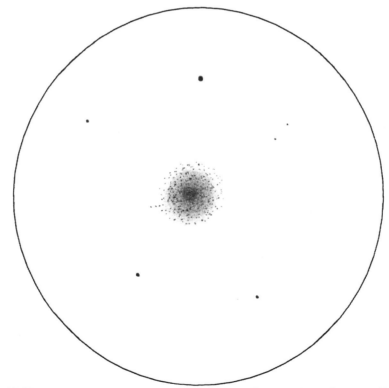

Figure 10.13c *M3. Drawing by the author using an 8-inch f/7 Newtonian and a 12-mm Tele Vue Nagler eyepiece (119x).*

western edge of the Keystone. M13 lies about one-third of the way from Eta to Zeta Herculis at the southwest corner. (See Chart 10.14a.)

Description: This is one of the finest globular clusters visible from mid-northern latitudes. Like M3 in the Spring sky, M13 (Figures 10.14b and 10.14c) requires a 6-inch telescope for partial resolution of its estimated 100,000 stars. Again, medium- and high-power oculars should produce the best results, but avoid the temptation to overpower your telescope and spoil the image.

As you gain experience, be sure to revisit M13 often to look for some fine structural detail. Many eighteenth- and nineteenth-century astronomers commented on how many of the cluster's stars form chains or rows; they remind me of the legs of a spider. This effect is quite apparent in 8-inch and larger telescopes but not as clear in smaller instruments (though some have recorded them in telescopes as small as 6 inches). At the same time, look for three peculiar dark lanes forming what looks like a propeller set off-center in the cluster. Once again, they grow more obvious as aperture increases. Can you see either of these unusual features?

M8 (NGC 6523)   Bright Nebula in Sagittarius
R.A. $18^h 03^m.8$ Dec. $-24° 23'$ Mag. = 5.8   Diameter = $90' \times 40'$

Where to look: With your eyes alone, draw an imaginary line between Sigma Sagittarii and Lambda Sagittarii, stars in the handle and lid of the teapot, respectively. Again, using your eyes alone, extend that line an equal distance toward the west-northwest and aim your finderscope toward that imaginary point. The finder should reveal the pair of 5th-magnitude stars 4 and 7 Sagittarii. Aiming at the latter will put the western half of M8 in your telescope's view. (See Chart 10.15a.)

Description: M8 (Figures 10.15b and 10.15c) appears as a great cloud through amateur telescopes. Telescopes up to 6 inches in aperture unveil a soft glow of remarkable complexity sliced in half by a dark lane, described once as a "lagoon." Though most modern-day observers fail to see the similarity to a lagoon (it strikes me more like a canal or channel), the name stuck, dubbing this the Lagoon Nebula. Eight- to 10-inch instruments uncover some of the fainter portions of the nebula, while larger telescopes unleash a tumultuous cloud. Always begin a visit to M8 with a low-power, wide-field ocular. This way, most of its full $90' \times 40'$ girth will squeeze inside a single field of view. Then switch to a higher power to zero in on areas that appear particularly interesting.

Engulfed in the clouds of M8 is the open cluster NGC 6530, an attractive congregation of about two dozen stars ranging from 7th to 9th magnitude and scattered across 15 arc-minutes. This isn't a bad deal: two objects for the price of one!

M57 (NGC 6720)   Planetary Nebula in Lyra
R.A. $18^h 53^m.6$ Dec. $+33° 02'$ Mag. = 9.7   Diameter = $70'' \times 150''$

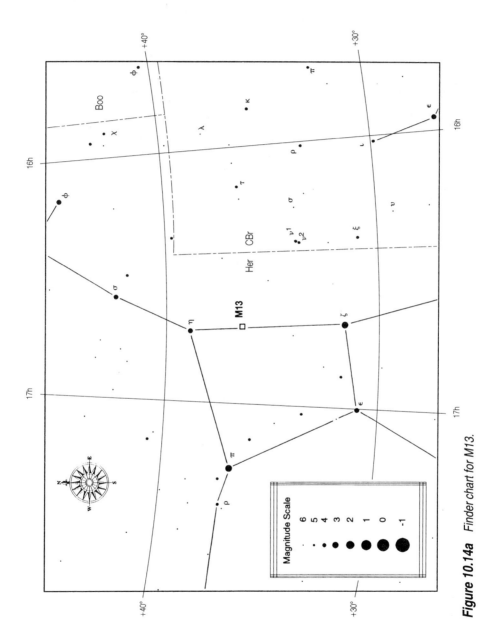

Figure 10.14a Finder chart for M13.

Figure 10.14b   *M13, the Great Globular Cluster in Hercules. Note the 12th-magnitude galaxy NGC 6207 in the lower right corner. Photograph by George Viscome (14.5-inch f/6 Newtonian, Tri-X film in cold camera, 30-minute exposure). South is up.*

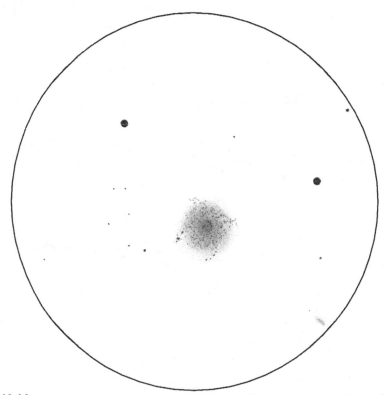

Figure 10.14c   *M13. Drawing by the author using an 8-inch f/7 Newtonian and a 26-mm Tele Vue Plössl eyepiece (55×).*

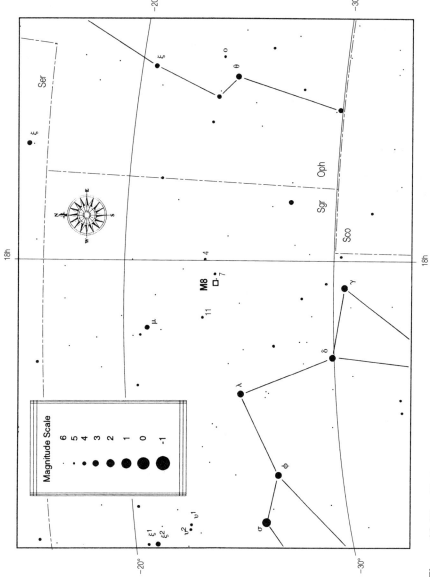

Figure 10.15a *Finder chart for M8.*

Figure 10.15b *M8, the Lagoon Nebula. Photograph by George Viscome (8-inch f/5.6 Newtonian, hypered Tech Pan 2415, 46-minute exposure). South is up.*

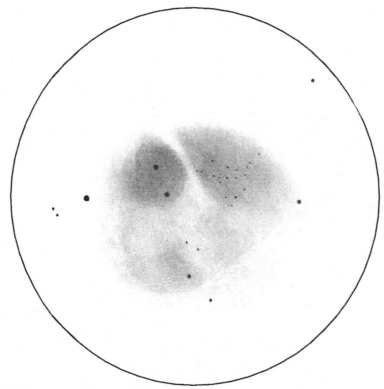

Figure 10.15c *M8. Drawing by the author using a 13.1-inch f/4.5 Newtonian, a 24-mm Tele Vue Wide Field eyepiece and a DayStar LPR filter (62×).*

Where to look: M57 lies almost exactly halfway between Beta Lyrae (Sheliak) and Gamma Lyrae (Sulafat) along the southern edge of Lyra's parallelogram. (See Chart 10.16a.)

Description: This is the famous Ring Nebula, an outstanding little smokering of stellar debris. Visible in binoculars as a faint point of light, M57 (Figures 10.16b and 10.16c) displays its annular shape in telescopes as small as 2 inches in aperture. A 6-inch instrument adds to the prominence of the ring effect, while 8-inch and larger scopes begin to show a subtle fraying along the edge of the nebula's east-west axis. Conditions permitting, try an eyepiece yielding at least 100× for the best view.

Can you spy the Ring's elusive 15th-magnitude central star? The star's visibility is greatly hampered by its inherent dimness against the intrinsic brightness of the surrounding nebulosity. Though some observers claim to have seen it in 10-inch telescopes, I have never seen it through my 13.1-inch Newtonian. Perhaps I'm not trying hard enough?

M27 (NGC 6853)   Planetary Nebula in Vulpecula
R.A. 19ʰ 59ᵐ.6   Dec. +22° 43′   Mag. = 8.1   Diameter = 480″×240″

Where to look: Center your view on Gamma Sagittae, the easternmost star in the Arrow, then turn north. About 3.5° later, watch for 6th-magnitude 14 Vulpeculae. M27 lies just to its southeast, near a 9th-magnitude sun. (See Chart 10.17a.)

Description: M27 (Figures 10.17b and 10.17c), nicknamed the Dumbbell Nebula, has one of the highest surface brightnesses of just about any planetary nebula in the northern sky. Even binoculars will show its fuzzy disk settled among a rich portion of the summer Milky Way. While the nickname comes from its resemblance to an old-fashioned dumbbell, it strikes me as more reminiscent of an hour- glass (indeed, another pet name for M27 is the Hourglass Nebula). Viewing through an 8-inch telescope reveals some blue-green coloring to the Dumbbell and begins to fill in the hourglass curve with fainter nebulous extensions, especially when viewed with a narrow-band LPR filter in place. These extensions grow in brightness as aperture increases, impairing the hourglass analogy but adding to the overall majesty. Large amateur scopes will also reveal the nebula's faint central star.

Autumn

M2 (NGC 7089)   Globular Cluster in Aquarius
R.A. 21ʰ 33ᵐ.5   Dec. −00° 49′   Mag. = 6.5   Diameter = 13′

Where to look: Begin at the star Alpha Aquarii (Sadalmelik). Orient yourself to the finder's field by identifying the fainter stars 28, 32, and Omicron Aquarii. Next, extend a line from Alpha to 28, then beyond. Four degrees to the northwest of 28 lies 11 Pegasi (identified on some star charts as 27 Aquarii). This star forms the tip of an equilateral triangle with 25 and 26 Aquarii. Center this triangle in your finder, then head southwest 2° toward a lone 6th-magnitude

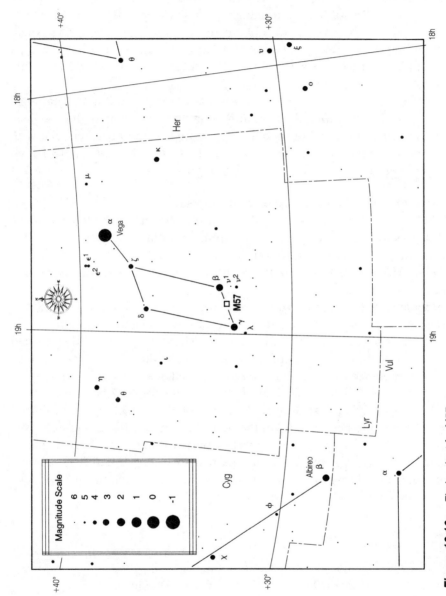

Figure 10.16a *Finder chart for M57.*

Figure 10.16b   *M57, the Ring Nebula. Photograph by George Viscome (14.5-inch f/6 Newtonian, hypered Tech Pan 2415 film, 15-minute exposure). South is up.*

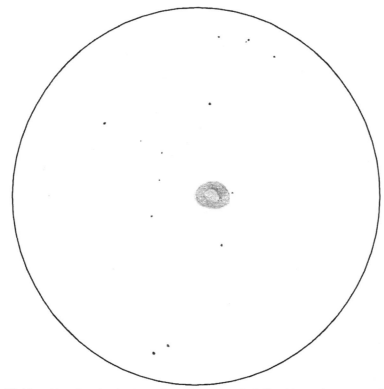

Figure 10.16c   *M57. Drawing by the author using a 13.1-inch f/4.5 Newtonian, a 7-mm Tele Vue Nagler eyepiece, and a Daystar LPR filter (214×).*

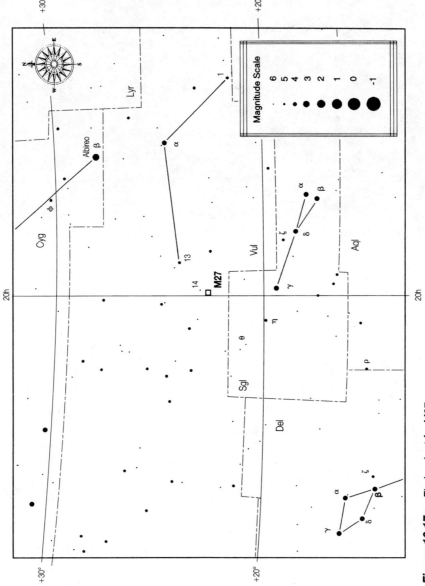

Figure 10.17a *Finder chart for M27.*

Figure 10.17b *M27, the Dumbbell Nebula. Photograph by George Viscome (14.5-inch f/6 Newtonian, hypered Tech Pan 2415 film, 26-minute exposure). South is up.*

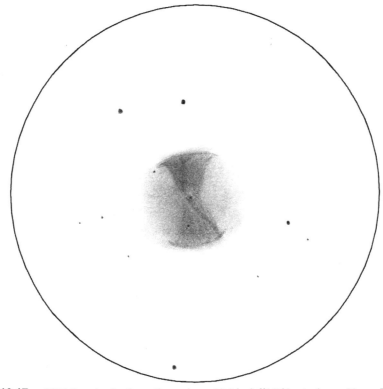

Figure 10.17c *M27. Drawing by the author using a 13.1-inch f/4.5 Newtonian, a 12-mm Tele Vue Nagler eyepiece, and a Daystar LPR filter (125×).*

star. From here, M2 lies about 1° farther west-southwest and should be visible in finderscopes as a faint smudge. (See Chart 10.18a.)

Description: M2 (Figures 10.18b and 10.18c) is one of autumn's finest glob-ular clusters. Small telescopes reveal it as a round puff of cotton marked by a bright center and diffuse edges. Partial resolution of M2 is possible in a 6-inch telescope, with some of the myriad of 13th-magnitude cluster stars visible around its outer edge. Twelve-inch and larger scopes smash the core of M2 by revealing uncountable stars teeming throughout.

M31 (NGC 224)   Galaxy in Andromeda
R.A. 00^h 42^m.7   Dec. +41° 16'   Mag. = 3.5   Diameter = 160' × 40'

Where to look: As outlined in Chapter 9, begin your quest for M31 at Beta Andromedae (Mirach). Head 4° northwest to the naked-eye star Mu Andro-medae, then northwest again for 3° to Nu Andromedae. You might have some trouble spotting these latter two stars, especially Nu, without optical aid if you are a captive of light pollution, but your finder will show them easily. From Nu, nudge your telescope due east for 1.33° for M31, which should be visible in the finder as a knockwurst-shaped smudge. (See Chart 10.19a.)

Description: M31, the Andromeda Galaxy, (Figures 10.19b and 10.19c) is a delight to behold in all instruments regardless of size. Under dark skies, low-power binoculars, and even the eyes alone, reveal the galaxy's huge span. Be-cause of its size, most telescopes can only squeeze portions of the galaxy into single fields of view. A well-planned tour of M31 begins at the galaxy's center. The bright, oval galactic core reaches a visual crescendo as it draws toward an intense stellar nucleus. Stretching out toward the northeast and southwest from the core is the comparatively faint glow of the spiral arms. A 6-inch tele-scope will reveal one, possibly two, bands of dark nebulosity along the north-west side of the great oval. These dark lanes grow more obvious as aperture increases. Also coming into view with larger telescopes is NGC 206, a rectan-gular star cloud found about 0.66° southwest of the core.

Two smaller companion galaxies stay close by the side of M31. The brighter of the pair is M32 (NGC 221), an elliptical galaxy found 24' due south of M31's core. Telescopes reveal M32 as an oval glow also punctuated by a bright stellar nucleus. The third member of this galactic family is M110, also known as NGC 205. Though twice the apparent size of M32, M110 appears much fainter because of its lower surface brightness. Look for its nondescript oval glow about 35' northwest of the center of M31. In place of the bright nucleus found in M32, M110 reveals only a homogeneous glow across its face.

NGC 457 Open Cluster in Cassiopeia
R.A. 01^h 19^m.1   Dec. +58° 20'   Mag. = 6.4   Diameter = 13'

Where to look: Extend the line along the eastern leg of the W shape of Cas-siopeia from Epsilon to Delta Cassiopeiae and then, looking through your finder, about one-third farther to 5th-magnitude Phi Cassiopeiae. Center Phi

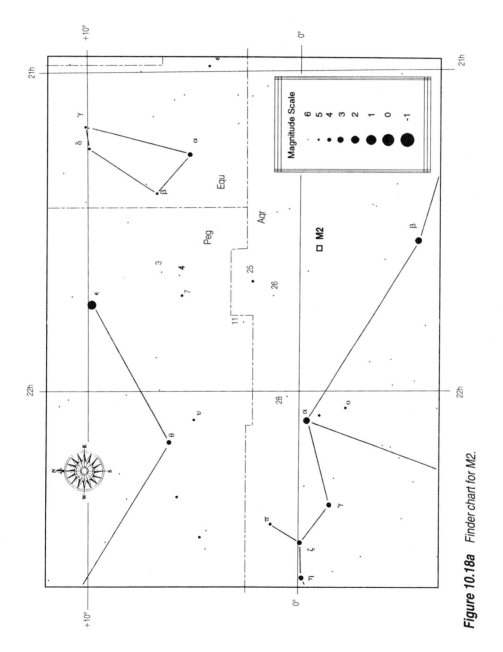

Figure 10.18a *Finder chart for M2.*

Figure 10.18b *M2. Photograph by George Viscome (14.5-inch f/6 Newtonian, Tri-X film, 60-minute exposure). South is up.*

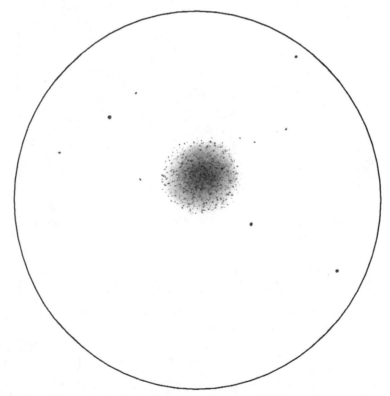

Figure 10.18c *M2. Drawing by the author using a 13.1-inch f/4.5 Newtonian and a 12-mm Tele Vue Nagler eyepiece (125×).*

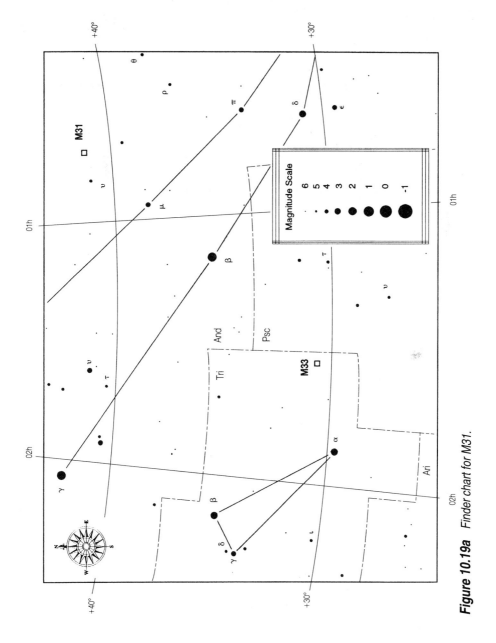

Figure 10.19a Finder chart for M31.

Figure 10.19b   *M31, the Andromeda Galaxy. Photograph by George Viscome (3-inch f/6.6 refractor, hypered Tech Pan 2415 film, 86-minute exposure). South is up.*

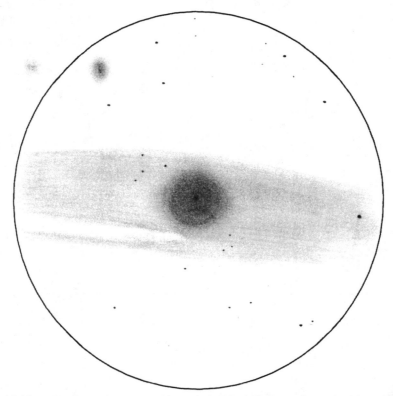

Figure 10.19c   *M31. Drawing by the author using an 8-inch f/7 Newtonian and a 24-mm Tele Vue Wide Field eyepiece (59×).*

in the field and switch to your telescope to see NGC 457 bordering Phi to the north. (See Chart 10.20a.)

Description: Although Phi appears to lie within the cluster's boundaries, some studies indicate that it is only a foreground star that happens to lie along the same line of sight as the cluster. True cluster member or not, Phi combines with a second foreground star, 7th-magnitude HD 7902, to create a pair of eyes staring back at us (Figure 10.20b). The cluster's true members shine between 8th and 11th magnitude and fall into a pattern that resembles a bird in flight. The bird's body is created from about a dozen stars of magnitudes 9 to 11. A pair of 10th-magnitude suns mark its tail feathers. The wings are comprised of about a half dozen stars each, set in long, graceful arcs. The east wing is highlighted by a distinctive 8th-magnitude orange star, which is the cluster's brightest star. Overall, the pattern resembles an owl, and so NGC 457 became known as the Owl Cluster.

NGC 869 Open Clusters in Perseus
R.A. 02^{h}19^m.0   Dec. +57° 09'   Mag. = 4.3   Diameter = 30'
NGC 884
R.A. 02^{h}22^m.4   Dec. +57° 07'   Mag. = 4.4   Diameter = 30'

Where to look: These open clusters, known collectively as the Double Cluster, are also easily found by using the W of Cassiopeia. Referring back to Chart 10.20a, extend the line from Gamma to Delta Cassiopeiae toward the stars of neighboring Perseus to the southeast. Center your attention about halfway between Delta Cassiopeiae and Gamma Persei. Even from suburbia, most observers can spot a dim smudge of light at this point. Aim your finderscope there to see two tight clumps of stars scattered in a rich starry field. Those clumps are NGC 869 and NGC 884.

Description: The Double Cluster (Figure 10.21) creates one of the most striking views to be had through a telescope or binoculars. Though attractive in all instruments, they are best seen through low-power short-focal-length telescopes and giant binoculars. These display a field of stardust of unparalleled beauty. Countless stars are strewn across the view, clustering together into two tight knots. Look toward the center of each group to see several tiny triangles and other geometric patterns created by the cluster stars. Most of the suns appear blue-white and white, though with some effort a few shining with subtle hues of yellow and red can be detected.

Winter

M45 Open Cluster in Taurus
R.A. 03^{h}47^m.0   Dec. +24° 07'   Mag. = 1.2   Diameter = 110'

Where to look: M45, better known as the Pleiades or Seven Sisters, is visible to the naked eye as a tiny dipper-shaped pattern of stars riding on the back of Taurus the Bull, west of the Bull's head and its bright star Aldebaran. (See Chart 10.22a.)

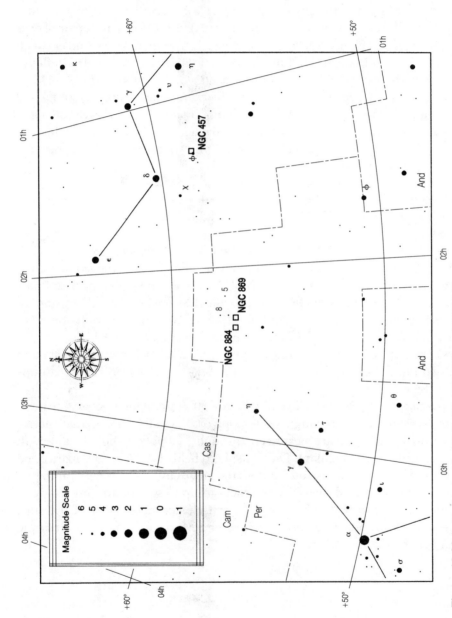

Figure 10.20a Finder chart for NGC 457, NGC 869, and NGC 884.

Figure 10.20b   NGC 457, the Owl Cluster. Photograph by George Viscome (14.5-inch f/6 Newtonian, hypered Tech Pan 2415 film, 45-minute exposure). South is up.

Figure 10.21   NGC 869 and NGC 884, the Double Cluster. Photograph by George Viscome (14.5-inch f/6 Newtonian, hypered Tech Pan 2415 film, 45-minute exposure). South is up.

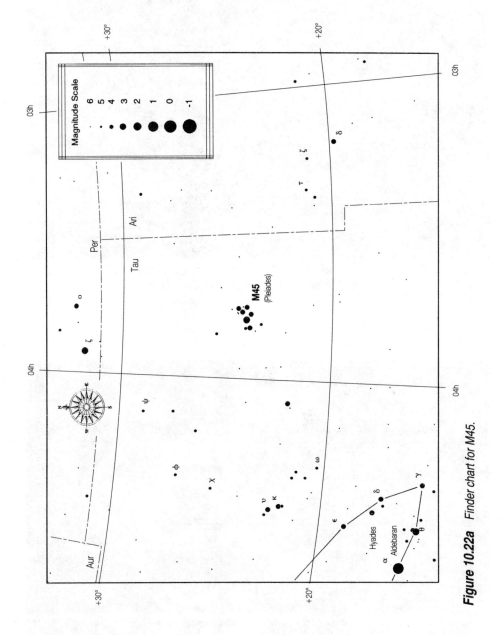

Figure 10.22a Finder chart for M45.

Description: I would like to nominate this as the grandest open star cluster of all for binoculars. Even the smallest glasses will cause a population explosion in the family of the Seven Sisters by revealing dozens of fainter stellar siblings (Figure 10.22b). The half dozen stars visible to the naked eye increases by a factor of ten or more when M45 is seen through 7× binoculars. An outstanding vista awaits observers viewing with giant binoculars, as dozens of brilliant blue-white stellar sapphires fill the field. Current estimates place at least 100 stars within the Pleiades.

Under very dark, very clear conditions, it is also possible to glimpse some patches of reflection nebulosity that engulfs some of the cluster stars. The brightest patch is labeled NGC 1435 and surrounds Merope, the southernmost star in the Pleiades' bowl-shaped grouping. Look for it extending to the south of the star. (Here's a tip: Many first-time observers looking at the Pleiades immediately think they are seeing the nebulosity around all of the bright stars in the cluster. The Pleiades nebulosity is *not* that easy to see. To confirm your observation, turn your telescope toward the Hyades cluster, which forms the head of Taurus. If you see nebulosity there, too, then what you are witnessing is light-scatter in the telescope—the Hyades are nebula-free. If, on the other hand, the glow disappears, then you are probably seeing a portion of the Pleiades' nebulosity.

M1 (NGC 1952)   Supernova remnant in Taurus
R.A. 05h 34m.5   Dec. +22° 01′   Mag. = 8.2   Diameter = 6′×4′

Figure 10.22b *M45, the Pleiades or Seven Sisters. Photograph by George Viscome (8-inch f/5.6 Newtonian, hypered Tech Pan 2415 film, 80-minute exposure). North is up.*

Where to look: Begin at Zeta Tauri, which marks the tip of Taurus's south-eastern horn. Looking through your finder, locate two 6th-magnitude stars just to its north that combine with Zeta to form a westward-pointing right triangle. Extend the leg of the triangle formed by the 6th-magnitude stars farther west for about half their distance. Through a low-power eyepiece, M1 should be visible in your telescope field just to the south of this point. (See Chart 10.23a.)

Description: M1 (Figures 10.23b and 10.23c) appears as an amorphous gray oval disk through 3- to 6-inch telescopes, while 8- to 12-inch instruments begin to show some of its irregularities. Larger apertures increase the mottled look of M1, with the biggest amateur telescopes revealing some of the crablike appendages that led to this object's nickname: the Crab Nebula.

M42 (NGC 1976)   Bright Nebula in Orion
R.A. 05^{h}35^m.4   Dec. $-05°$ 27'   Mag. = 2.9   Diameter = 66' $\times$ 60'

Where to look: M42 may be seen with the naked eye as the middle star in Orion's sword, directly below the three stars in the Hunter's belt. (See Chart 10.24a.)

Description: The Great Nebula in Orion, M42, is the finest deep-sky object visible from the Northern Hemisphere. All magnifications work well, with each offering a different perspective. Low powers (Figures 10.24b and 10.24c) are best for seeing the BIG picture; medium magnification reveals the nebula's complex structure and its varying colors and contrasts; high power works best for spying the intricate area in and around the nebula's center. On one especially transparent night a few years ago, my 13.1-inch f/4.5 reflector displayed a tremendous cloud with many stars embedded within. Tenuous curved fingers of glowing gas reached from the main body of the nebula to grasp many of the neighboring stars.

Even the smallest telescopes will show the Trapezium, four young, hot, blue-white stars buried within the center of M42. Their energy combines with that from other, fainter stars within the cloud to excite the Orion Nebula's hydrogen into luminescence. This results in the reddish color that is so vivid in photographs but tough to see otherwise. Hints of red can be seen along the misty fringes of M42 through large amateur instruments, but the overall coloring of the cloud appears blue-green.

Just north of M42 is a second, much smaller tuft of nebulosity catalogued separately as M43. Interestingly, although M42 was discovered telescopically in 1610, M43 was not recognized until 1731.

M35 (NGC 2168)   Open Cluster in Gemini
R.A. 06^{h}08^m.9   Dec. $+24°$ 20'   Mag. = 5.3   Diameter = 28'

Where to look: Chart 10.25a shows how M35 forms one corner of a triangle with Eta and Mu Geminorum, two stars in the foot of the twin brother Castor in Gemini. Recreate this triangle by aiming at the invisible third corner, where you should spot M35 through your finder. In fact, this cluster is so bright that

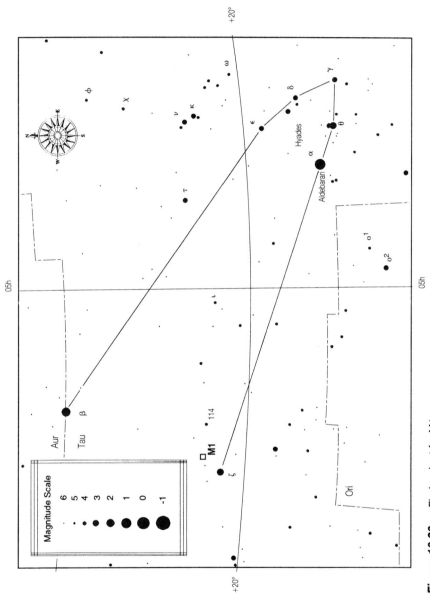

Figure 10.23a *Finder chart for M1.*

Figure 10.23b   *M1, the Crab Nebula. Photograph by George Viscome (8-inch f/5.6 Newtonian, hypered Kodak Tech Pan 2415, 70-minute exposure). South is up.*

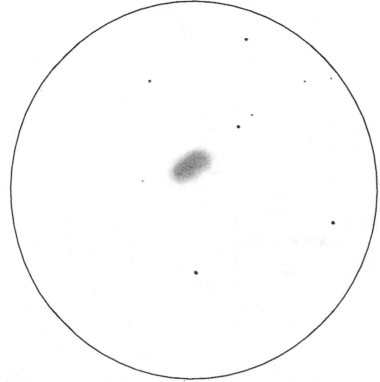

Figure 10.23c   *M1. Drawing by the author using an 8-inch f/7 Newtonian and a 12-mm Tele Vue Nagler eyepiece (119×).*

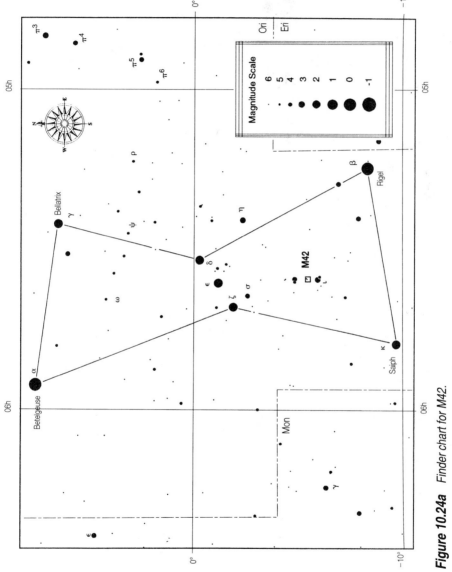

Figure 10.24a *Finder chart for M42.*

Figure 10.24b *M42, the Orion Nebula. Photograph by George Viscome (6-inch f/8 Newtonian, Tri-X film, 40-minute exposure). South is up.*

it might be visible faintly to the naked eye under extremely dark skies. (See Chart 10.25a.)

Description: M35 (Figure 10.25b) is one of winter's finest open clusters. Binoculars begin to show some individual stars, but at least a 3-inch scope is needed for resolution of the cluster. Eight-inch and larger instruments resolve just about all of the 200 stars that call this group home. Most appear blue-

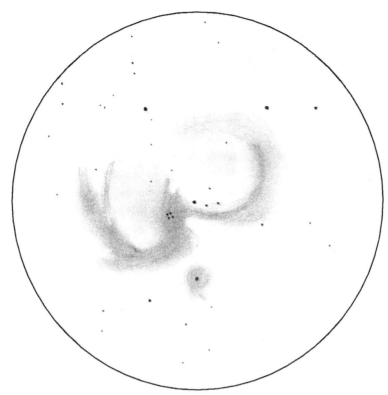

Figure 10.24c   *M42. Drawing by the author using an 8-inch f/7 Newtonian, 24-mm Tele Vue Wide Field eyepiece, and a DayStar LPR filter (59×).*

white, though some shine yellow and orange. Your lowest-power eyepiece will provide the best view.

Half a degree southwest of M35 is a second, much fainter, more distant open cluster: NGC 2158. Visible in a 3-inch, NGC 2158 requires at least a 6-inch telescope for partial resolution.

Astrophotography 101

One of the most popular pastimes for amateur astronomers is trying to capture the beauty of the universe on film. Just look at the superb photographs that highlight this book. Outstanding, aren't they? Though they rival the best photos from professional observatories, all were taken by amateur astronomers!

Yet in spite of its popularity, astrophotography can be one of the most time-consuming and frustrating aspects of our hobby. For those who have dabbled in it before, how often have you sent out a roll of film to be processed, confident at your success, only to get back pictures that are either out-of-focus, blurred, overexposed, underexposed, or just not right? An astrophotographer

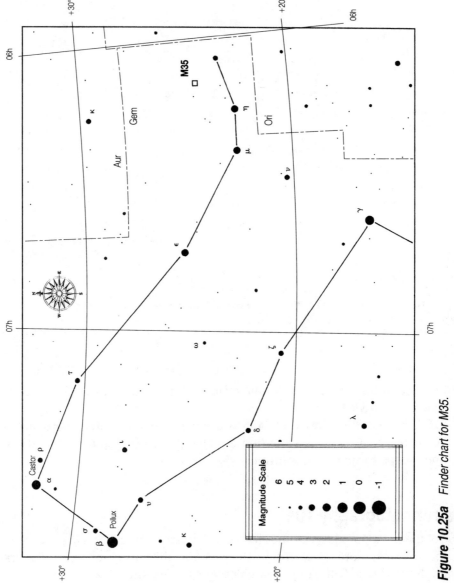

Figure 10.25a *Finder chart for M35.*

Figure 10.25b *M35. Photograph by George Viscome (8-inch f/5.6 Newtonian, hypered Tech Pan 2415 film, 40-minute exposure). South is up.*

friend once stated that even after years of perfecting his craft, he expects about five bad photographs for every one high-quality picture.

What makes astrophotography so difficult? Today, we live in a point-and-shoot world. Modern cameras are capable of determining exposure, setting the correct lens opening, focusing the lens, advancing the film to the next frame, and even telling the camera's light meter what kind of film is being used . . . all automatically! All of this is amazing, indeed, but applies only to terrestrial, not celestial, photography. By comparison, astrophotography is still in the Dark Ages, with good results coming only after much initial trial and error. Although it can be frustrating at times, perhaps this is also part of its appeal.

For the sake of this brief introduction, I have chosen to break the subject down into four broad categories, as shown below:

1. Fixed-camera
2. Short exposures through the telescope
3. Guided exposures
4. Long exposures through the telescope

This order was chosen as an approximate indicator of difficulty, ranging from low to high. It might be argued by some that items 2 and 3 could be reversed, but I will leave that judgment to you. For the moment, let's discuss the specifics.

Fixed-camera

Fixed-camera photography requires four things: a manually adjustable camera and lens, a tripod on which to set it, a roll of fast (>ISO 200) film, and a locking cable release. That's all—no expensive telescope or elaborate clock-driven mounting is required. To take a photograph, aim the camera at the desired area of sky, set the camera's focus at infinity (∞), close the aperture one f-stop from the maximum (to lessen edge distortion), and set the shutter speed to B (bulb) or T (time). Open the shutter for anywhere from a few seconds to a half hour or more. The net result is an accurate record of the night sky to at least the naked-eye limit.

Even though the camera is attached to a sturdy tripod, it is always moving because of the Earth's rotation. As a result, all of the stars in the photograph will be recorded as trails rather than points—the longer the exposure, the longer the trails. If star trails are the desired result, then exposure duration depends only on sky darkness. Under dark, rural conditions, exposures up to 30 minutes or more are possible, whereas suburban and urban photographs must probably be limited to no more than five or ten minutes because of light pollution.

If you want to use a fixed camera to photograph the stars as points, then the exposure time must be limited even more. Just how limited depends on the focal length of the camera lens being used. Table 10.2 compares lens focal length to maximum exposure before trailing will become evident. This table also shows how wide a slice of sky is covered by a given focal-length lens, an important consideration for proper framing.

Figure 10.26 shows the effects of exposure duration on the constellation Orion. Using a 50-mm lens, a ten-minute exposure was taken. Five minutes later, a 20-second exposure was made on the same frame. The stars were re-

Table 10.2 **Lens Focal Length versus Maximum Time Exposure**

(35-mm–format film)		
Focal Length of Lens (millimeters)	**Maximum Exposure[1] (seconds)**	**Sky Coverage[2] (degrees)**
28	25	49 x 74
50	14	28 x 41
85	8	16 x 24
105	7	13 x 20
200	4	7 x 10
300	2	5 x 7

Notes:

1. *The maximum duration of an exposure before stars begin to trail. These values are for stars at 0° declination; longer exposures are possible if the target area is above or below the celestial equator. Increase exposure by up to 50% at 45° declination, 100% above 60° declination.*

2. *Area covered by a frame of 35-mm film using a lens of the given focal length.*

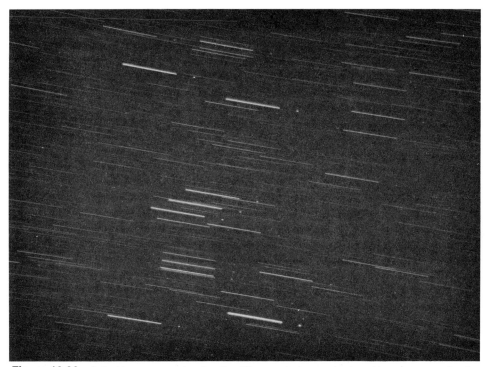

Figure 10.26 *A double exposure showing the difference between short- and long-exposure fixed-camera photography. Five minutes elapsed between the long 10-minute exposure and shorter 20-second exposure of the constellation Orion. (T-Max 400 film, 50-mm f/1.8 lens set at f/2.8).*

corded as points during the 20-second exposure but as trails in the ten-minute exposure.

Through the Telescope I: Short Exposures

The simplest method for photographing through a telescope, short-exposure telescopic work can be done with just about any type of telescope on just about any type of mounting (yes, even Dobsonian mounts, if balanced correctly). Subject matter appropriate for short exposures is restricted to the Sun, Moon, and the brighter planets, with exposure times ranging from $1/1000$th second to maybe four or five seconds. Other subjects and longer exposures are covered under a separate heading later on.

Just about any type and speed of film can be used to photograph the Sun, Moon, and planets, though slower films (for example, <ISO 200) are usually preferred. These offer finer grain structure than high-speed emulsions, an important consideration when trying to record subtle detail on a planet's surface or atmosphere.

As Figure 10.27 shows, there are four basic camera-telescope combinations available to the astrophotographer. Some are more appropriate than oth-

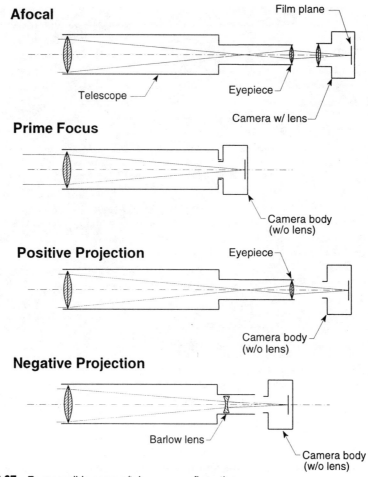

Figure 10.27 *Four possible camera/telescope configurations.*

ers for capturing certain objects, but no single setup is best for everything. Table 10.3 compares all four options illustrated in Figure 10.27.

Afocal. The afocal system uses both a camera's lens and a telescope's eyepiece. This allows just about any single-lens reflex (SLR) camera to be used for celestial photography, regardless of whether its lens is removable or not. Recall from the discussion in Chapter 6 that the viewfinder of an SLR camera looks through the camera lens and therefore shows exactly what the film will see. This ability makes life a lot easier when checking the camera's alignment with the telescope. While through-the-telescope astrophotography is possible with nonreflex cameras, it is not recommended.

Achieving a sharp focus is the biggest challenge facing the afocal photographer (as it is with the other methods as well). Place the camera on a separate tripod and align it to the telescope's eyepiece. With the camera's lens wide open

Table 10.3 **Camera-Telescope Configurations**

System name	Camera lens?	Telescope eyepiece?	Best for
Afocal	Yes	Yes	Moon, Sun
Prime focus	No	No	Moon, Sun (whole–disk)
Positive projection	No	Yes	Moon, Sun (close–ups), planets
Negative projection	No	Yes (Barlow Lens)	Moon, Sun (close–ups)

and focused at infinity, look through the viewfinder. Slowly turn the telescope's focusing knob in and out until the image is at its sharpest.

What exposure should be used? Over the years, many articles and books have been written on how to calculate proper exposure for the Moon, Sun, and planets. Unfortunately, most recommendations vary from one source to another, attesting to the fact that astrophotography is more art than science. However, as a guide, Table 10.4 offers suggested exposures for different subjects. Be sure to bracket the exposures *at least* one shutter speed either side of the suggestion.

Notice that these charts are based on knowing the *effective focal ratio* (EFR) of the camera-telescope combination. You DON'T know that value? Yes, you do! For the afocal method, the EFR can be calculated using the following formula:

$$EFR = \frac{(\text{camera lens focal length} \times \text{telescope magnification})}{\text{aperture of telescope}}$$

For example, consider taking a photograph of the first-quarter Moon using the afocal method, an 8-inch f/10 telescope, a 25-mm eyepiece, and a 50-mm f/1.8 camera lens. First, convert inches to millimeters or vice versa, so that the units of measure are all the same. For the sake of this discussion, we will convert the telescope's 8-inch aperture to 203 millimeters (8 inches $\times$ 25.4 millimeters per inch). From the formula in Chapter 1, the telescope's magnification is $80\times$. Plugging these values into the formula above yields:

$$EFR = \frac{(50\text{mm} \times 80)}{203 \text{ mm}} = f/19.7$$

With the f-ratio known, an estimation for proper exposure may now be picked off the chart. Assuming ISO 100 film, the suggested exposure is $\frac{1}{15}$ of a second. But take more than one shot, varying the exposure of each.

Prime focus. Prime focus is the cleanest camera-telescope combination, but it requires a single-lens reflex camera with a removable lens. This method

Table 10.4 Suggested Exposures for Selected Objects

Film Speeds

3200	1600	800	400	200	100	50	25
1	2	4	8	16	32	64	128
1/2	1	2	4	8	16	32	64
1/4	1/2	1	2	4	8	16	32
1/8	1/4	1/2	1	2	4	8	16
1/15	1/8	1/4	1/2	1	2	4	8
1/30	1/15	1/8	1/4	1/2	1	2	4
1/60	1/30	1/15	1/8	1/4	1/2	1	2
1/125	1/60	1/30	1/15	1/8	1/4	1/2	1
1/250	1/125	1/60	1/30	1/15	1/8	1/4	1/2
1/500	1/250	1/125	1/60	1/30	1/15	1/8	1/4
1/1000	1/500	1/250	1/125	1/60	1/30	1/15	1/8
	1/1000	1/500	1/250	1/125	1/60	1/30	1/15
		1/1000	1/500	1/250	1/125	1/60	1/30
			1/1000	1/500	1/250	1/125	1/60
				1/1000	1/500	1/250	1/125
					1/1000	1/500	1/250
						1/1000	1/500
							1/1000

Focal ratio (f-number) of camera/telescope combination or camera lens only

1.4 2 2.8 4 5.6 8 11 16 22 32 45 64 90 122 180 256

Target lines: Saturn · Crescent Moon, Mars · Quarter Moon · Sun (with Neutral Density 5 Filter) · Jupiter · Gibbous Moon · Full Moon · Venus

Recommended exposures for brighter sky targets. To read the table, find your camera/telescope focal ratio and your selected target. Follow the boxes to the left until you line up with your film's ISO column. The suggested exposure lies at their intersection.

332

couples the lensless camera body to the eyepieceless telescope, in effect making the telescope itself a large telephoto lens. Determining the effective focal ratio requires no calculation; it is simply the telescope's own focal ratio. For example, the prime-focus EFR of an 8-inch f/10 telescope is f/10, while the EFR of a 14-inch f/5 telescope is f/5, and so on.

To take prime-focus photos through a telescope, the camera is attached directly to the telescope using a two-piece camera-to-telescope adapter. Focusing is done directly through the camera's viewfinder by turning the focusing knob(s) in and out until a sharp image is seen. Unfortunately, this is largely a hit-or-miss technique. To take the guesswork out of focusing, use one of the focusing aids reviewed in Chapter 6. Though they may seem an unnecessary frill at first, there is no worse feeling than getting back a perfectly exposed, perfectly framed photograph that is out of focus!

Just about all modern refractors and catadioptrics are designed for prime-focus photography, but not so with some Newtonian reflectors. Many Newtonians are constructed in such a way that the prime-focus point lies too far down in the focusing mount to be accessible by a camera. In cases like this, the only alternative is to move the main mirror up the tube to reduce its distance to the eyepiece. Reducing this distance will push the prime focus out toward the end of the focusing mount.

If your Newtonian telescope will not permit prime-focus photography, measure the distance from the primary to the diagonal mirror. Add to this the distance from the diagonal's center to about the upper edge of the eyepiece holder's base. To access the prime focus, this total distance should be about an inch or so less than the telescope's focal length. If it is more, then subtract the instrument's focal length from your overall measurement plus 1 inch. Moving the primary up the tube by that difference (probably no more than a couple of inches) will push the prime focus farther up the focusing tube and make it accessible to the camera. For example, an 8-inch f/7 telescope has a focal length of 56 inches. Measuring from the primary to the diagonal and then from the diagonal to the eyepiece holder indicates a length of 58 inches. From the discussion above, the primary must be moved [(58 + 1) − 56], or 3 inches, up the tube to access the prime focus with a 35-mm single-lens reflex camera. (The extra inch is to allow for the distance from the eyepiece holder to the film plane in the back of the camera.) Do not move the mirror too far up, or the telescope may not focus when used visually. If this latter problem occurs, then an extension tube made from a 1.25-inch brass drainpipe (used in bathroom sinks) will be required. Already mentioned in Chapter 8, these drainpipes decrease from a 1.25-inch inside diameter to a 1.25-inch outside diameter. Note also that the diagonal mirror may also be too small to reflect all the light coming from the primary if its distance is reduced too much. In addition, when you move the mirror up, you will probably have to rebalance the telescope tube in its mount.

Positive projection. Sometimes called *eyepiece projection*, this arrangement uses an eyepiece to enlarge whatever is being photographed, in effect stretching a telescope's focal length. Like prime focus, a camera-to-telescope adapter is used, but this time with an extension tube between the telescope adapter and T-ring. An eyepiece is placed inside the extension tube (don't forget to lock the eyepiece in place using the side-mounted screw) to project the image from the telescope into the camera. Positive projection is ideal for solar and lunar close-ups as well as detailed photographs of the planets.

As with the afocal and prime-focus systems, the effective focal ratio of the camera-telescope team must be known before proper exposure can be estimated. The following equation may be used for this calculation:

$$\text{EFR} = \frac{f_t \times (L_e - f_e)}{f_e}$$

where f_t = telescope focal ratio
L_e = projection distance from eyepiece
f_e = eyepiece focal length

To illustrate this, consider taking a photograph of, say, Jupiter with an 8-inch (200-mm) f/10 telescope. Because of the planet's small apparent size, the positive-projection method will be used. A 17-mm eyepiece is selected for the task, with an eyepiece-to-film projection distance of 3 inches (75-mm). Plugging these values into the positive-projection equation above produces

$$\text{EFR} = \frac{\text{f/10} \times (75 - 12)}{12} = \text{f/10} \times 5.25 = \text{f/52.5}$$

Thus, the overall system has an effective focal ratio of f/52.5, yielding an effective focal length of 10,500 mm (2000 mm × 5.25). An exposure estimate may now be gleaned from the accompanying tables.

Negative Projection. Similar to positive projection, negative projection puts a negative, or concave, lens between the telescope and the lensless camera body. Negative projection does not extend a telescope's focal length to the great extent that positive projection does, making this method ideal for enlarged shots of the Moon and Sun but not as useful for photographing the planets.

The most commonly available negative lenses are the Barlow Lens and the photographic teleconverter, with many photographers preferring the latter. To use a teleconverter on a telescope, simply connect it between the camera body and the T-ring/telescope adapter used for prime-focus work. A Barlow is used as a projection lens by inserting it into the eyepiece holder and affixing the camera to it using a telescope adapter/T-ring. The resulting increase in magnification will be equal to the power of the teleconverter or Barlow, usually either 2× or 3×. Therefore, using either a 2× teleconverter or a 2× Barlow with, for example, an f/10 telescope will double both the instrument's focal length and effective focal ratio (in this case, to f/20).

Telecompression. Though not shown separately in Figure 10.27, telecompression may be thought of as reverse projection. Rather than enlarging an image, placing a telecompressor or focal reducer between the telescope and lensless camera will lower the effective focal ratio. The net result is a negative magnification effect and a wider field of view. This is especially useful when trying to photograph the full disk of either the Moon or Sun through a long-focal-length telescope.

Most telecompressors/focal reducers state their *deflation factor* right on them. For instance, the Celestron Reducer-Corrector cuts the effective focal ratio of an f/10 telescope by 37% to f/6.3 (and, therefore, the effective focal length from 80 inches to 50 inches, in the case of an 8-inch f/10). That same telecompressor will reduce an f/6.3 instrument (such as some of Meade's Schmidt-Cassegrains) to f/4, and so on. Not only does lowering a telescope's focal ratio increase the field covered in a photograph, but it also decreases the exposure time needed to take the picture. To see how dramatic this effect is, look at any of the exposure charts in this chapter and compare the straight telescope's f-ratio with the same instrument telecompressed.

If you are considering using a telecompressor for photography, be sure to use a reducer/corrector. A telecompressor without the corrective optics causes much more distortion around the edges of the film frame.

Guided-camera

Sometimes called *piggyback astrophotography*, this next step up requires the camera to be placed on an equatorial mounting and tracked with the stars. Frequently, the camera is mounted sidesaddle on an equatorially mounted telescope, giving rise to the piggyback nickname.

Both Meade and Celestron (as well as many after-market companies such as Orion) sell piggyback brackets that mount directly to their Schmidt-Cassegrain telescopes. Any camera with a standard tripod socket on its baseplate may then be attached to the bracket. Some equatorially mounted refractors and reflectors also come with provisions for attaching a camera for guided exposures. For instance, Celestron's C6 Newtonian has a built-in ¼-20 stud sticking out of one of its tube rings, letting the owner thread on a ball-and-socket tripod head.

An expensive clock-driven telescope mounting is not needed for guided-camera astrophotography. For about $20, you can build a Scotch mount, a hand-driven camera platform that will track the stars as accurately as a mounting costing 50 times as much. Plans for making a Scotch mount are found in Chapter 7.

Preferred films for guided-camera astrophotography are any with an ISO rating of at least 200, with many photographers favoring films with ISO values in excess of 800. These allow the maximum amount of starlight to be recorded in the minimal amount of time.

Before taking the first picture, align the mounting with the celestial pole. If you are using an equatorially mounted telescope, follow the directions given in the last chapter; to align a Scotch mount, follow the instructions in Chapter 7.

With the camera firmly attached to the polar-aligned mounting and aimed at the desired area, begin the exposure. Follow the same advice given under the fixed-camera section earlier. Set the lens focus at infinity, the shutter to either B or T, and move the lens down one stop. Before opening the shutter, lock the camera's mirror up (if so equipped). Failing that, hold a black card in front of the lens with one hand while you trigger the shutter with the other. Count several seconds to let any mirror-slap-induced vibrations dampen out, then pull the card away. Finally, choose the exposure duration. Begin with 1 minute and double the exposure on each of the next five frames to 16 minutes. If the mount's polar axis is aimed properly at the celestial pole, the clock drive should track the stars accurately.

Through the Telescope II: Long Exposures

The most challenging of all astrophotos to take are those magnificent pictures of deep-sky objects that adorn the pages of every astronomy magazine and book (including those found in this one). How do these photographers do it?

To begin with, long-exposure telescopic photography employs either the prime-focus or telecompression methods highlighted in the short-exposure section. To take successful long exposures, however, requires more than a camera, telescope, and adapter. With exposure times that can extend for an hour or more, long-exposure astrophotography requires a rock-steady equatorial mounting precisely aligned to the celestial pole, sophisticated (in other words, *expensive*) accessories and an overabundance of patience.

Unlike guided exposures, where the clock drive can run unattended during the exposure, through-the-telescope photography requires constant monitoring to make certain that the telescope is following the stars as it should. (Though modern-day clock drives are amazingly accurate—especially those featuring periodic-error-correction circuits—they still experience tracking errors.) To control the telescope's tracking rate, the photographer must place and keep a star in the center of an illuminated-reticle eyepiece.

But how can the photographer watch a guide star to make sure the telescope follows it correctly if the camera is looking through the eyepiece holder? There are two alternatives. Many astrophotographers opt for off-axis guiders, while others prefer side-mount, long-focus refractors. Both methods work the same way: The observer keeps his or her eye trained on that star to make sure it does not deviate from the eyepiece's cross-hairs during the photograph. Any minor adjustments in the tracking rate of the clock drive are made using a handheld control box plugged into either a built-in or external dual-axis drive corrector. (Note that an increasing number of astrophotographers are using the autoguider capability of CCD cameras. This way, the telescope automatically compensates for any tracking errors while the photographer relaxes. Many observe side by side with binoculars or another telescope, but others read or watch television!) Choice of film and exposure must be based on what is being photographed. In general, beginning astrophotographers will do best

Astrophotography Log

Date	Frame #	Film Type	Object	Exposure	f/#	Lens/Telescope	Conditions	Comments

Figure 10.28   *Suggested log form for recording photographic details (exposures, film, etc).*

by selecting a fast (high ISO) film. Take a look back at Chapter 6 for some suitable choices. Also widely used today is hypersensitized Kodak Technical Pan 2415 black-and-white film. Normally operating at about ISO 25, TP 2415 jumps to ISO 200 when hypersensitized. Best of all, hypersensitizing the film significantly reduces reciprocity failure. This makes it ideal for the long exposures required for faint deep-sky objects.

Logging Off

Regardless of what kinds of astrophotos you take, it pays to record all the particulars in a permanent logbook. Be sure to include such items as date, subject, equipment (camera, lens, and/or telescope), film type and ISO value, frame number, length of exposure, and f-number used. One possible log format is included as Figure 10.28. By knowing all of this information, it will be easy to compare technique with the actual results once the film is returned. All the books in the world will not teach you as much about astrophotography as will learning from your mistakes.

Appendix A
The Astronomical Yellow Pages

Manufacturers

ARO Instrument Company
1245 South 6th Street
Coshocton, Ohio 43812
Phone: (614) 622-8895
Main product(s): INTES Maksutov
 telescope

Astronomical Innovations
P.O. Box 14853
Lenexa, Kansas 66285
Phone: (800) 473-7382
Main product(s): Astronomical
 accessories

Astro-Physics, Inc.
11250 Forest Hills Road
Rockford, Illinois 61111
Phone: (815) 282-1513
Main product(s): Refractors

AstroSystems, Inc.
1109 Limbark Street
Longmount, Colorado 80501
Phone: (303) 678-5339
Main product(s): Telescope parts and
 accessories

Bean, L.L.
Freeport, Maine 04033-0001
Phone: (800) 341-4341
Main product(s): Clothing and outdoor
 accessories

Beattie Systems, Inc.
P.O. Box 3142
Cleveland, Tennessee 37311
Phone: (800) 251-6333
Main product(s): Intenscreen camera
 focus screens

Bike Nashbar
4111 Simon Road
Youngstown, Ohio 44512
Phone: (800) 627-4227
Main product(s): Cold-weather
 clothing and accessories (bike stuff,
 too!)

Bausch and Lomb
9200 Cody
Overland Park, Kansas 66214
Phone: (800) 423-3537
Main product(s): Binoculars

Byers, Edward R., Company
29001 West Highway 58
Barstow, California 92311
Phone: (619) 256-2377
Main product(s): Telescope mountings
 and drive systems

Campmor
810 Route 17 North, P.O. Box 997A
Paramus, New Jersey 07653
Phone: (800) 526-4784
Main product(s): Clothing and outdoor
 accessories

Carton Optics
1037 Enderby Way
Sunnyvale, California 94087
Phone: (408) 245-4818 (TAD
 International)
Main products: Adlerblick binoculars

Celestron International
2835 Columbia Street
Torrance, California 90503
Phone: (800) 421-1526
Main product(s): All types of
 telescopes, binoculars, eyepieces,
 and accessories

Ceravolo Optical Systems
Box 1427
Ogdensburg, New York 13669
Phone: (613) 258-4480
Main product(s): Maksutov-Newtonian
   telescopes

Clausing, P.A., Inc.
8038 Monticello Avenue
Skokie, Illinois 60076
Phone: (312) 267-3399
Main product(s): Mirror coatings

Coulter Optical, Inc.
P.O. Box K
Idyllwild, California 92349
Phone: (714) 659-4621
Main product(s): Dobsonian-style
   Newtonian reflectors and optics

D&G Optical
6490 Lemon Street
East Petersburg, Pennsylvania 17520
Phone: (717) 560-1519
Main product(s): Refractors

Damart
3 Front Street, Department 0A640
Rollinsford, New Hampshire 03805
Phone: (800) 258-7300
Main product(s): Cold-weather
   clothing and accessories

DayStar Filter Corporation
P.O. Box 1290
Pomona, California 91769
Phone: (714) 591-4673
Main product(s): Solar and LPR filters

Edmund Scientific Company
Department 10B1, N937 Edscorp
   Building
Barrington, New Jersey 08007
Phone: (609) 573-6250
Main product(s): Telescopes and
   accessories

Equatorial Platforms
11065 Peaceful Valley Road
Nevada City, California 95959
Phone: (916) 265-3183
Main product(s): Equatorial platforms
   for alt-azimuth mounts

Evaporated Metal Films
701 Spencer Road
Ithaca, New York 14850
Phone: (800) 456-7070
Main product(s): Mirror coatings

Fujinon, Inc.
10 High Point Drive
Wayne, New Jersey 07470
Phone: (201) 633-5600
Main product(s): Binoculars

Galaxy Optics
P.O. Box 2045
Buena Vista, Colorado 81211
Phone: (719) 395-8242
Main product(s): Telescope mirrors
   and optics

GrandView Instruments
P.O. Box 278
Concord, California 94522
Phone: (510) 825-3019
Main product(s): GrandView
   Binocular Mount

Hollywood General Machining, Inc.
   (Losmandy)
1033 North Sycamore Avenue
Los Angeles, California 90038
Phone: (213) 462-2855
Main product(s): Equatorial mounts
   and attachments

Jim's Mobile Industries (JMI)
810 Quail Street, Unit E
Lakewood, Colorado 80215
Phone: (303) 233-5353
Main product(s): Newtonian reflectors
   and peripherals

Jupiter Telescope Company
810 Saturn Street, Suite 16
Jupiter, Florida 33477
Phone: (407) 694-1154
Main product(s): Dobsonian-style
   Newtonian reflectors and equatorial
   platforms

Kalmbach Publishing Company
21027 Crossroads Circle, P.O. Box
   1612
Waukesha, Wisconsin 53187
Phone: (800) 446-5489
Main product(s): Astronomy magazine,
   books, and related products

Kufeld, Steve
P.O. Box 6780
Pine Mountain Club, California 93222
Phone: (805) 242-5421
Main product(s): Telrad aiming device

Lumicon
2111 Research Drive, Suites 4 & 5
Livermore, California 94550
Phone: (800) 767-9576
Main product(s): Filters, drive
  correctors, finders, and hypered film

Meade Instruments Corporation
16542 Millikan Avenue
Irvine, California 92714
Phone: (714) 756-2291
Main product(s): All types of
  telescopes, binoculars, eyepieces,
  and accessories

Minolta Corporation
101 Williams Drive
Ramsey, New Jersey 07446
Phone: (201) 825-4000
Main product(s): Binoculars

Mountain Instruments
5050 Laguna Blvd., Suite 112
Elk Grove, California 95758
Phone: (916) 422-0962
Main product(s): Telescope mounts
  and drive systems

Nikon, Incorporated
623 Stewart Avenue
Garden City, New York 11530
Phone: (516) 222-0200
Main product(s): Binoculars

Northern Lites
640 Cains Way, R.R. 1
Sooke, British Columbia, CANADA
  V0S 1N0
Phone: (604) 642-6601
Main product(s): Astrophotography
  accessories

Novak, Kenneth F., & Co.
Box 69
Ladysmith, Wisconsin 54848
Phone: (715) 532-5102
Main product(s): Telescope
  components

Obsession Telescopes
P.O. Box 804
Lake Mills, Wisconsin 53551
Phone: (414) 648-2328
Main product(s): Dobsonian-style
  Newtonian reflectors

Optic-Craft Machining
33918 Macomb
Farmington, Michigan 48335
Phone: (313) 476-5893
Main product(s): Telescope mounts

Optica b/c Company
4100 MacArthur Blvd.
Oakland, California 94619
Phone: (415) 530-1234
Main product(s): Astronomical
  accessories and publications

Optical Guidance Systems
2450 Huntingdon Pike
Huntingdon Valley, Pennsylvania
  19006
Phone: (215) 947-5571
Main product(s): Newtonian and
  Cassegrain telescopes

Opto-Data
1755 East Bayshore Road, Unit 24B
Redwood City, California 94063
Phone: (415) 599-9075
Main product(s): Starport
  computerized telescope-aiming
  system

Orion Telescope Center
P.O. Box 1158
Santa Cruz, California 95062
Phone: (800) 447-1001 / (800) 443-1001
  (in California only)
Main product(s): Telescopes,
  binoculars, eyepieces, filters, and
  other accessories

P & S Skyproducts
R.R. #1 20095 Con. 7
Mount Albert, Ontario, CANADA L0G
  1M0
Phone: (416) 473-1627
Main product(s): Kwik Focus focusing
  device

P.A.P. Coating Services
1112 Chateau Avenue
Anaheim, California 92802
Phone: (714) 778-2525
Main product(s): Mirror coatings

Parallax Instruments
8318 Pineville-Matthews Road
Suite 708, Box 192
Charlotte, North Carolina 28226
Phone: (704) 542-4817
Main product(s): Newtonian reflectors

Parks Optical
270 Easy Street
Simi Valley, California 93065
Phone: (805) 522-6722
Main product(s): Reflecting telescopes,
  binoculars, and accessories

Pentax Corporation
35 Inverness Drive East
Englewood, Colorado 80112
Phone: (303) 799-8000
Main product(s): Refractors and
  binoculars

Performance Bicycle Shop
P.O. Box 2741
Chapel Hill, North Carolina 27514
Phone: (800) 727-2453
Main product(s): Cold-weather
  clothing and accessories (bike stuff,
  too!)

Questar Corporation
P.O. Box 59
New Hope, Pennsylvania 18938
Phone: (215) 862-5277
Main product(s): Maksutov-Cassegrain
  catadioptric telescopes and
  accessories

Rigel Systems
26850 Basswood Avenue
Rancho Palos Verde, California 90274
Phone: (310) 375-4149
Main product(s): Starlite LED
  flashlight and other astronomical
  accessories

Safari Telescopes
12400 Skyline Drive
Albuquerque, New Mexico 87123
Phone: (505) 293-0117
Main product(s): Dobsonian-style
  telescopes and binocular mounts

**Santa Barbara Instrument Group
  (SBIG)**
1482 East Valley Road, Suite #J601
Santa Barbara, California 93108
Phone: (805) 969-1851
Main product(s): ST-4 and ST-6 CCD
  cameras and accessories

Sky Scientific
28578 Highway 18, P.O. Box 184
Skyforest, California 92385
Phone: (909) 337-3440
Main product(s): Astrophotographic
  accessories

Sky Designs
4100 Felps, #C
Colleyville, Texas 76034
Phone: (817) 581-9878
Main product(s): Dobsonian-style
  Newtonian reflectors

Sky Publishing Corporation
P.O. Box 9111
Belmont, Massachusetts 02178-9111
Phone: (800) 253-0245
Main product(s): Sky & Telescope
  magazine, books, related products

Small Parts, Inc.
13980 N.W. 58th Court, P.O. Box 4650
Miami Lakes, Florida 33014-0650
Phone: (305) 557-7955
Main product(s): Nuts, bolts, fasteners,
  knobs, and other assorted hardware

Spectra Astro Systems
6631 Wilbur Avenue, Suite 30
Reseda, California 91335
Phone: (800) 735-1352
Main product(s): Focusing devices,
  astrophoto accessories (also see
  listing under "Dealers and
  Distributors")

SpectraSource Instruments
31324 Via Colinas, Suite 114
Westlake Village, California 91362
Phone: (818) 707-2655
Main product(s): Lynxx CCD cameras
   and accessories

Star Bound
68 Klaum Avenue
North Tonawanda, New York 14120
Phone: (716) 692-3671
Main product(s): Binocular mounts

Star Instruments
P.O. Box 597
Flagstaff, Arizona 86002
Phone: (602) 774-9177
Main product(s): Mirrors and other
   telescope optics

Star-Liner Company
1106 South Columbus Blvd.
Tucson, Arizona 85711
Phone: (602) 795-3361
Main product(s): Newtonian and
   Cassegrain reflectors

Starsplitter Telescopes
3228 Rikkard Drive
Thousand Oaks, California 91362
Phone: (805) 493-2489
Main product(s): Dobsonian-style
   Newtonian reflectors

Starry Messenger
P.O. Box 6552
Ithaca, New York 14851
Phone: (201) 992-6865
Main product(s): Monthly used-
   equipment newsletter

Swift Instruments, Inc.
952 Dorchester Avenue
Boston, Massachusetts 02125
Phone: (617) 436-2960
Main product(s): Binoculars

Tectron Telescopes
2111 Whitfield Park Avenue
Sarasota, Florida 34243
Phone: (813) 758-9890
Main product(s): Dobsonian-style
   Newtonian reflectors and
   collimation devices

Tele Vue, Inc.
100 Route 59
Suffern, New York 10901
Phone: (914) 357-9522
Main product(s): Refractors, eyepieces,
   and attachments

Thousand Oaks Optical
Box 5044-289
Thousand Oaks, California 91359
Phone: (805) 491-3642
Main product(s): Glass solar filters

Torus Optical
67 Bon-Aire
Iowa City, Iowa 52240
Phone: (319) 339-0524
Main product(s): Dobsonian-style
   Newtonian reflectors and optics

Tuthill, Roger W., Inc.
Box 1086
Mountainside, New Jersey 07092
Phone: (800) 223-1063
Main product(s): Mylar solar filters
   and other astronomical accessories

Unitron, Inc.
170 Wilbur Place, P.O. Box 469
Bohemia, New York 11716
Phone: (516) 589-6666
Main product(s): Refractors and
   binoculars

VERNONscope and Company
5 Ithaca Road
Candor, New York 13743
Phone: (607) 659-7000
Main product(s): Eyepieces, filters,
   and other accessories

Vista Instrument Company
307 East Tunnell Street
Santa Maria, California 93454
Phone: (805) 925-1240
Main product(s): Drive correctors and
   other astronomical accessories

Vogel Enterprises
38W150 Hickory Court
Batavia, Illinois 60510
Phone: (800) 457-8725
Main product(s): Drive correctors

Zeiss, Carl, Inc.
1015 Commerce Street
Petersburg, Virginia 23803
Phone: (800) 338-2984
Main product(s): Binoculars,
    telescopes (not readily available in
    North America)

Dealers and Distributors

An important warning: If you are shopping by mail, ALWAYS ask about shipping charges BEFORE ordering. Some
offer exceptionally low prices only to
charge the consumer exorbitant (and
unpublished) shipping and handling
costs. Not only will these hidden costs
offset any savings, but these dealers may
also end up being more expensive!

Arizona

Stellar Vision and Astronomy Shop
1835 South Alvernon, #208
Tucson, Arizona 85711
Phone: (602) 571-0877
Product line(s): Tele Vue, Takahashi,
    Celestron, Meade, Questar, Unitron,
    and more

California

Los Angeles Optical Company
12129 Magnolia Boulevard
North Hollywood, California 91607
Phone: (818) 762-2206
Product line(s): Celestron, Meade, Tele
    Vue, Lumicon, and Fujinon

Lumicon
2111 Research Drive, Suites 4 & 5
Livermore, California 94550
Phone: (800) 767-9576
Product line(s): Celestron, Meade,
    Takahashi, and Lumicon (also see
    previous listing)

Orion Telescope Center
P.O. Box 1158
Santa Cruz, California 95062
Phone: (800) 447-1001 / (800) 443-1001
    (in California only)
Product line(s): Celestron, Tele Vue,
    Orion (also see previous listing)

Scope City
679 Easy Street
Simi Valley, California 93065
Phone: (800) 235-3344
Product line(s): Parks, Celestron,
    Edmund, Meade, Questar, Tele Vue,
    JMI, and more

Spectra Astro Systems
6631 Wilbur Avenue, Suite 30
Reseda, California 91335
Phone: (800) 735-1352
Product line(s): Celestron, Tele Vue,
    Takahashi, and Fujinon (also see
    previous listing)

Colorado

Jim's Mobile Industries (JMI)
810 Quail Street, Unit E
Lakewood, Colorado 80215
Phone: (303) 233-5353
Product line(s): JMI, Meade, Celestron,
    Tele Vue, and Astrophysics (also see
    previous listing)

S & S Optika
5174 South Broadway
Englewood, Colorado 80110
Phone: (303) 789-1089
Main product(s): Takahashi, Pentax,
    Zeiss, DayStar, Bausch & Lomb,
    Celestron, and more

Connecticut

Pauli's Wholesale Optics
Professional Building #401
57 North Street
Danbury, Connecticut 06810
Phone: (203) 746-3579
Product line(s): Meade, Celestron, Tele
    Vue, Fujinon, Bausch and Lomb,
    Parks, and more

Florida

Sarasota Camera Exchange
1055 South Tamiami Trail
Sarasota, Florida 34237
Phone: (813) 366-7484
Product line(s): Meade, Celestron, Tele
  Vue, Nikon, Minolta, and Pentax

Georgia

Coke's Camera Center
735 Cherry Street
Macon, Georgia 31213
Phone: (800) 768-2653
Product line(s): Celestron, Meade, and
  JMI

Illinois

Cosmic Connections
P.O. Box 7
North Aurora, Illinois 60542
Phone: (800) 634-7702
Product line(s): Tele Vue, Celestron,
  Meade, JMI, Lumicon, Vogel, and
  Edmund; repair

Shutan Camera and Video
312 West Randolph Street
Chicago, Illinois 60606
Phone: (800) 621-2248
Product line(s): Meade, Celestron, Tele
  Vue, and JMI

Kansas

Science Education Center
125 South Hillside
Wichita, Kansas 67211
Phone: (316) 682-1921
Product line(s): Celestron, Tele Vue,
  Minolta, and Zeiss; telescope repair

Maryland

Company Seven
Box 2587
Montpelier, Maryland 20708
Phone: (301) 953-2000
Product line(s): Astro-Physics, Questar,
  JMI, SBIG, Celestron, Tele Vue, and
  more

Massachusetts

Meischner, F.C., Company, Inc.
182 Lincoln Street
Boston, Massachusetts 02111
Phone: (800) 321-8439
Product line(s): Takahashi, Tele Vue,
  Questar, Celestron, Meade, and
  more; plus repair

Michigan

City Camera
15336 West Warren
Dearborn, Michigan 48126
Phone: (800) 359-5085
Product line(s): Celestron and Vixen

New Hampshire

Rivers Camera Shop
454 Central Avenue
Dover, New Hampshire 03820
Phone: (603) 742-4888
Product line(s): Meade, Celestron, Tele
  Vue, Questar, Parks, and more; used
  equipment

New Jersey

Dover Photo Supply
25 East Blackwell Street
Dover, New Jersey 07801
Phone: (201) 366-0994
Product line(s): Celestron, Meade, Tele
  Vue, Nikon, Swift, and Steiner

Tuthill, Roger W., Inc.
Box 1086
Mountainside, New Jersey 07092
Phone: (800) 223-1063
Product line(s): Meade, Celestron, Tele
  Vue, and Coulter

New York

Adorama
42 West 18th Street
New York, New York 10011
Phone: (800) 223-2500
Product line(s): Celestron, Meade, Tele
  Vue, Thousand Oaks, and Edmund

Berger Brothers Camera Exchange
209 Broadway
Amityville, New York 11701
Phone: (800) 262-4160
Product line(s): Celestron, Meade, and
    Tele Vue

Focus Camera, Inc.
4419 13th Avenue
Brooklyn, New York 11219
Phone: (800) 221-0828
Product line(s): Celestron, Meade, JMI,
    and Fujinon

Hirsch, Edwin
29 Lakeview Drive
Tomkins Cove, New York 10986
Phone: (914) 786-3738
Product line(s): DayStar filters and
    Tele Vue's Solaris telescope

Ohio

Eastern Hills Camera
7875 Montgomery Road
Cincinatti, Ohio 45236
Phone: (513) 791-2140
Product line(s): Celestron, Swift,
    Minolta, Vixen, and Pentax

Oklahoma

Astronomics
2401 Tee Circle, Suites 105/106
Norman, Oklahoma 73069
Phone: (800) 422-7876
Product line(s): Celestron, Edmund,
    Meade, Questar, Tele Vue, Lumicon,
    and DayStar

Pennsylvania

Pocono Mountain Optics
R.R. #6, Box 6329
North Pocono Village
Moscow, Pennsylvania 18444
Phone: (800) 569-4323
Product line(s): Celestron, Meade, Tele
    Vue, JMI, Telrad, Vogel, Edmund,
    and Questar

South Carolina

Wonder Works
280 West Coleman Boulevard
Mount Pleasant, South Carolina 29464
Phone: (800) 352-2316
Product line(s): Celestron, Meade, Tele
    Vue, and Nikon

Texas

Fort Davis Astronomical Supply
P.O. Box 922
Fort Davis, Texas 79734
Phone: (915) 426-3008
Product line(s): Parks, JMI, Tele Vue,
    and Lumicon

**Land, Sea and Sky (Texas Nautical
    Repair)**
3110 South Shepherd
Houston, Texas 77098
Phone: (713) 529-3551
Product line(s): Takahashi

Wisconsin

Eagle Optics
716 South Whitney Way
Madison, Wisconsin 53711
Phone: (608) 271-4751
Product line(s): Questar, Meade,
    Celestron, Tele Vue, Fujinon, and
    Nikon

Canada

Cosmic Connection
32 Ashgrove Boulevard
Brandon, Manitoba R7B 1C2
Phone: (204) 727-3111
Product line(s): Celestron, Lumicon,
    and Vixen

Khan Scope Centre
3243 Dufferin Street
Toronto, Ontario M6A 2T2
Phone: (416) 783-4140
Product line(s): Celestron, Meade, Tele
    Vue, Lumicon, and Thousand Oaks

Lire La Nature, Inc.
100 Goyer Street, Store 110
LaPrairie, Quebec J5R 5G5
Phone: (514) 659-3578
Product line(s): Meade, Celestron,
    Lumicon, and Bausch & Lomb

United Kingdom

Beacon Hill Telescopes
112 Mill Road
Cleethorpes, South Humberside
DN35 8JD
Phone: (0472) 692959
Product line(s): Vixen, Clavé, and
    more

Broadhurst, Clarkson & Fuller, Ltd.
63 Farrington Road
London EC1M 3JB
Phone: (071) 405 2156
Product line(s): Meade, Takahashi,
    Vixen, Tele Vue, Celestron, JMI, and
    more

Countryside Optics
Lower Wood End Farm
Wood End, Near Marston Moretaine,
    Beds. MK43 0PA
Phone: (0234) 765151
Product line(s): Celestron, Nikon,
    Zeiss, Bausch & Lomb, Fujinon, and
    custom-designed instruments

Dark Star Telescopes
6 Pinewood Drive, Ashley Heath
Market Drayton, Salop TF9 4PA
Phone: (0630) 672958
Product line(s): Celestron, JMI, Vixen,
    and Lumicon

Osborne Optics
139 Dean House, Eastfield Avenue
Walker, Newcastle Upon Tyne
NE6 4UU
Phone: (091) 263 8826
Product line(s): Tele Vue, Parks, and
    Celestron

Orion Optics
Unit 12, Quakers Coppice
Crewe Gates Industrial Estate
Crewe, Cheshire CW1 1FA
Phone: (0270) 500089
Product line(s): Celestron, Vixen

Australia

Astro Optical Supplies
9B Clarke Street
Crows Nest, NSW 2065
Phone: +61 2 436 4360
Product line(s): Celestron and
    Australian-made AOS reflectors

Astro Optical Supplies
13 Lower Plaza, 131 Exhibition Street
Melbourne, VIC 3000
Phone: +61 3 650 8072
Product line(s): Celestron and
    Australian-made AOS reflectors

Astronomy & Electronics Centre
P.O. Box 45
Cleve, SA 5640
Phone: +61 86 282 435
Product line(s): Meade, Astro-Physics,
    Lumicon, Parks, ARO, and
    Lumicon.

Binocular and Telescope Shop
310 George Street
Sydney, NSW 2000
Phone: +61 2 235 3344
Product line(s): Parks, Tele Vue,
    Lumicon, Daystar, and Thousand
    Oaks

Elken Ridge
155 Ridge Road
Mount Dandenong, VIC 3767
Phone: +61 3 751 2176
Product line(s): JMI and SBIG

York Optical and Scientific
316 St. Paul's Terrace
Fortitude Valley, QLD 4006
Phone: +61 7 252 2061
Product line(s): Meade, Questar,
    Bausch & Lomb, Fujinon, and Vixen

York Optical and Scientific
7/270 Flinders Street
Melbourne, VIC 3000
Phone: +61 3 654 7212
Product line(s): Meade, Questar,
    Bausch & Lomb, Fujinon, and Vixen

York Optical and Scientific
939 Hay Street
Perth, WA 6000
Phone: +61 9 322 4410
Product line(s): Meade, Questar,
    Bausch & Lomb, Fujinon, and Vixen

Carl Zeiss Jena
114 Pyrmont Bridge Road
Camperdown, NSW 0666
Phone: +61 2 516 1333
Product line(s): Zeiss telescopes and
    binoculars

Carl Zeiss Jena
396 Neerim Road
Carnegie, VIC 3163
Phone: +61 3 568 3355
Product line(s): Zeiss refractors and
    binoculars

New Zealand

Blaxhall Science Company
P.O. Box 25094
Christchurch
Phone: +64 3 366 2828
Product line(s): Celestron

Skylab
172 St. Asaph Street
Christchurch
Phone: +64 3 366 2827
Product line(s): Celestron, Meade, and
    more

Telescope & Optics
Otaraoa Road, Waitara
Taranaki
Phone +64 6 754 6434
Product line(s): Locally-made
    Newtonian reflectors

Carl Zeiss Jena
5 Wakefield Street
Lower Hutt
Phone: +64 4 566 7601
Product line(s): Zeiss refractors and
    binoculars

Appendix B
An Astronomer's Survival Guide

It's not always easy being an amateur astronomer. We are nocturnal creatures by nature, going outdoors when the rest of the world sleeps, braving cold, heat, bugs, and things that go bump in the night, always looking up when most everyone else is looking down (astronomers are the eternal optimists).

Here is a checklist of things that I like to bring along for a night under the stars. They make the experience much more pleasurable and the cold night a little warmer.

Astronomical:

Telescope ____   Binoculars ____   Eyepieces ____   LPR filter(s) ____

Color filter(s) ____   Star atlas ____   List of things to look at ____

Clipboard ____   Pen/pencil ____   Flashlight (red) ____   Flashlight (white) ____

Etc. _____

_____

Photographic:

Camera(s) ____   Auxiliary lenses ____   Film ____   Tripod ____

Camera-to-telescope adapters ____   Scotch mount ____   Drive corrector ____

Cable releases (always bring two, in case one breaks) _____

Etc. _____

_____

Miscellaneous:

Sweatshirt ____   Long underwear ____   Heavy socks ____   Jacket ____

Winter coat ____   Gloves/mittens ____   Boots ____   Foot/hand warmers ____

Hat ____   Folding table ____   Chair ____   Insect repellent ____

Something warm to drink (nonalcoholic) ____   Food/snacks ____   Radio ____

Etc. _____

_____

Appendix C
Upcoming Eclipses, 1994–2000

Lunar

Date (U.T.)	Type of Eclipse	First Contact (U.T.)*	Mid-eclipse (U.T.)*	Last Contact (U.T.)*
25 May 1994	Partial	02:39	03:31	04:23
15 Apr 1995	Partial	11:42	12:18	12:54
3–4 Apr 1996	Total	22:22	00:10	01:58
27 Sep 1996	Total	01:13	02:54	04:35
24 Mar 1997	Partial	02:59	04:40	06:21
16 Sep 1997	Total	17:08	18:46	20:24
28 Sep 1999	Partial	10:22	11:33	12:44
21 Jan 2000	Total	03:03	04:44	06:25
16 Jul 2000	Total	11:58	13:56	15:54

Solar

Date (U.T.)*	Type of eclipse	Zone of maximum eclipse
10 May 1994	Annular	Pacific Ocean, northwestern Mexico, North America (from Texas through New England and into Ontario and the Maritime Provinces)
3 Nov 1994	Total	Uruguay, South Atlantic Ocean
29 Apr 1995	Annular	Peru/Ecuador border
24 Oct 1995	Total	India, Vietnam, Brunei, South China Sea
9 Mar 1997	Total	Siberia, Arctic Ocean
26 Feb 1998	Total	Pacific Ocean, Colombia, Venezuela, Atlantic Ocean
22 Aug 1998	Annular	North of New Guinea
16 Feb 1999	Annular	South Indian Ocean
11 Aug 1999	Total	Atlantic Ocean, England, France, Germany, Austria, Hungary, Rumania, Turkey, Iran, Pakistan, India

* A Timely Note:

In order to standardize when a particular celestial event is due to occur, astronomers have chosen to express the time of day according to Universal Time*, or simply* U.T. *Universal Time (aka Greenwich Mean Time or Coordinated Universal Time) is based on the local time at the Greenwich, England, prime meridian. To find out when a celestial event is to take place from your location, its U.T. must be converted to your civil clock time. In general, there is an hour's difference for every 15° of longitude traveled away from the Earth's prime meridian. The following table shows how to convert Universal Time to various time zones around the world. Remember to change both the time* and *date if necessary.*

Converting Universal Time to Local Time

Time Zone	Local Standard Time*
North America:	
Atlantic Standard Time	U.T. minus 4 hours
Eastern Standard Time	U.T. minus 5 hours
Central Standard Time	U.T. minus 6 hours
Mountain Standard Time	U.T. minus 7 hours
Pacific Standard Time	U.T. minus 8 hours
Yukon Standard Time	U.T. minus 9 hours
Alaska-Hawaii Standard Time	U.T. minus 10 hours
Great Britain	U.T.
Europe	U.T. plus 1 hour
Australia (west coast)	U.T. plus 8 hours
Australia (east coast)	U.T. plus 10 hours
New Zealand	U.T. plus 12 hours

Note:

If you are converting to Daylight Savings Time, be sure to add an extra hour (or subtract an hour less, depending how you look at it) to the standard time.

As an example, look at the total lunar eclipse of January 21, 2000. The eclipse is due to begin at 3:03 U.T. An observer in the Eastern Standard Time (EST) zone of the United States will see the eclipse start at 10:03 P.M. on January 20, while the event will begin at 7:03 P.M. on January 20 for observers in Pacific Standard Time (PST).

Appendix D
Visibility of the Planets, 1994–2000

Which bright planets are visible tonight? This table will tell you where to look for Venus, Mars, Jupiter, and Saturn. The positions of the planets are listed for the middle of each month. To make the table a little more concise, each constellation name is with its standard three-letter abbreviation. (A translation table is given in Appendix F.) An *italicized* constellation means the planet will be visible in the early evening sky.

		Venus	Mars	Jupiter	Saturn
1994	January	Sgr	Sgr	Lib	Aqr
	February	Aqr	Cap	Lib	Aqr
	March	*Psc*	Aqr	Lib	Aqr
	April	*Ari*	Psc	Lib	Aqr
	May	*Tau*	Ari	*Vir*	Aqr
	June	*Cnc*	Ari	*Vir*	Aqr
	July	*Leo*	Tau	*Vir*	Aqr
	August	*Vir*	Tau	*Vir*	Aqr
	September	*Vir*	Gem	*Lib*	Aqr
	October	*Lib*	Cnc	*Lib*	Aqr
	November	Vir	Leo	Lib	Aqr
	December	Vir	Leo	Sco	Aqr
1995	January	Lib	*Leo*	Sco	*Aqr*
	February	Sgr	*Leo*	Sco	*Aqr*
	March	Cap	*Leo*	Oph	*Aqr*
	April	Aqr	*Leo*	Oph	*Aqr*
	May	Psc	*Leo*	*Oph*	Psc
	June	Tau	*Leo*	*Sco*	Psc
	July	Gem	*Vir*	Sco	Psc
	August	Leo	*Vir*	Sco	*Psc*
	September	Vir	*Vir*	Sco	*Psc*
	October	*Vir*	*Lib*	Oph	Aqr
	November	*Oph*	*Oph*	Oph	Aqr
	December	*Sgr*	*Sgr*	Sgr	Aqr
1996	January	*Cap*	Sgr	Sgr	Aqr
	February	*Psc*	Aqr	Sgr	*Psc*

		Venus	Mars	Jupiter	Saturn
	March	*Ari*	Psc	Sgr	Psc
	April	*Tau*	Psc	Sgr	Psc
	May	*Tau*	Ari	Sgr	Psc
	June	Tau	Tau	*Sgr*	Psc
	July	Tau	Tau	*Sgr*	Psc
	August	Gem	Gem	*Sgr*	*Psc*
	September	Cnc	Cnc	*Sgr*	*Psc*
	October	Leo	Leo	*Sgr*	*Psc*
	November	Vir	Leo	Sgr	*Psc*
	December	Lib	Vir	Sgr	*Psc*
1997	January	Oph	*Vir*	Sgr	*Psc*
	February	Cap	*Vir*	Cap	*Psc*
	March	Aqr	*Vir*	Cap	Psc
	April	Psc	*Leo*	Cap	Psc
	May	*Tau*	*Leo*	Cap	Psc
	June	*Gem*	*Vir*	Cap	Psc
	July	*Cnc*	*Vir*	Cap	Psc
	August	*Vir*	*Vir*	Cap	Psc
	September	*Lib*	*Lib*	*Cap*	Psc
	October	*Oph*	*Oph*	*Cap*	Psc
	November	*Sgr*	*Sgr*	*Cap*	Psc
	December	*Cap*	*Sgr*	*Cap*	*Psc*
1998	January	Sgr	*Cap*	Cap	*Psc*
	February	Sgr	*Aqr*	Aqr	*Psc*
	March	Cap	*Psc*	Aqr	*Psc*
	April	Aqr	Ari	Aqr	Psc
	May	Psc	Tau	Aqr	Psc
	June	Ari	Tau	Psc	Psc
	July	Tau	Gem	Psc	Psc
	August	Cnc	Gem	Psc	Ari
	September	Leo	Cnc	*Aqr*	Ari
	October	Vir	Leo	*Aqr*	*Ari*
	November	Lib	Leo	*Aqr*	Saturn
	December	Sgr	Vir	*Aqr*	Ari
1999	January	Cap	Vir	*Psc*	Ari
	February	Psc	Lib	*Psc*	Ari
	March	Psc	Lib	Psc	Ari
	April	*Tau*	Lib	Psc	Ari
	May	*Gem*	*Vir*	Psc	Ari
	June	*Cnc*	*Vir*	Ari	Ari
	July	*Leo*	*Lib*	Ari	Ari
	August	*Leo*	*Lib*	Ari	Ari
	September	Leo	*Oph*	*Ari*	Ari
	October	Leo	*Sgr*	Ari	Ari

		Venus	Mars	Jupiter	Saturn
	November	Vir	*Sgr*	*Ari*	*Ari*
	December	Lib	*Cap*	Psc	*Ari*
2000	January	Oph	*Aqr*	*Psc*	*Ari*
	February	Sgr	Psc	Psc	Ari
	March	Aqr	*Psc*	*Ari*	*Ari*
	April	Psc	*Ari*	*Ari*	*Ari*
	May	Ari	*Tau*	Ari	Ari
	June	Tau	Tau	Tau	Tau
	July	*Cnc*	Gem	Tau	Tau
	August	*Leo*	Cnc	Tau	Tau
	September	*Vir*	Leo	Tau	Tau
	October	*Lib*	Leo	Tau	Tau
	November	*Sgr*	Vir	*Tau*	*Tau*
	December	*Cap*	Vir	*Tau*	*Tau*

Appendix E
The Messier Catalogue Plus

Here's a listing of 175 of the finest deep-sky objects visible in amateur telescopes.

M #		NGC #	Con[1]	Season	Type[2]	R.A. (2000) h	m	Dec. (2000) °	'	Mag.	Size[3]	Remarks
1		1952	Tau	Winter	SNR	05	34.5	+22	01	8.2	6'×4'	Crab Nebula
2		7089	Aqr	Autumn	GC	21	33.5	−00	49	6.5	13'	
3		5272	CVn	Spring	GC	13	42.2	+28	23	6.4	16'	
4		6121	Sco	Summer	GC	16	23.6	−26	32	6.0	26'	
5		5904	Ser	Summer	GC	15	18.6	+02	05	5.8	17'	
6		6405	Sco	Summer	OC	17	40.1	−32	13	4.2	15'	Butterfly Cluster
7		6475	Sco	Summer	OC	17	53.9	−34	49	3.3	80'	
8		6523	Sgr	Summer	EN	18	03.8	−24	23	5.8	90'×40'	Lagoon Nebula
9		6333	Oph	Summer	GC	17	19.2	−18	31	7.9	9'	
10		6254	Oph	Summer	GC	16	57.1	−04	06	6.6	15'	
11		6705	Sct	Summer	OC	18	51.1	−06	16	5.8	14'	Wild Duck Cluster
12		6218	Oph	Summer	GC	16	47.2	−01	57	6.6	15'	
13		6205	Her	Summer	GC	16	41.7	+36	28	5.9	16'	
14		6402	Oph	Summer	GC	17	37.6	−03	15	7.6	12'	
15		7078	Peg	Autumn	GC	21	30.0	+12	10	6.4	12'	
16		6611	Ser	Summer	EN+OC	18	18.8	−13	47	6.0	35'	Eagle Nebula
17		6618	Sgr	Summer	EN	18	20.8	−16	11	7.0	46'×37'	Omega Nebula
18		6613	Sgr	Summer	OC	18	19.9	−17	08	6.9	9'	
19		6273	Oph	Summer	GC	17	02.6	−26	16	7.1	14'	
20		6514	Sgr	Summer	EN+RN	18	02.6	−23	02	8.5	29'×27'	Trifid Nebula
21		6531	Sgr	Summer	OC	18	04.6	−22	30	5.9	13'	
22		6656	Sgr	Summer	GC	18	36.4	−23	54	5.1	24'	
23		6494	Sgr	Summer	OC	17	56.8	−19	01	5.5	27'	
24		—	Sgr	Summer	OC	18	16.9	−18	29	4.5	90'	Small Sgr Star cloud
25	IC	4725	Sgr	Summer	OC	18	31.6	−19	15	4.6	32'	
26		6694	Sct	Summer	OC	18	45.2	−09	24	8.0	15'	
27		6853	Vul	Summer	PN	19	59.6	+22	43	8.1	8'×4'	Dumbbell Nebula
28		6626	Sgr	Summer	GC	18	24.5	−24	52	6.9	11'	
29		6913	Cyg	Summer	OC	20	23.9	+38	22	6.6	7'	
30		7099	Cap	Autumn	GC	21	40.4	−23	11	7.5	11'	
31		224	And	Autumn	Gx	00	42.7	+41	16	3.5	160'×40'	Andromeda Galaxy
32		221	And	Autumn	Gx	00	42.7	+40	52	8.2	3'×2'	

M #	NGC #	Con[1]	Season	Type[2]	R.A. (2000) h	m	Dec. (2000) °	'	Mag.	Size[3]	Remarks
33	598	Tri	Autumn	Gx	01	33.9	+30	39	6.3	60'×35'	
34	1039	Per	Autumn	OC	02	42.0	+42	47	5.5	35'	
35	2168	Gem	Winter	OC	06	08.9	+24	20	5.3	28'	
36	1960	Aur	Winter	OC	05	36.1	+34	08	6.0	12'	
37	2099	Aur	Winter	OC	05	52.4	+32	33	5.6	24'	
38	1912	Aur	Winter	OC	05	28.7	+35	50	6.4	21'	
39	7092	Cyg	Summer	OC	21	32.2	+48	26	4.6	32'	
40	—	UMa	Spring	**	12	22.4	+58	05	9.0,9.3	50"	Also known as Winnecki 4
41	2287	CMa	Winter	OC	06	46.0	−20	44	4.6	38'	
42	1976	Ori	Winter	EN	05	35.4	−05	27	2.9	66'×60'	Orion Nebula
43	1982	Ori	Winter	EN	05	35.6	−05	16	6.9	20'×15'	
44	2632	Cnc	Spring	OC	08	40.1	+19	59	3.1	95'	Beehive or Praesepe
45	—	Tau	Winter	OC	03	47.0	+24	07	1.2	110'	Pleiades
46	2437	Pup	Winter	OC	07	41.8	−14	49	6.1	27'	
47	2422	Pup	Winter	OC	07	36.6	−14	30	4.5	30'	
48	2548	Hya	Spring	OC	08	13.8	−05	48	5.8	55'	
49	4472	Vir	Spring	Gx	12	29.8	+08	00	8.4	9'×7'	
50	2323	Mon	Winter	OC	07	03.2	−08	20	5.9	16'	
51	5194	CVn	Spring	Gx	13	29.9	+47	12	8.4	11'×8'	Whirlpool Galaxy
52	7654	Cas	Autumn	OC	23	24.2	+61	35	6.9	13'	
53	5024	Com	Spring	GC	13	12.9	+18	10	7.7	13'	
54	6715	Sgr	Summer	GC	18	55.1	−30	29	7.7	9'	
55	6809	Sgr	Summer	GC	19	40.0	−30	58	7.0	19'	
56	6779	Lyr	Summer	GC	19	16.6	+30	11	8.2	7'	
57	6720	Lyr	Summer	PN	18	53.6	+33	02	9.7	70"×150"	Ring Nebula
58	4579	Vir	Spring	Gx	12	37.7	+11	49	9.8	5'×4'	
59	4621	Vir	Spring	Gx	12	42.0	+11	39	9.8	5'×3'	
60	4649	Vir	Spring	Gx	12	43.7	+11	33	8.8	7'×6'	
61	4303	Vir	Spring	Gx	12	21.9	+04	28	9.7	6'×5'	
62	6266	Oph	Summer	GC	17	01.2	−30	07	6.6	14'	
63	5055	CVn	Winter	Gx	13	15.8	+42	02	8.6	12'×7'	
64	4826	Com	Spring	Gx	12	56.7	+21	41	8.5	9'×5'	Black-Eye Galaxy
65	3623	Leo	Spring	Gx	11	18.9	+13	05	9.3	10'×3'	
66	3627	Leo	Spring	Gx	11	20.2	+12	59	9.0	9'×4'	
68	4590	Hya	Spring	GC	12	39.5	−26	45	8.2	12'	
69	6637	Sgr	Summer	GC	18	31.4	−32	21	7.7	7'	
70	6681	Sgr	Summer	GC	18	43.2	−32	18	8.1	8'	
71	6838	Sge	Summer	GC	19	53.8	+18	47	8.3	7'	
72	6981	Aqr	Autumn	GC	20	53.5	−12	32	9.4	6'	
73	6994	Aqr	Autumn	OC	20	58.9	−12	38	9.0	3'	
75	6864	Sgr	Summer	GC	20	06.1	−21	55	8.6	6'	
76	650	Per	Autumn	PN	01	42.2	+51	34	11.4	3'×1'	Little Dumbbell Nebula
77	1068	Cet	Autumn	Gx	02	42.7	−00	01	8.9	6'×5'	
78	2068	Ori	Winter	RN	05	46.7	+00	03	8	8'×6'	
79	1904	Lep	Winter	GC	05	24.5	−24	33	8.4	3'	

M #	NGC #	Con[1]	Season	Type[2]	R.A. (2000) h	m	Dec. (2000) °	′	Mag.	Size[3]	Remarks
80	6093	Sco	Summer	GC	16	17.0	−22	59	7.2	9′	
81	3031	UMa	Spring	Gx	09	55.6	+69	04	7.0	26′×14′	
82	3034	UMa	Spring	Gx	09	55.8	+69	41	8.4	11′×5′	
83	5236	Hya	Spring	Gx	13	37.0	−29	52	7.6	11′×10′	
84	4374	Vir	Spring	Gx	12	25.1	+12	53	9.3	5′×4′	
85	4382	Com	Spring	Gx	12	25.4	+18	11	9.2	7′×5′	
86	4406	Vir	Spring	Gx	12	26.2	+12	57	9.2	7′×6′	
87	4486	Vir	Spring	Gx	12	30.8	+12	24	8.6	7′	
88	4501	Com	Spring	Gx	12	32.0	+14	25	9.5	7′×4′	
89	4552	Vir	Spring	Gx	12	35.7	+12	33	9.8	4′	
90	4569	Vir	Spring	Gx	12	36.8	+13	10	9.5	9′×5′	
91	4548	Com	Spring	Gx	12	35.4	+14	30	10.2	4′×3′	
92	6341	Her	Summer	GC	17	17.1	+43	08	6.5	11′	
93	2447	Pup	Winter	OC	07	44.6	−23	52	6.2	22′	
94	4736	CVn	Spring	Gx	12	50.9	+41	07	8.2	11′×9′	
95	3351	Leo	Spring	Gx	10	44.0	+11	42	9.7	7′×5′	
96	3368	Leo	Spring	Gx	10	46.8	+11	49	9.2	7′×5′	
97	3587	UMa	Spring	PN	11	14.8	+55	01	11.2	3′	
98	4192	Com	Spring	Gx	12	13.8	+14	54	10.1	10′×3′	
99	4254	Com	Spring	Gx	12	18.8	+14	25	9.8	5′×4′	
100	4321	Com	Spring	Gx	12	22.9	+15	49	9.4	7′×6′	
101	5457	UMa	Spring	Gx	14	03.2	+54	21	7.7	27′×26′	Pinwheel Galaxy
102	(mistaken duplicate observation of M101)										
103	581	Cas	Autumn	OC	01	33.2	+60	42	7.4	6′	
104	4594	Vir	Spring	Gx	12	40.0	−11	37	8.3	9′×4′	Sombrero Galaxy
105	3379	Leo	Spring	Gx	10	47.8	+12	35	9.3	5′×4′	
106	4258	CVn	Spring	Gx	12	19.0	+47	18	8.3	18′×8′	
107	6171	Oph	Summer	GC	16	32.5	−13	03	8.1	10′	
108	3556	UMa	Spring	Gx	11	11.5	+55	40	10.1	8′×3′	
109	3992	UMa	Spring	Gx	11	57.6	+53	23	9.8	8′×5′	
110	205	And	Autumn	Gx	00	40.4	+41	41	8.0	10′×5′	
	Iota	Cnc	Spring	**	08	46.7	+28	46	4.0,6.6	30″	Yellow/blue
	2516	Car	Spring	OC	07	58.3	−60	52	3.8	30′	
	2808	Car	Spring	GC	09	12.0	−64	52	6.3	14′	
	3114	Car	Spring	OC	10	02.7	−60	07	4.2	35′	
	3115	Sex	Spring	Gx	10	05.2	−07	43	9.1	8′×3′	Spindle Galaxy
	3132	Vel	Spring	PN	10	07.7	−40	26	8.2p	84″×53″	Eight-Burst Nebula
	3242	Hya	Spring	PN	10	24.8	−18	38	8.6p	16″	Ghost of Jupiter Nebula
	3293	Car	Spring	OC	10	35.8	−58	14	4.7	40′	
	3324	Car	Spring	OC+EN	10	37.3	−58	38	6.7	6′	
	Zeta	UMa	Spring	**	13	23.9	+54	56	2.3,4.0	14″	Alcor
	3372	Car	Spring	EN	10	43.8	−59	52	5.0	120′	Eta Carinae Nebula
	3532	Car	Spring	OC	11	06.4	−58	40	3.0	55′	
	3572	Car	Spring	OC	11	10.4	−60	14	6.6	20′	
	3766	Cen	Spring	OC	11	36.1	−61	37	5.3	12′	

M #	NGC #	Con[1]	Season	Type[2]	R.A. (2000) h	m	Dec. (2000) °	′	Mag.	Size[3]	Remarks
Melotte	111	Com	Spring	OC	12	25	+26		1.8	275′	Coma star cluster
		Cru	Spring	Dk	12	53	−63		—	7°×5°	Coalsack
	4755	Cru	Spring	OC	12	53.6	−60	20	4.2	10′	Jewel Box Cluster
	5128	Cen	Spring	Gx	13	25.5	−43	01	7	18′×14′	Centaurus A
	5139	Cen	Spring	GC	13	26.8	−47	29	3.7	36′	Omega Centauri
	6210	Her	Summer	PN	16	44.5	+23	49	9.2	15″	
	6231	Sco	Summer	OC	16	54.0	−41	48	2.6	15′	
	6281	Sco	Summer	OC	17	04.8	−37	54	5.4	8′	
B59,65–7		Oph	Summer	Dk	17	21	−27			5°×3°	Pipe Nebula
B78		Oph	Summer	Dk	17	33	−26			3°×2°	
	6369	Oph	Summer	PN	17	29.3	−23	46	10.4	30″	Little Ghost Nebula
	6388	Sco	Summer	GC	17	36.3	−44	44	6.9	9′	
IC 4665		Oph	Summer	OC	17	46.3	+05	43	4.2	41′	
	6572	Oph	Summer	PN	18	12.1	+06	51	9.0	7″	
B92		Sgr	Summer	Dk	18	15.5	−18	11		12′×6′	See M24
	6712	Sct	Summer	GC	18	53.1	−08	42	8.3	7′	
Collinder	399	Vul	Summer	OC	19	25.4	+20	11	3.6	60′	Coathanger Cluster
	Beta	Cyg	Summer	**	19	30.7	+27	58	3.2,5.4	34″	Albireo, yellow/blue
	6826	Cyg	Summer	PN	19	44.8	+50	31	9.8	30″	Blinking Planetary
	6939	Lac	Summer	OC	20	31.4	+60	38	7.8	8′	
	6960	Cyg	Summer	SNR	20	45.7	+30	43		70′×6′	West half of Veil Nebula
	6992	Cyg	Summer	SNR	20	56.4	+31	43		60′×8′	East half of Veil Nebula
	7000	Cyg	Summer	EN	20	58.8	+44	20		2°	North America Nebula
	7009	Aqr	Autumn	PN	21	04.2	−11	22	8.4	26″	Saturn Nebula
	7293	Aqr	Autumn	PN	22	29.6	−20	48	6.5	15′×12′	Helix Nebula
	7662	And	Autumn	PN	23	25.9	+42	33	8.9p	32″×28″	
	55	Scl	Autumn	Gx	00	14.9	−39	11	8.0	32′×6′	
	246	Cet	Autumn	PN	00	47.0	−11	53	8.5	240″×210″	
	253	Scl	Autumn	Gx	00	47.6	−25	17	7.1	22′×6′	
	288	Scl	Autumn	GC	00	52.8	−26	35	8.1	14′	
	457	Cas	Autumn	OC	01	19.1	+58	20	6.4	13′	
Gamma		Ari	Autumn	**	01	53.5	+19	18	4.6,4.7	8″	Orange/green; outstanding
	752	And	Autumn	OC	01	57.8	+37	41	5.7	50′	
Gamma		And	Autumn	**	02	03.9	+42	20	2.1,5.1	10″	Orange/yellow;beautiful
	869	Per	Autumn	OC	02	19.0	+57	09	4.3	30′	Double Cluster (h Per)
	884	Per	Autumn	OC	02	22.4	+57	07	4.4	30′	Double Cluster (Chi Per)
	891	And	Autumn	Gx	02	22.6	+42	21	10.0	14′×3′	
Melotte	20	Per	Autumn	OC	03	22	+49		1.2	185′	Alpha Per Cluster
	1316	For	Autumn	Gx	03	22.7	−37	12	8.9	4′×3′	
	1360	For	Autumn	PN	03	33.3	−25	51		390″	
	1514	Tau	Winter	PN	04	09.2	+30	47	10.9	2′	
	1973	Ori	Winter	EN+RN	05	35.1	−04	44		40′×25′	Also NGC 1975,1977
	2194	Ori	Winter	OC	06	13.8	+12	48	8.5	10′	
	Beta	Mon	Winter	**	06	28.8	−07	02	4.6,5.1	7″	

M #	NGC #	Con[1]	Season	Type[2]	R.A. (2000) h	m	Dec. (2000) °	'	Mag.	Size[3]	Remarks
	2237	Mon	Winter	EN	06	32.3	+05	03		80'×60'	Rosette Nebula
	2244	Mon	Winter	OC	06	32.4	+04	52	4.8	24'	Rosette Nebula Cluster
	2261	Mon	Winter	EN+RN	06	39.2	+08	44	10.0	2'	Hubble's Variable Nebula
	2264	Mon	Winter	OC	06	41.1	+09	53	3.9	20'	Christmas Tree Cluster
	2359	CMa	Winter	EN	07	18.6	−13	12		8'×6'	
	2392	Gem	Winter	PN	07	29.2	+20	55	8.3	13"	Eskimo Nebula
	2403	Cam	Winter	Gx	07	36.9	+65	36	8.4	18'×11'	
	2539	Pup	Winter	OC	08	10.7	−12	50	6.5	22'	

Notes:

1. *Constellation—see Appendix F.*
2. *Object type:*

**	= *Double Star*
OC	= *Open Cluster*
GC	= *Globular Cluster*
EN	= *Emission Nebula*
RN	= *Reflection Nebula*
Dk	= *Dark Nebula*
PN	= *Planetary Nebula*
SNR	= *Supernova Remnant*
Gx	= *Galaxy*

3. *Apparent size of object in either minutes of arc, seconds of arc, or degrees. Most measurements were made from photographs; visual appearance may be smaller. For double stars, this number is a measure of the stars' separation from one another.*

Appendix F
The Constellations

Constellation	Abbr.	Genitive Form	Meaning
Andromeda	And	Andromedae	The Daughter of Queen Cassiopeia
Antlia	Ant	Antliae	The Air Pump
Apus	Aps	Apodis	The Bird of Paradise
Aquarius	Aqr	Aquarii	The Water-bearer
Aquila	Aql	Aquilae	The Eagle
Ara	Ara	Arae	The Altar
Aries	Ari	Arietis	The Ram
Auriga	Aur	Aurigae	The Charioteer
Boötes	Boo	Boötis	The Herdsman
Caelum	Cae	Caeli	The Chisel
Camelopardalis	Cam	Camelopardalis	The Giraffe
Cancer	Cnc	Cancri	The Crab
Canes Venatici	CVn	Canum Venaticorum	The Hunting Dogs
Canis Major	CMa	Canis Majoris	The Big Dog
Canis Minor	CMi	Canis Minoris	The Little Dog
Capricornus	Cap	Capricorni	The Sea-goat
Carina	Car	Carinae	The Keel (of the mythical ship Argo)
Cassiopeia	Cas	Cassiopeiae	The Queen
Centaurus	Cen	Centauri	The Centaur
Cepheus	Cep	Cephei	The King
Cetus	Cet	Ceti	The Whale
Chamaeleon	Cha	Chamaeleontis	The Chameleon
Circinus	Cir	Circini	The Compasses
Columba	Col	Columbae	The Dove
Coma Berenices	Com	Comae Berenices	The Queen Berenice's Hair
Corona Australis	CrA	Coronae Australis	The Southern Crown
Corona Borealis	CrB	Coronae Borealis	The Northern Crown
Corvus	Crv	Corvi	The Crow
Crater	Crt	Crateris	The Cup
Crux	Cru	Crucis	The Southern Cross
Cygnus	Cyg	Cygni	The Swan
Delphinus	De	Delphini	The Dolphin
Dorado	Dor	Doradus	The Swordfish
Draco	Dra	Draconis	The Dragon
Equuleus	Equ	Equulei	The Little Horse
Eridanus	Eri	Eridani	The River

Constellation	Abbr.	Genitive Form	Meaning
Fornax	For	Fornacis	The Furnace
Gemini	Gem	Geminorum	The Twins
Hercules	Her	Herculis	The Giant
Horologium	Hor	Horologii	The Clock
Hydra	Hya	Hydrae	The Water Snake (male)
Hydrus	Hyi	Hydri	The Water Snake (female)
Indus	Ind	Indi	The Indian
Lacerta	Lac	Lacertae	The Lizard
Leo	Leo	Leonis	The Lion
Leo Minor	LMi	Leonis Minoris	The Little Lion
Lepus	Lep	Leporis	The Hare
Libra	Lib	Librae	The Scales of Justice
Lupus	Lup	Lupi	The Wolf
Lynx	Lyn	Lyncis	The Lynx
Lyra	Lyr	Lyrae	The Lyre
Mensa	Men	Mensae	The Table
Microscopium	Mic	Microscopii	The Microscope
Monoceros	Mon	Monocerotis	The Unicorn
Musca	Mus	Muscae	The Fly
Norma	Nor	Normae	The Square
Octans	Oct	Octantis	The Octant
Ophiuchus	Oph	Ophiuchi	The Serpent-bearer
Orion	Ori	Orionis	The Hunter
Pavo	Pav	Pavonis	The Peacock
Pegasus	Peg	Pegasi	The Flying Horse
Perseus	Per	Persei	The Warrior
Phoenix	Phe	Phoenicis	The Phoenix
Pictor	Pic	Pictoris	The Painter
Pisces	Psc	Piscium	The Fishes
Piscis Austrinus	PsA	Piscis Austrini	The Southern Fish
Puppis	Pup	Puppis	The Stern (of the mythical ship Argo)
Pyxis	Pyx	Pyxidis	The Compass
Reticulum	Ret	Reticuli	The Reticle
Sagitta	Sge	Sagittae	The Arrow
Sagittarius	Sgr	Sagittarii	The Archer
Scorpius	Sco	Scorpii	The Scorpion
Sculptor	Scl	Sculptoris	The Sculptor
Scutum	Sct	Scuti	The Shield
Serpens	Ser	Serpentis	The Serpent
Sextans	Sex	Sextantis	The Sextant
Taurus	Tau	Tauri	The Bull
Telescopium	Tel	Telescopii	The Telescope
Triangulum	Tri	Trianguli	The Triangle
Triangulum Australe	TrA	Trianguli Australis	The Southern Triangle
Tucana	Tuc	Tucanae	The Toucan
Ursa Major	UMa	Ursae Majoris	The Great Bear
Ursa Minor	UMi	Ursae Minoris	The Little Bear

Constellation	Abbr.	Genitive Form	Meaning
Vela	Vel	Velorum	The Sails (of the mythical ship Argo)
Virgo	Vir	Virginis	The Maiden
Volans	Vol	Volantis	The Flying Fish
Vulpecula	Vul	Vulpeculae	The Fox

Appendix G
English/Metric Conversion

Most amateur astronomers in the United States will speak of a telescope's aperture in terms of inches, while the rest of world use centimeters. The table below acts as a translation table to help convert telescope apertures from one system to another. Recall that there are 2.54 centimeters per inch.

English (inches)	Metric (centimeters)
2	5
3.1	8
4	10
6	15
8	20
10	25
12	30
14	35
16	40
18	45
20	50
24	60
30	75
32	80
36	90

Index

Abbe eyepiece (see eyepieces)
aberrations, optical
    astigmatism, 21, 35, 232, 235
    chromatic, 13, 113
    coma, 21, 35
    curvature of field, 113
    distortion, 113
    optical, 35
    pinched optics, 232, 234
    rough optical surface, 233, 235
    secondary spectrum, 29
    spherical, 13, 19, 35, 113, 232–233, 235
achromatic lens, 14
achromatic refractor (see telescopes)
adapters, camera-to-telescope, 164–165
addresses, manufacturers, 339–348
Adlerblick binoculars, 54
airy disk, 6
altitude-azimuth telescope mountings (see mountings, telescope)
American Association of Variable Star Observers, 285
Andromeda Galaxy (see M31)
aperture, 3
apochromatic refractor (see telescopes)
astigmatism (see aberrations)
Astro-Physics refractors, 61–63
Astrofest, 249
Astronomy Book Club, 143
astrophotography, 325, 327–338
    fixed camera, 328–329
    guided-camera, 335–336
    through the telescope, 329–335, 336, 338
autocollimator eyepiece, 141
averted vision, 260–261

Barlow Lens, 114, 120
barn-door camera mount (see Scotch mount)
batteries, rechargeable, 154–155
Bausch and Lomb binoculars, 54–55
Bayer, Johannes, 286
Beehive Cluster (see M44)
binocular viewing attachments (for telescopes), 141–142
binoculars (see also brand names), 54–60
    consumer considerations, 27–29
books, 142–150

annuals, 145–146
astrophotography, 148–149
deep-sky catalogues, 149–150
introductory, 146
miscellaneous, 150
observing guides, 146–148
optics, 149
periodicals, 144–145
star atlases, 143–144
telescope making, 149
Bouwers, A., 22
Brandon eyepiece (see eyepieces)

cameras, 35mm single-lens reflex, 159–160
cameras, CCD, 167–169
Carton binoculars (see Adlerblick binoculars)
Cassegrain reflector (see telescopes)
Cassegrain, Sieur, 18
celestial coordinates, 250–251
Celestron International
    binoculars, 55
    eyepieces (see eyepieces)
    reflectors, 70
    refractors, 63
    Schmidt-Cassegrain telescopes, 87–91
Ceravolo Optical Maksutov-Newtonian telescopes, 91
chair, observing
    commercial, 175–176
    homemade, 187–190
Cheshire eyepiece, 140
chromatic aberration (see aberrations)
Clark, Alvan, 15
clipboard, illuminated, homemade, 190–193
cloaking device, light-pollution, 261
clock-drive correctors (see correctors, clock-drive)
clothing, cold-weather, 169–171
coatings, enhanced versus standard, 32
coatings, lens, 113
collimation, 35–36, 224–231
collimation tools, 139–141
color filters (see filters lunar and filters, planetary)
coma (see aberrations)
coma corrector, 121–122
computer software, 150–151

computerized telescope-aiming systems, 156–159

constellations (table), 363–365

conventions, astronomy, 247–249

correctors, clock-drive, 151–154

Coulter Optical reflectors, 70–72

Crab Nebula (see M1)

curvature of field (see aberrations)

D&G Optical Company refractors, 64

Dall-Kirkham reflector (see telescopes)

Dawes' Limit, 7

dew cap
    commercial, 172–175
    homemade, 195–198

dew guns, 175

distortion (see aberrations)

dog biscuit (see aberrations)

Dollond, John, 15

Double Cluster (see NGC 869 and NGC 884)

Dumbbell Nebula (see M27)

ED glass (see extra-low dispersion glass)

Edmund Scientific
    binoculars, 55–56
    reflectors, 72–73

Erfle eyepiece (see eyepieces)

exit pupil, 109–110

extra-low dispersion glass, 31

eye, human, 258–261

eye lens, 105

eye relief, 112

eyecups, rubber, 141

eyepieces, 105–126
    alternatives (table), 125
    recommendations (table), 124
    types of (see also Barlow Lens and
        Telecompressors), 114–120
        Abbe (see orthoscopic)
        Brandon, 117–118
        Celestron International Ultima, 119
        Erfle, 114, 116
        Huygens, 114–115
        Kellner, 114, 115–116
        Lanthanum LV, 118
        Meade Instruments
            Super Plössl, 119
            Super Wide Field, 119
            Ultra Wide Field, 119
        Orion Telescope Center
            MegaVista, 119–120
            Ultrascopic, 119–120
        orthoscopic, 114, 116
        Plössl, 114, 117
        Ramsden, 114, 116
        reticle, 122–123
        RKE, 114, 116
        Tele Vue

            Nagler, 114, 119
            Panoptic, 119
            Wide Field, 119
        zoom, 116–117

f-ratio (see focal ratio)

field lens, 105

field of view
    apparent, 110–112
    real, 110–112

film, 161–163

filters, 130–139
    color (see filters, lunar and filters,
        planetary)
    light-pollution reduction, 131–134
    LPR (see filters, light-pollution reduction)
    lunar, 135, 136
    nebula (see filters, light-pollution
        reduction)
    planetary, 135, 136
    solar, 135, 137–139
        hydrogen-alpha, 138–139

finderscopes, 128–130

Flamsteed, John, 286

flashlight
    commercial, 172
    homemade, 193–195

fluorite, 31

focal ratio, 3

Focal Reducer (see telecompressor)

focusing aids, 165–167

focusing device, motorized
    commercial, 155–156
    homemade, 210–213

Foucault, Jean, 20

Fraunhofer, Joseph von, 15

Fujinon binoculars, 56–57

Galileo, 11–13

goggles, light-pollution, 261

Greek alphabet, 286, 287

Gregorian reflector (see telescopes)

Gregory, James, 17

Gregory, John, 23

Gregory-Maksutov telescopes (see telescopes)

Guinard, Pierre Louis, 15

Hadley, John, 19

Hall, Chester, 14

Herschel, William, 20

Herschelian reflector (see telescopes)

Hidden Hollow, 249

Huygens eyepiece (see eyepieces)

insect repellent, 171

International Dark-Sky Association, 242

INTES Maksutov telescopes, 92

Jim's Mobile Industries reflectors, 73–75

JMI reflectors (see Jim's Mobile Industries reflectors)
Jupiter Telescope Company reflectors, 75–76

Kellner eyepiece (see eyepieces)
Kepler, Johannes, 13

Lagoon Nebula (see M8)
Lanthanum LV eyepiece (see eyepieces)
lenses, camera, 160–161
light loss, 21
light pollution (see also observing site selection), 241–242, 261
light-gathering ability, 4–5
light-pollution reduction filters (see filters, light-pollution)
light-pollution shields, homemade, 183–187
limiting magnitude (see magnitude, limiting)
Lippershey, Jan, 11
locating sky objects, 250–258
    setting circles, 254–258
    star hopping, 251–254
LPR filters (see filters, light-pollution reduction)

M-objects (see Messier catalogue)
M1, 319–320, 321–322
M2, 305, 310, 311–312
M3, 296–297, 298–299
M8, 300, 303–304
M13, 297, 300, 301–302
M27, 305, 308–309
M31, 252, 253
M31, 310, 313–314
M32 (see M31)
M33, 252–254
M35, 320, 324–325, 326–327
M42, 320, 323, 324, 325
M44, 288, 290, 291
M45, 315, 318–319
M51, 291, 294–296, 297
M57, 300, 305, 306–307
M81, 288, 291, 293, 294
M82, 288, 291, 293, 294
M110 (see M31)
magnification, 4, 107–108
magnitude, limiting (table), 5
mail fraud, 100–102
Maksutov telescopes (see telescopes)
Maksutov, D., 22
Meade Instruments Corporation
    eyepieces (see eyepieces)
    reflectors, 33, 76–77
    refractors, 64–65
    Schmidt-Cassegrain telescopes, 39, 92–94
Messier catalogue, 287
metric conversion, telescope aperture (table), 367

Minolta binoculars, 57
Miyauchi binoculars, 57–58
moon filters (see filters, lunar)
mountings, binocular
    commercial, 177–179
    homemade, 198–202
mountings, telescope, 43–49, 176–177
    altitude-azimuth, 43, 45–46
        Dobsonian, 43, 45
    equatorial, 46–49
        fork, 48–49
        German, 47–48
        polar alignment, 254–257

nebula filters (see filters, light-pollution reduction)
New General Catalogue, 287
Newton, Isaac, 18
Newtonian reflector (see telescopes)
NGC (see New General Catalogue)
NGC 205 (see M110)
NGC 206 (see M31)
NGC 221 (see M32)
NGC 224 (see M31)
NGC 457, 310, 315, 316–317
NGC 869, 315, 316–317
NGC 884, 315, 316–317
NGC 1952 (see M1)
NGC 1976 (see M42)
NGC 2158 (see M35)
NGC 2168 (see M35)
NGC 2632 (see M44)
NGC 3031 (see M81)
NGC 3034 (see M82)
NGC 5194 (see M51)
NGC 5195 (see M51)
NGC 5272 (see M3)
NGC 6205 (see M13)
NGC 6523 (see M8)
NGC 6530 (see M8)
NGC 6720 (see M57)
NGC 6853 (see M27)
NGC 7089 (see M2)
night vision, 260
Nikon binoculars, 58
Nova East, 249

observatory, homemade, 213–216
observing, 267–325
    asteroids, 279
    checklist, 349
    comets, 279–281
    deep-sky objects (see also specific entries), 283–325
    site selection, 245–247
        double stars, 284
        galaxies, 286
        nebulae, 285–286

star clusters, 285
    variable stars, 284–285
    Messier Catalogue Plus (table), 357–361
eclipses
    lunar, 271
    solar, 282
    upcoming (table), 351
lunar occultations, 270
Moon, 267–271
planets, 271–279
    Jupiter, 275–277
    locations (table), 353–355
    Mars, 274–275
    Mercury, 272
    Neptune, 278–279
    Pluto, 279
    Saturn, 277–278
    Uranus, 278
    Venus, 272–273
Sun, 281–283
Obsession reflectors, 78–79
oculars (see eyepieces)
off-axis guiders, 165
Okie-Tex Star Party, 249
Optical Guidance Systems reflectors, 79–80
optical quality (see also star test), 26–27
Orion Nebula (see M42)
Orion Telescope Center
    binoculars, 58–59
    eyepieces (see eyepieces)
    reflectors, 80–81
    refractors, 66
orthoscopic eyepiece (see eyepieces)
Owl Cluster (see NGC 457)

Parallax Instruments reflectors, 81–82
parfocal, 112
Parks Optical
    binoculars, 59
    reflectors, 37, 82–83
Pentax refractors, 66
pinched optics (see aberrations)
Pleiades (see M45)
Plössl eyepiece (see eyepieces)
polar alignment, 254–257
power (see magnification)
power packs (see batteries, rechargable)
projects, homemade, 183–216

Questar Maksutov telescopes, 94–95

Ramsden eyepiece (see eyepieces)
record keeping, 262, 263
reflecting telescopes (see telescopes)
refracting telescopes (see telescopes)
resolving power, 5-9
Ring Nebula (see M57)
Ritchey-Kirkham reflector (see telescopes)

Riverside Telescope Maker's Conference, 248
RKE eyepiece (see eyepieces)
Rosse, Lord, 20
rough optical surface (see aberrations)

Safari reflectors, 83
Schmidt, Bernard, 22
Schmidt telescope (see telescopes)
Schmidt-Cassegrain telescope (see telescopes)
Schmidt-Newtonian telescope (see telescopes)
scintillation (see sky conditions)
Scotch mount, homemade, 202–206
secondary spectrum (see aberrations)
seeing (see sky conditions)
setting circles, 254–258
Seven Sisters (see M45)
Short, James, 20
sight tube, 140
sketching, 262–264
sky conditions, 241–245
Sky Designs reflectors, 84
solar filters (see filters, solar)
solar viewing device, homemade, 206–210
spherical aberration (see aberrations)
star hopping, 251–254
star parties, 247–249
star test, 231–236
Star-Liner reflectors, 84–85
Starbeam one-power finder, 130
Starfest, 249
Starsplitter reflectors, 85
Steinheil, Karl, 20
Stellafane, 247–248
sun filters (see filters, solar)
Swift binoculars, 59–60

Takahashi
    refractors, 66–67
    Schmidt-Cassegrain telescopes, 95–96
Tectron reflectors, 85–86
Tele Vue
    eyepieces (see eyepieces)
    refractors, 30, 67–68
telecompressors, 120–121
telescope care, 219–224
    cleaning corrector plates, 219–221
    cleaning lenses, 219–221
    cleaning mirrors, 221–223
telescope mountings (see mountings,
    telescope)
telescope pop quiz, 49–51
telescope shopping, 96–100
telescope storage, 217–219
telescope, traveling with, 236–239
telescopes
    consumer considerations
        catadioptric, 38–43
        reflectors, 32–38

refractor, 29–32
history of, 11–23
point-counterpoint comparison (table),
    44–45
types of (see also specific brand names),
    1–2
    catadioptric, 22–23
    reflectors, 17–22, 34–35
    refractors, 12–15, 31–32, 60–70
    used, 102–104
Telrad one-power finder, 129–130
Texas Star Party, 248–249
Torus Optical reflectors, 86–87
transparency (see sky conditions)
tripods, camera, 163–164

Unitron
    binoculars, 60
    refractors, 68–69

Velcro, 179–180
vibration dampers, 179
Vixen Optical Industries
    reflectors, 87
    refractors, 69–70

Winter Star Party, 249

Zeiss binoculars, 60
zoom eyepiece (see eyepieces)